Basic Biology

Basic Biology

Khushboo Chaudhary and
Pankaj Kumar Saraswat

Delve
PUBLISHING
www.delvepublishing.com

Basic Biology

Khushboo Chaudhary and Pankaj Kumar Saraswat

Delve Publishing
224 Shoreacres Road
Burlington, ON L7L 2H2
Canada
www.delvepublishing.com

Email: orders@arclereducation.com

ISBN: 978-1-77469-155-7 (Hardcover)

ABOUT THE AUTHORS

Dr Khushboo Chaudhary is presently working as a Research Associate in NRCE, Hisar Haryana, India and having one year of teaching experience and seven years of research experience. Previously, she worked on "Improvement of Phytoremediation efficiency of Fluoride". She has published several research papers in international and national journals. She has published five international text books. She has got seven best paper and poster presentation awardsfromthe Indian Society of Genetics and Biotechnology Research and Development and received the president appreciation awards also in the International conference. She has got the best poster award by ISSGPU Central Institute Research on Goats, Makhdoom. She has published several gene banks in NCBI Pubmed. She has also published a research article in the virology journal. She is likely to be a co-author in several of the publications and coauthor in J. Virological Methods.

Dr.Pankaj Kumar Saraswat graduated from R.B.S. College Bichpuri (Agra University) Agra in 1996, completed post-graduate and Ph.D. in Soil Science & Agricultural Chemistry from Banaras Hindu University Varanasi in 1999 and 2004 respectively. Dr. Saraswat started his career at H.N.B.G. University Srinagar Garhwal Uttrakhand in 2005 as a lecturer in Soil Science and moved as Subject Matter Specialist (Soil Science) to KVK Banasthali Vidyapith Tonk Rajasthan. In October 2015 Dr. Saraswat joined ICAR-RC for N.E.H Region Umiam as Sr. Scientist & Head at KVK Tamenglong Manipur. Dr. Saraswat as P.I. completed two research projects externally funded from SERC-DST Govt. of India New Delhi and DST-Govt. of Rajasthan Jaipur. At the same time, he also conducted four INSPIRE-Internship Science Camps under the SEAT program of DST for 10th pass top 1% students of Rajasthan. He has published three international text books Presently Dr. Saraswat has been working for agricultural technology assessment and demonstrations for its wider application at farmers' field along with capacity development programs for farmers, farm women, rural youth, line departments and other stakeholders and also working as P.I. in one of NEC Shillong funded demonstration based project in Tamenglong district Imphal, Manipur.

TABLE OF CONTENTS

PREFACE

Basic Biology has always been a fundamental area of science. The exponential enhance in the quantity of scientific information and the rate at which new discoveries are made, require much elaborated, interdisciplinary and up-to date information and their understanding. The book "Basic Biology" has been written to serve as a guide book for undergraduates and postgraduates of different universities in the world. This guide book has covered a range of syllabus and the student is an easy way to go through this book content. Books are available on the subject of Basic Biology but still not up to syllabus run in the university so the students face problem to find out notes. While this book is for students to acquire or gain fair knowledge of Basic Biology and easy presentation of language has written for proper understanding of each and every topic. The chapters have been so arranged to give a sequential knowledge of the introductory basic biochemistry, enzymology, cell biology and molecular biology, techniques in biochemistry, biotechnology, plant and animal tissue culture in this guide book. For phe preparing this book, I have been greatly helped by several books and presentationsto which I express my acknowledgment in the references and if I have forgotten any references in my notice, please ignore them. This book would serve not only for undergraduate but postgraduate students of all life science students, and also research scholars of various universities. The numbers of scientists and authors have been quoted or fully acknowledge and bibliography atthe end of the book. We claim no originality for the matter or text presented in this book but the method of presentation and illustration is our own. Last but not least I would be thankful to readers for pointing out errors and most welcome for suggestions in the next book.

Khushboo Chaudhary

Pankaj Kumar Saraswat

CHAPTER 1

BASIC BIOCHEMISTRY

1. BIOCHEMISTRY: DEFINITION, SCOPE AND IMPORTANCE IN AGRICULTURE.

Introduction

The term biochemistry was first introduced by a German chemist Carl Neuberg in 1903. Biochemistry is the division of science that explores the chemical processes within and related to living organisms, it is a science that brings together biology and chemistry. Biochemistry plays an important role in nutrition and health. Similarly, biochemistry in general deals with body substances like enzymes, carbohydrates, amino acids, fats, proteins, hormones, DNA, RNA, pigments, etc. It is used in clinical diagnosis, manufacturing of various biological products, treatment of diseases, in nutrition, agriculture, etc. Carbohydrates, lipids, proteins and nucleic acids are the most common biological molecules studied by biochemists. It also encompasses the science of molecular biology. This includes immunochemistry, neurochemistry and bioinorganic, bioorganic and biophysical chemistry.

In terms of history, biochemistry is a young science. It started largely as an off shoot of organic chemistry and later incorporated ideas obtained from physical chemistry. In fact, the science of biochemistry may be regarded to have begun with the writings of Theophrastus Bombastus von Hohenheim, better known as Phiippus Aureolus Paracelsus (Lifetime or LT, 1493-154),a swedish physician and alchemist who laid the foundation of chemotherapy as a method of treating diseases and also the promoter of doctrine of signatures. He first acquired the knowledge of chemistry of his time and then entered the field of medicine to apply his knowledge of chemistry. The proclaimed, "Life processes are essential of chemical nature and diseases can be cured by mclicines", Later, his: followers notably Jan Baptist van Helmont (LT, 1577-164) amalgamated the science of chemistry with medicine which emerged under the name of 'medical chemistry' (or Iatrochemistry). " The basis of biochemistry was in fact, laid down by chemists like Scheele and Lavoisier. Karl Wilhelm LT. 1742-1786), a swedish pharmacist, discovered the chemical composition of various drugs and the plant and animal materials. He also isolated a number of substances having discovered even a single one of malic acid from apple and uric acid from urine. Scheele, thus, laid the foundation of descriptive biochemistry.

Similarly, Antoine Lavoisier (LT, 1743-1794), a French chemist, who studied the composition of air and propounded the theory of conservation of matter. put dynamic biochemistry on firm stand footings. He developed the concept of oxidation and also clarified the nature of animal respiration. He had concluded that respiration could be equated with combustion and that it was slower but not essentially different from the combustion of charcoal. Lavoiser is often spoken of as "father of modern biochemistry"

Biochemistry is applied in various areas, including medicine, dentistry, industry, and agriculture and food science. Keeping in view the importance of biochemistry in a sustainable way, it is proposed to commercialize all the fields of biochemistry in the world as it is the most powerful and sustainable tool for reducing global poverty and hunger in the world. The importance of biochemistry can be seen from the fact that it is used in many day-to-day activities like in clinical diagnosis, manufacture of various biological products; treatment of diseases, in nutrition, agriculture, etc. It is a branch of science which deals with the chemical basis of life. It is concerned about the internal chemistry of biological systems viz. animals and plants. It is also taught as a branch of biology such as Bio-chemistry. The study of biochemistry helps one understand the actual concepts of the functioning of various body processes and physiology of plants and animals. Biochemistry in general deals with enzymes, hormones, carbohydrates, amino acids, fats, proteins, DNA, RNA, pigments ,etc. It describes their origin, formation, function, deficiency

Agricultural biochemistry is the study of both chemistry and biochemistry which are important in agricultural production, the processing of raw products into foods and beverages, and in environmental monitoring and remediation. These studies emphasize the relationships between plants, animals and bacteria and their environment.

2. BIOTECHNOLOGY CONCEPT OF LIFE AND LIVING PROCESSES: INTRODUCTION, NATURE OF BIOLOGICAL MATERIAL, PROPERTIES, THE IDENTIFYING CHARACTERISTICS OF A LIVING MATTER, MOLECULAR LOGIC OF LIFE.

Introduction

The origin of life on earth is one of the great mysteries. Where and how did life on earth originate. It is difficult to determine because it began at least 3.8 billion years ago and the earth is slightly more than 4.5 billion years old. Life is generally characterized by three functions.

- ***Compartmentalization***: The ability to keep its components together and distinguish itself from the environment.
- ***Replication***: The ability to process and transmit heritable information to progeny.
- ***Metabolism***: The ability to capture the energy and material resources, staying away from thermodynamic equilibrium.

All these functions are operated by biopolymers such as DNA, RNA, proteins and phospholipids. It is typically assumed that the precursors of these vital components were synthesized abiotically, accumulated somewhere, condensed into polymers, interacted mutually and eventually evolved into a self-sustaining system through natural phenomena on the primitive earth. Many rescarchers and scientists are trying to find the answers such as from where the raw materials came from, how organic molecules such as RNA formed and how the first cell arose. Haldane and Oparin attempted to answer this question. They proposed that life had arisen from simpler molecules on the lifeless earth under much different atmospheric conditions that exist today. However, instead of life arising suddenly, as previous spontaneous generation theories proposed. All living things have life processes. Life processes are essential functions of a living organism in order to remain in the living condition.

1. Supporting life processes is a part of managing wildlife populations.
2. When life processes stop, an organism dies.
3. There are eight life processes used in wildlife science.

Life activities depends on

- Growth and Repair
- Food
- Circulation
- Locomotion
- Reproduction
- Respiration
- Secretion
- Sensation

Growth is the process of an organism increasing in size by adding cells, by the cells getting larger, or by replacing cells. With young organisms, growth is primarily adding cells to increase size. With mature organisms, growing new cells is needed to replace those that are lost. Growth requires food with essential nutrients.

Food is any substance that nourishes an organism. Animals get their food by swallowing, otherwise known as ingestion. Once ingested, digestion, which is the process that changes food into forms that can be absorbed by the animal. Plants make their food by the process of photosynthesis.

Circulation is the movement of substances within an organism. Food nutrients, digested food, and other substances are involved in circulation. Animals have a circulatory system that involves the heart, and blood vessels.

Respiration is the process by which an organism provides its cells with food and oxygen. Oxygen and carbon dioxide are the two primary substances important in respiration. Animals use structures, such as lungs and gills, for respiration.

Secretion is the production within an organism of substances needed for life processes. Specialized body structures, known as glands, are often responsible for secretion.

Sensation is the ability of an organism to feel or otherwise gather information and respond to its environment. Animal's sensation is easier to understand than plants. Animals typically have touch, taste, sight, smell, and hearing. All organisms have some type of movement. Internal movements are needed for an organism to carry out life processes.

Locomotion is the movement of an animal from one place to another.

Reproduction is the process by which organisms give rise to new organisms of the same species. This is not a life process that is essential for living; however it is essential for a species to continue itself. Reproduction may involve sexual or asexual reproduction.

3. THE CELL - A BRIEF INTRODUCTION: STRUCTURE OF PROKARYOTIC AND EUKARYOTIC CELLS, CELL MEMBRANE SYSTEM AND CELL WALL: CELL MEMBRANE AND ITS ORGANIZATION; ELEMENTARY IDEA OF CELLULAR CONSTITUENTS: NUCLEUS, MITOCHONDRIA, GOLGI BODIES, ENDOPLASMIC RETICULUM, LYSOSOMES AND MICROBODIES; BACTERIAL AND PLANT CELL WALLS.

Introduction

The basic structural and functional unit of cellular organisms is the cell. It is an aqueous compartment bound by a cell membrane that is capable of independent existence and performing the essential functions of life. All organisms are more complex than viruses consist of cells. Viruses are non-cellular organisms because they lack cell and cell-like structures. In the year 1665, Robert Hooke first discovered cells in a piece of cork and also coined the word cell. The word cell is derived from the Latin word cellula, which means a small compartment. Hooke only observed cell walls because cork cells are dead and without cytoplasmic contents. Anton van Leeuwenhoek was the first person who observed living cells under the microscope and named them animalcules, meaning little animals. On the basis of the internal architecture, all cells can be subdivided into two major classes, prokaryotic cells and eukaryotic cells.

Cells that have unit membrane-bound nuclei are called eukaryotic, whereas cells that lack a membrane-bound nucleus are prokaryotic. Eukaryotic cells have much more complex intracellular organization with internal membranes as compared to prokaryotic cells. Besides the nucleus, eukaryotic cells have other membrane-bound organelles like the endoplasmic reticulum, mitochondria, Golgi complex, lysosomes, microbodies and vacuoles. The region of the cell lying between the plasma membrane and the nucleus is the cytoplasm, comprising the cytosol and the organelles. The prokaryotic cells lack such unit membrane-bound organelles. The cell theory states that all organisms are composed of similar units of organization called cells.

There are 3 major parts of the cell theory:

1. All organisms are made of cells.
2. All existing cells are produced by other living cells.
3. Cell is the most basic unit of life.

Contributors

There are **5** contributors to tell about cell theory.

1. Robert Hooke
2. Anton van Leeuwenhoek
3. Matthias Schleiden
4. Theodor Schwann
5. Rudolf Virchow

Each of the above contributors produced evidence through investigations and experiments that led to today's cell theory. Before the cell theory, people believed in spontaneous generation which means living things come from non-living things. Two scientists disproved spontaneous generation. Francisco Redi proved that macroorganisms do not spontaneously generate. People thought that maggots came from meat. Redi was disproved this in a controlled experiment.

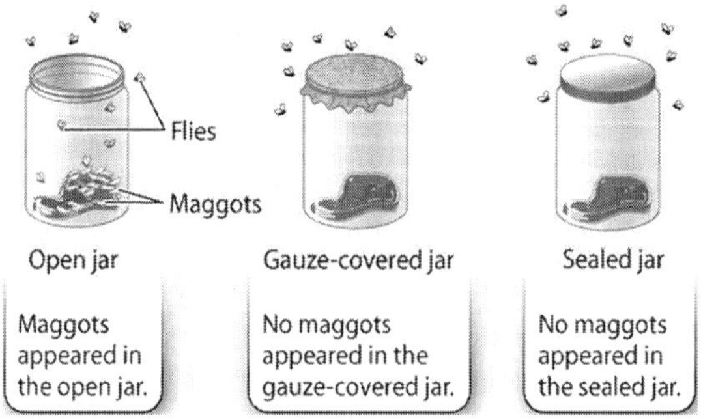

Pasture was proved that microoganisms do not spontaneously generate.

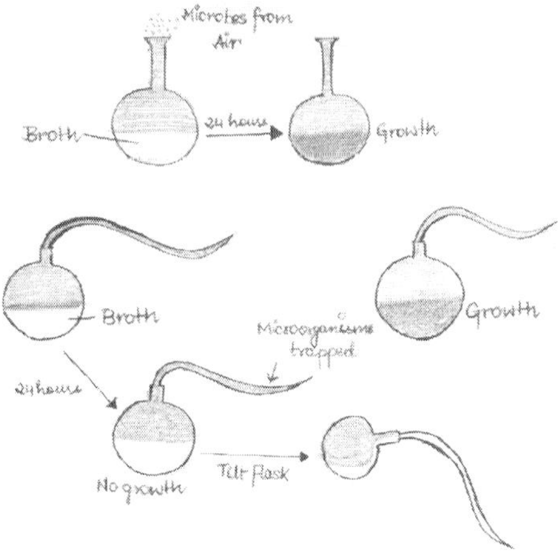

Theodor Schwann and Matthias Schleiden was credited with coming up with the cell theory.

Schwann was discovered animal cells.

Schleiden was discovered plant cells. The cell theory would not exist if it wasn't for the development of the microscope and the work of others.

Janssen was invented the first compound microscope.

Robert Hooke is an English scientist, improved Janssen's invention in 1663. He used the compound microscope to observe cork. Hooke was observed that cork is composed of small and hollow compartments. Cells was investigated through cork experimenting with the compound microscope and originated with the name cells. Hooke was discovered the cell in 1665, which started formulating the cell theory.

Anton van Leeuwenhoek was study new approaches for creating lenses to observe cloth. Leeuwenhoek's microscope was more powerful than Hooke's compound microscope. From investigating and experimenting with his microscope, Leeuwenhoek became one of the first scientists to refer to living cells when he observed an abundant number of single-celled organisms, which he called animalcules (plant & animal), swimming in a drop of pond water.

Matthias Schleiden (a German Scientist) was involved with plant cells, Schleiden used the compound microscope and studied about plant cells. He was discussed what he observed with his friend Theodor Schwann from investigating and experimenting on plant parts are made of cells.

Theodor Schwann: He was studied about plant & animal cells, and fascinated by the similarities between the plant and animal cells. Schwann was able to determine that all animals are made of cells. Schwann published the 1st statement of the cell theory: all living things are made of cells and cell products.

Rudolf Virchow was investigations and experiments and he stated that all cells come from preexisting cells, which is the 2nd part of the cell theory: all existing cells are produced by other living cells. After Virchow's completed his work, the cell theory was finally formulated in 1838.

Types of cells

Prokaryotic cells are mainly single-celled organisms. Prokaryotes have no membrane-bound organelles. Example – Bacteria cell

Eukaryotic cells are mainly multicellular organisms and although some are single-celled (amoeba, algae and yeast). Eukaryotes have contain membrane-bound organelles.

Prokaryotic cells are single-celled entities that are primitive in structure and function as they lack a membrane-bound nucleus and other organelles. The term "prokaryote" is derived from two Greek words, 'pro' meaning 'before' and 'karyon' meaning 'nucleus'. Prokaryotes are considered to be the first living organisms of the earth as they are the simplest form of life. Single-celled. Reproduce by binary fission (another copy by dividing). No cell nucleus or any other membrane-bound organelles. DNA travels openly around the cell. All bacteria are prokaryotes. Eukaryotic cells are mainly multicellular organisms are dervived from two Greek words **Eu = "true"; karyon = "nucleus".** Most organisms that we can see, such as trees, grass, worms, flies, mice, humans, mushrooms and yeast are eukaryotes. It can either be single-celled or multi-celled. It can reproduce in one of several ways (eg. meiosis and mitosis). It has a cell nucleus within containing its DNA. Nucleus is the most evident difference between these cell types. A plasma membrane: an outer covering that separates the cell's interior from its surrounding environment. Cytoplasm: a jelly-like cytosol within the cell in which other cellular components are found

- DNA: the genetic material of the cell.
- Ribosomes: where protein synthesis occurs.

However, prokaryotes differ from eukaryotic cells in several ways. A prokaryote is a simple, single-celled (unicellular) organism that lacks an organized nucleus or any other membrane-bound organelle. Prokaryotic DNA is found in a central part of the cell: the nucleoid. Most prokaryotes have a peptidoglycan cell wall and many have a polysaccharide capsule. The cell wall acts as an extra layer of protection, helps the cell maintain its shape and prevents dehydration. The capsule enables the cell to attach to surfaces in its environment. Some prokaryotes have flagella, pili, or fimbriae. Flagella are used for locomotion. Pili are used to exchange genetic material during a type of reproduction called conjugation. Fimbriae are used by bacteria to attach to a host cell.

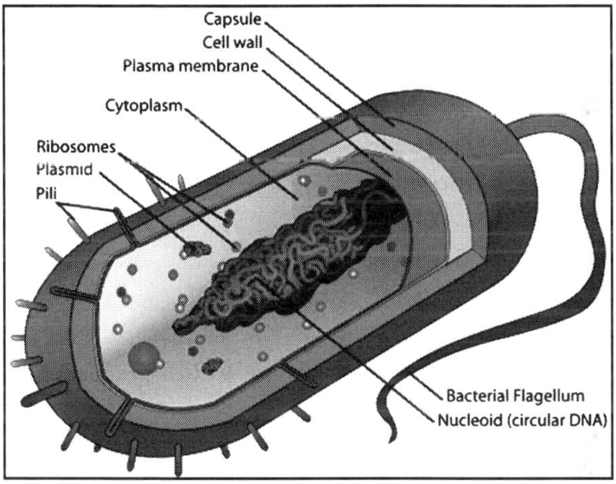

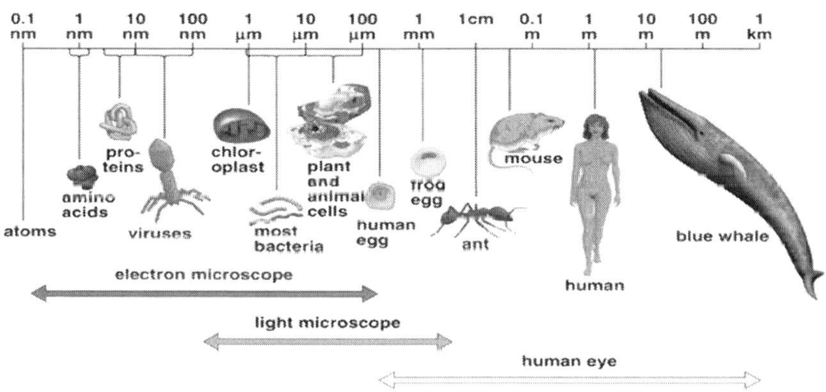

Prokaryote

Nucleoid

The region of cytoplasm where prokaryote's genome (DNA) is located. Usually a singular and circular chromosome.

Plasmid

A small extra piece of chromosome or genetic material. It has 5 - 100 genes. It can provide genetic information to promote.

- Antibiotic resistance
- Virulence factors

The molecules produced by a pathogen that specifically influencehost's function to allow the pathogen to thrive. It promote conjugation (transfer of genetic material between bacteria through cell- to-cell contact).

Cytoplasm

It is also known as protoplasm. It is gel-like matrix of water, enzymes, nutrients, wastes, and gases and contains cell structures. It has also promote the location of growth, metabolism, and replication.

Granules

The way of storing nutrients by bacteria and staining of some granules aids in identification.

Cytoskeleton

Cellular "scaffolding" or "skeleton" within the cytoplasm.

Ribosomes

It is found within the cytoplasm or attached to the plasma membrane. It is made of protein & rRNA. It is composed of two subunits.The cell may contain thousands of ribosomes.

Prokaryotes plasma membrane

It is a outer membrane of cell which protect from environement . Phospholipid molecules are oriented so that hydrophilic water-loving heads are directed outward and hydrophobic water-hating tails are directed inward. Proteins are embedded in two layers of lipids (lipid bilayer). The membrane is semi-permeable.

Plasma membrane as a barrier

Osmosis

The diffusion of water across a semi-permeable membrane. The environment surrounding cells may contain amounts of dissolved substances (solutes).

Tonicity and osmosis

Isotonic: It is equal concentration of a solute inside and outside of the cell. Hypertonic: A higher concentration of solute. Hypotonic: A lower concentration of solute. Water will always move toward a hypertonic environment.

Plasma Membrane As A Barrier

Active transport

The molecules are move across the plasma membrane. Analogous to pump moving water uphill. Active transport are classified by type of energy used to drive molecules across membranes. ATP Driven active transport. It build up energy from adenosine triphosphate (ATP) drives substances across the plasma membrane with the aid of carrier molecules.

Cell wall

The peptidoglycan inwards all bacteria are very similar. It is going further out, the bacterial world divides into two major classes gram positive and gram negative.

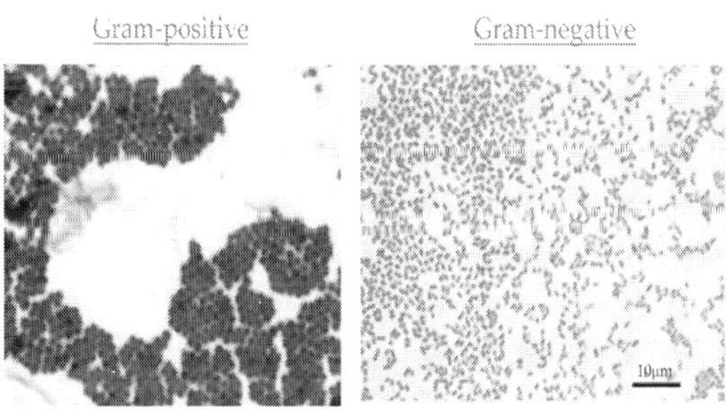

Bacterial Cell Wall

Peptidoglycan is a huge polymer of interlocking chains of alternating monomers. It provides rigid support while freely permeable to solutes. The backbone of peptidoglycan molecule composed of two amino sugar derivatives of glucose. The "glycan" part of peptidoglycan.

- N-acetylglucosamine (NAG)
- N-acetlymuramic acid (NAM)

NAG / NAM strands are connected by interlocking peptide bridges. The "peptide" part of peptidoglycan; see in figure 3.1.

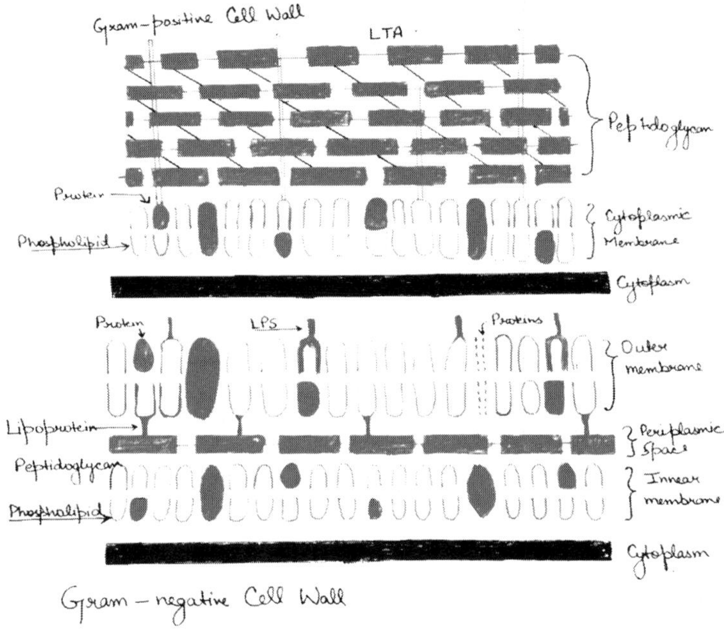

Figure 3.1 Prokaryotes - Cell wall

Glycocalyx

Some bacteria have an additional layer outside of the cell wall called the glycocalyx. This additional layer can come in one of two forms:

1. Slime layer

Glycoproteins are loosely associated with the cell wall. The slime layer causes bacteria to adhere to solid surfaces and helps prevent the cell from drying out. The slime layer of Gram+ *Streptococcus mutans* allows it to accumulate on tooth enamel (yuck mouth and one of the causes of cavities). Other bacteria in the mouth become trapped in the slime and form a biofilm & eventually a buildup of plaque.

2. Capsule

Polysaccharides are firmly attached to the cell wall. The capsules adhere to solid surfaces and to nutrients in the environment. Adhesive power of capsules is a major factor in the initiation of some bacterial diseases. The capsule also protects bacteria from being phagocytized by cells of the host's immune system.

Endospores

It is dormant, tough, non-reproductive structure produced by a small number of bacteria. It hase resistant to radiation, desiccation, lysozyme, temperature, starvation, and chemical disinfectants. Endospores are commonly found in soil and water, where they may survive for very long periods of time.

Surface appendages

Some prokaryotes have distinct appendages that allow them to move about or adhere to solid surfaces. It consists of delicate strands of proteins. Flagella: It is long, thin extensions that allow some bacteria to move about freely in aqueous environments. Axial filament (endoflagella): It has wind around bacteria, causing movement in waves. Fimbriae: The gram-negative bacteria have these short, fine appendages surrounding the cell and gram positive bacteria don't have. It has no role in motility. It helps to bacteria adhere to solid surfaces. The Major factor in virulence. Pili: Tubes that are longer than fimbriae, usually shorter than flagella and use for movement, like grappling hooks, and also use conjugation pili to transfer plasmids.

Cell Shapes

Most bacteria are classified according to shape:

- Bacillus (bacilli) is a rod-shaped
- Coccus (cocci – sounds like cox-eye) is a spherical
- Spiral shaped

- Spirillum (spirilla) is a spiral with a rigid cell wall and flagella.
- Spirochete (spirochetes) is a spiral with flexible cell wall and axial filament.

There are many more shapes

- Coccobacilli is elongated coccal form
- Filamentous is bacilli that occur in long threads
- Vibrios is short, slightly curved rods
- Fusiform is bacilli with tapered ends

Eukaryotic cells

The genome is made of DNA and it may include several to many linear chromosomes within a membrane-bound nucleus. Cytoplasm is a matrix in which the organelles are suspended. Inside the plasma membrane, the nucleus is surrounded by cytoplasm. A water-like substance that fills cells. The cytosol is made up of water, salts, organic molecules and many enzymes that catalyze reactions. Eukaryotic cells have many organelles. Prokaryotes only have ribosomes, which are not bound by a membrane. Membrane-bound eukaryotic organelles organize functions within the cell. The system of internal membranes within eukaryotic cells that divide the cell into compartments, or organelles. The moving molecules through the interior of a cell, as well as interactive surfaces for lipid and protein synthesis. The membranes of the endomembrane system are made of a lipid bilayer, with proteins.

Nucleus

It is control center and separates the genetic material (DNA) from the rest of the cell. DNA is the genetic material, is a blueprint, or code for making proteins. The plasma membrane and cell membrane is the flexible boundary of a cell and separates a cell from its surroundings. It allows nutrients to enter the cell and waste to be removed. This is referred to as selective permeability. Keeping a healthy balance of nutrients and water within the cell is called homeostasis. The largest organelle in the cell and it is the most inner compartment of the cell contains chromatin (DNA); genetic information on strands called chromosomes "control centre" for cell metabolism and reproduction.

Nucleolus

It is found inside nucleus and ribosomes are also made in nucleolus.

Ribosomes

It make proteins and made from RNA also called as "factories".

Cytoplasm

Clear gel-like fluid inside the cell, in which all the organelles are present inside the cytoplasm.

Endoplasmic reticulum

It is extensive network of membranes. It contain **rough ER:** with ribosomes and **Smooth ER:** with no visible ribosomes.

Golgi apparatus

It is sorts of protein made by the ribosomes and sends them to needed places in the cell.

Lysosomes

Organelles that are filled with digestive enzymes to remove waste and bacteria.

Mitochondria

It is also called "powerhouse" of the cell which release energy for the cell. It converts the energy stored in glucose into ATP for the cell.

Vacuoles

It is fluid-filled organelles which enclosed by a membrane and store materials such as food, sugar, water, and waste products.

Eukaryotic plant cell

Plant cells are also eukaryotic cells, but plant cells contain some organelles that are not found in animal cells.

Plant cell organelles

Cell wall

It contains a rigid wall outside the plasma membrane. It provides the cell with extra support.

Chloroplasts

It captures light and energy and converts it into chemical energy.

Chlorophyll

It has green pigment found inside the chloroplast which have major role in photosynthesis process and also very important role to make food.

Plastids

Plastid has major role to store things such as food in the plant cell.

4. BIOMOLECULES: CARBON AND ITS BRIEF CHEMISTRY, GENERAL PROPERTIES OF BIOMOLECULES, WATER: STRUCTURE AND UNIQUE PROPERTIES, ACID, BASE, BUFFERS, POLYPROTIC ACIDS.

Introduction

Carbon is the central element. All biomolecules contain a carbon chain or ring. Carbon has 4 outer shell electrons and valency is 4.Therefore it is bonding capacity is great. It forms covalent bonds hence, has strong bonds. Once bound to other elements (or to other Carbons), it is very stable. The unique ability to combine in many ways with itself and other elements, carbon has a central role in the chemistry of living organisms. Carbon atoms can form single, double or triple bonds with other carbon atoms. Carbon can form up to 4 bonds. This allows carbon atoms to form long chains, almost unlimited in length. Carbon can bond with other carbons, form straight chains, branched chains and rings.

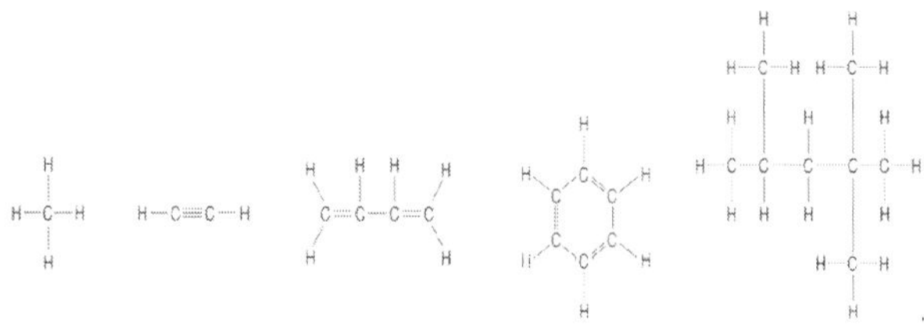

Chemistry of Carbon

It must contain at least one carbon. CH_4 is the simplest organic molecule. Carbon has 4 valence electrons. Therefore, carbon will always make 4 bonds with other atoms. They have ability to form millions of different compounds with other elements.

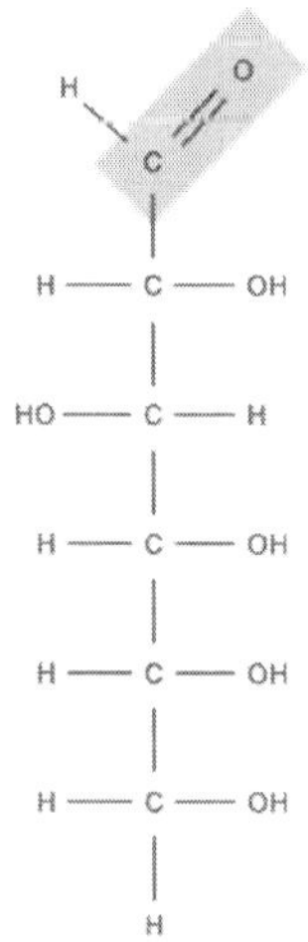

Glucose

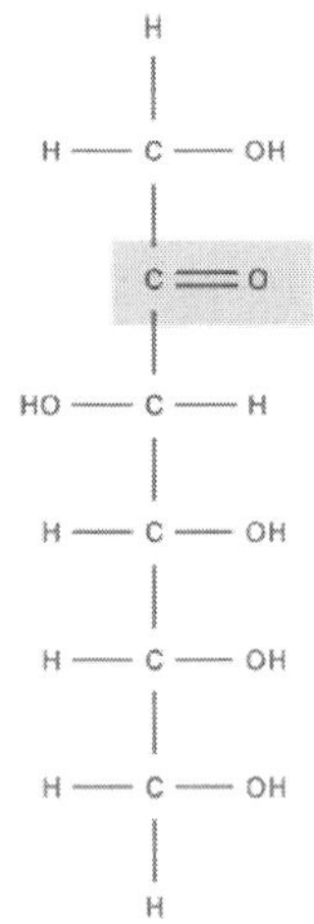

Fructose

Diamond, graphite, fullerenes and nanotubes are four forms of the element carbon. Diamond is a crystalline form of carbon in which each carbon atom is bonded strongly to four other carbon atoms. It is formed from high temperature and pressure. The melting point is more than 350° C. It can be made artificially and isused in industry as cutting tools.

Graphite: each carbon atom is bonded tightly to three other carbon atoms in flat layers. In this bonds are very weak. "Lead" in pencils is mostly graphite. It is used as a lubricant in machines.

Fullerenes consists of carbon atoms arranged in the shape of a hollow sphere. It is called buckyballs after an architect.

Nanotube is a carbon atoms which are arranged in the shape of a long hollow cylinder. Tiny, light, flexible and extremely strong. It is good conductors of heat and electricity.

Organic compounds are compounds that contain carbon. Many organic compounds have similar properties in terms of melting and boiling points, odor, electrical conductivity and solubility. Many organic compounds are gases at room temperature. It has a strong odor. It does not dissolve in water.

Hydrocarbons are compounds that contains only the elements carbon and hydrogen. Hydrocarbons mix poorly with water. All hydrocarbons are flammable; CH4 (methane), C_2H_6 (ethane), C_3H_8 (propane).

The carbon chains in the hydrocarbon may be straight, branched or ring-shaped. Structural formula shows the kind, number and arrangement of atoms in a molecule. Isomers are compounds that have the same chemical formula but different structural formulas whichmake them have different properties. Saturated hydrocarbons are only single bonds, has a maximum number hydrogen atoms attached. Unsaturated hydrocarbons has double or triple bonds, have fewer hydrogen than saturated hydrocarbons. Substituted hydrocarbon atoms of other elements replace one or more hydrogen atoms in a hydrocarbon. It is include halogen-containing compounds, alcohols, and organic compounds. Alcohol is a substituted hydrocarbon that contains one or more hydroxyl groups.

- Hydroxyl group –OH.
- It is dissolve well in water.
- They have higher boiling points than others.
- Hydrocarbons with similar numbers of carbon.

Life with Carbon

There are 4 classes of organic compounds required by all living things called macromolecules. Foods provide these organic compounds which cells of living things use, change, and store. These 4 classes are nutrients-substances that provide the energy and raw materials the body needs to grow, repair worn parts, and function properly. Macromolecule (polymer) is made by joining many monomers (single unit). Polymerization is a chemical reaction which joins monomers to make polymers.

The four main Classes Of Biological Molecules

1. Carbohydrates (such as sugar, starches and cellulose)
2. Lipids (such as wax, fats, oils and steroids)
3. Proteins (such as muscle, hair, hormones and enzymes)
4. Nucleic acids (DNA and RNA)

Carbohydrate is energy-rich organic compound made of the elements carbon, hydrogen and oxygen.

Simple carbohydrate is the simplest carbohydrate are sugars (glucose is in your body – $C_6H_{12}O_6$).

Complex carbohydrate is a polymer made of smaller molecules that are simple carbohydrate bonded to one another.

- It contains C, H, and O in a 1:2:1 ratio. Most end with "ose". An animal's main energy source. Carbohydrates are burned first in the body.
- Monosaccharides ($C_6H_{12}O_6$) such as Glucose, fructose and galactose
- Disaccharides such as Sucrose, lactose and maltose
- Polysaccharides are complex carbohydrates.

A) Glycogen (carbohydrate is stored in animal liver)
B) Starch (Stored in plants)
C) Cellulose (cell walls, cotton) "roughage"
D) Chitin (exoskeletons of arthropods)

Protein

Proteins are formed from smaller molecules called amino acids. Amino acid is a monomer that is a building block of proteins. Each amino acid molecule has a carboxyl group (–COOH) and an amino group ($–NH_3$). The body uses proteins from food to build and repair body parts and to regulate cell activities. Proteins are monomer and made from amino acids. It is essential to the structures and activities of life. It contain C, H, O, N (S, P). 50% of your dry weight (eg. groups of proteins).

1 Enzymes (amylase, sucrase, maltase, lactase)
2. Structural (collagen, elastin)
3. Contractile (actin, myosin)

4. Transport (hemoglobin, protein channels)
5. Hormones (insulin)

Amino acid

The amino acid has an amino group (-NH_2) and a carboxyl group (COOH). An R group, which distinguishes each of the 20 different amino acids. Each amino acid has specific properties based on the R-group. Peptide bonds link amino acids together to form polypeptide (protein).

Lipids

Lipids are energy rich compounds made of carbon, oxygen and hydrogen. Lipids are include fats, oils, waxes and cholesterol. Lipids are release twice as much energy in your body as do carbohydrates.

- Fatty acids are organic compound that is a monomer of a fat or oil.
- Cholesterol is a waxy lipid in animal cells.

It is mostly C and H atoms are linked by nonpolar covalent bonds. It is energy storage molecules (burned after carbohydrates are gone). It is insoluble in water (polar). It is soluble in nonpolar solvents (ether). It has more energy in lipids than in carbohydrate such as 9 cal/g lipid and 4 cal/g for carbohydrate.

Examples: triglycerides, phospholipids, steroids (cholesterol), waxes, oils, fats

Triglyceride has 3 fatty acids + 1 glycerol. Saturated Fats are all single bonds in the chain.

- Solid at room temp (ex: butter, lard).
- Unsaturated fats: one or more C=C bond in the chain.
- Liquid at room temp (ex: all oils).

Nucleic acids

Nucleic acids are very large organic molecules made up of carbon, oxygen, hydrogen, nitrogen and phosphorus. There are two types such as DNA and RNA. There are elements that make up all living things.

- C – Carbon

- H – Hydrogen
- N – Nitrogen
- Oxygen
- P – Phosphorus
- S – Sulfur

Nucleic acids (DNA and RNA) are store and transmit genetic information.

- DNA = Deoxyribonucleic acid.
- RNA = Ribonucleic acid.

The macromolecules are containing C, H, O, N, P. One nucleotide has 5-carbon sugar, phosphate (PO_{4-}), nitrogenous base. The sugars and phosphates are the backbone of the nucleic acid.

Vitamins

Vitamins are organic compounds that serve as helper molecules in a variety of chemical reactions in your body.

Minerals are elements in the form of ions in your body. Water makes up most of your body's fluids.

Water Structure

Water is essential for life. The most abundant substance of the living cell is water. It accounts for 70% or more of a cell's weight. It provides the medium in which most substances remain dissolved. Water is colorless, odorless and tasteless in nature. Water is formed when two hydrogen atoms bond to one oxygen atom. It is not symmetrical. Electrons spend more time near the (O) and less time near the (H); see figure **4.1.** Water molecule is polar.

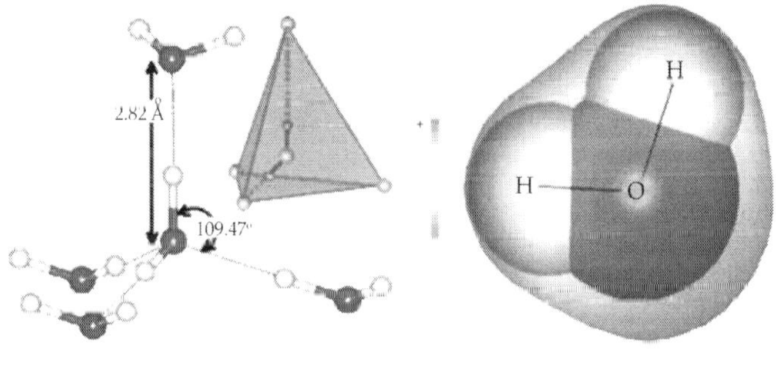

Figure 4.1 Water molecules are held together by a hydrogen bond.

There is a greater concentration of electrons around the nucleus of the oxygen than around the hydrogen. Therefore, the hydrogen end is slightly positive and the oxygen end is slightly negative.

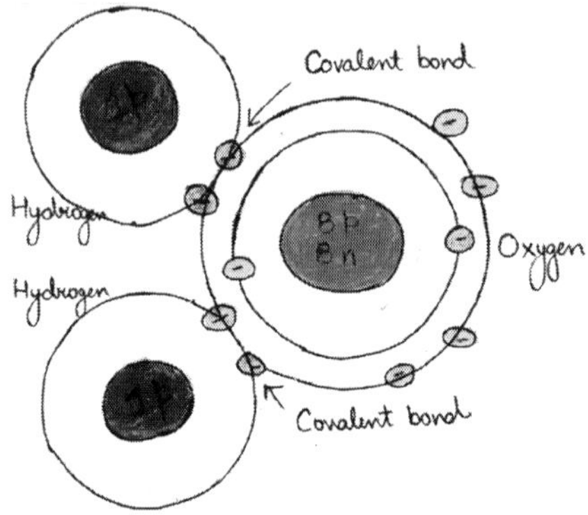

Bohr Model of H_2O

It is made up of two hydrogen atoms and one oxygen atom, with a total atomic mass of 18 daltons. The H-O-H bond angle is 104.5 which is smaller than the typical tetrahedral angle of 109. Since oxygen is more electronegative than hydrogen, the oxygen nucleus attracts electrons more strongly than does the hydrogen nucleus. It means that the shared electrons are more shifted towards the oxygen atom than of the hydrogen. The result of this unequal electron sharing is the formation of two electric dipoles in the water molecule, one along each of the H-O bonds. Each hydrogen atom bears a partial positive (+) and the oxygen atom carries a partial negative charge equal in magnitude to the sum of the two partial positives(-). When water molecules attract to one another, the hydrogen end of one molecule is attracted to the oxygen end of another. Water is the only substance found on earth in all three states (phases):

- Liquid
- Solid (Ice)
- Gas (Steam or vapor)

Physical States

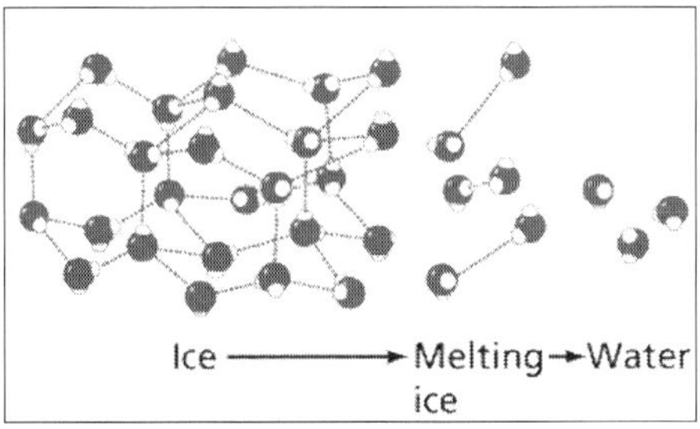

Water molecules are constantly moving. Temperature is increase so increase in movement. When water molecules move faster, they tend to break their hydrogen bonds. This is called evaporation. When gas or vapor molecules slow down, they clump or join together. This is called condensation. As water becomes cooler, it becomes less dense: 1 gram per cubic cm at 39.2° 0.95865 gram per cubic cm at 212°. As water changes from a liquid to a solid, molecules form crystals. In ice crystals, molecules are spaced further apart. Since molecules are spaced further apart, ice is less dense than water (it floats). Helpful for aquatic organisms; forms a "blanket" If ice was denser than water, lakes would freeze from the bottom upwards.

Thermal properties of water

Water has a high specific heat, or heat capacity (the amount of heat needed to raise 1g of a substance 1°C.). This quality of water is important for several reasons. It regulates the rate at which air temperature changes. It affects the global climate (milder temperatures in coastal regions). So, aquatic organisms face lesser temperature variation than terrestrial organisms. Therefore, water is used in cooling systems found in cars and industrial plants.

Surface tension

Water has the highest surface tension of all liquids. This is a result of the tendency of water molecules to attract to one another, or cohere, at the surface of any accumulation of water. The surface tension allows insects to walk on water. Interaction between hydrogen bonding and the earth's gravitational pull. **Cohesion** is the attraction between water molecules to each other through hydrogen bonds (H). **Adhesion** is the attraction of water molecules to another substance. **Capillarity** is the movement of water within the spaces of a porous material due to the forces of surface tension, adhesion, and cohesion.

pH: Alkalinity/Acidity

The measurement of the H^+ ions found in that particular substance. The scale goes from 0 to 14. When the pH is 7 ; it is neutral. Below pH 7 is acidic. Above pH 7 is alkaline (or basic). One pH unit represents a ten-fold change in H^+ concentration. Since water can dissolve more things than any other natural substance, it is known as the " Universal solvent"

Properties of Solvents

Interacts with other polar compounds. It is repelled by non-polar compounds. The small size allows it to saturate areas. It can convey other substances in solutions. The water is especially good at dissolving salts. The salts formed from the combination of particles with opposite electrical charges (or ions) such ex $\mathbf{Na^+ + Cl^- = NaCl}$. When salt is placed in water, the strongly charged salt ions attract to the weaker charged water molecules. Water molecules are surround each ion. Salt crystals are fall apart, or dissociates, and the salt dissolves.

Ionization of water

Although neutral water has a tendency to ionize.

$H_2O <-> H^+ + OH^-$

The free proton is associated with a water molecule to form the hydronium ion H_3O^+. High ionic mobility is due to proton jumping. Proton is jumping large proton and hydroxide mobility

$H_3O+ : 362.4 \times 10^{-5}\ cm^2{\bullet}V^{-1}{\bullet}s^{-1}$

$Na^+ : \quad 51.9 \times 10^{-5}\ cm^2{\bullet}V^{-1}{\bullet}s^{-1}$

Hydronium ion is migration hops by switching partners at 10^{12} per second.

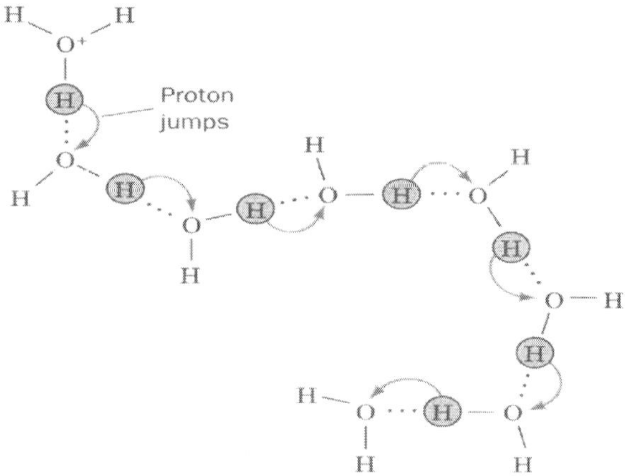

Equilibrium expression

- Described by:

$$K = \frac{[H^+][OH^-]}{[H_2O]}$$

- Where K is the dissociation constant
- Considering $[H_2O]$ constant yields

$$K_w = [H^+][OH^-]$$

$$K_w$$

$$K_w = [H^+][OH^-]$$

- Where K_w is the ionization constant of water
- For pure water ionization constant is

$$10^{-14}\ M^2 \text{ at } 25^o$$

- For pure water

$$[H^+] = [OH^-] = (K_w)^{1/2} = 10^{-7}\ M$$

Acids and bases

- For pure water (neutral)

  $[H^+] = [OH^-] = (K_w)^{1/2} = 10^{-7}$ M
- Acidic if $[H^+] > 10^{-7}$ M
- Basic if $[H^+] < 10^{-7}$ M
- Acid is a substance that can donate a proton.
- Base is a substance that can accept a proton.

$HA + H_2O$ $H_3O^+ + A^-/OH^-$

Acid Base Conjugate Acid Conjugate Base

or

HA $A^- + H^+$

Acid Conjugate Base Conjugate Acid

- If you establish equilibrium, changes in $[H^+]$ will shift the ratio of HA and A^-
- By adding more H^+, A^- will be consumed forming HA.
- If there is sufficient $[A^-]$, the extra H^+ will also be consumed and the $[H^+]$ will not change.

Buffer

A buffer can resist pH changes if the pH is at or near a weak acid pK value. **Buffer range:** the pH range where maximum resistance to pH change occurs when adding acid or base. It is $^+$1 pH from the weak acid pK. If pK is 4.8 the buffering range is 3.8 5.8.

Blood buffering system

Bicarbonate is most significant buffer. It is formed from gaseous CO_2.

$CO_2 + H_2O <-> H_2CO_3$

$H_2CO_3 <-> H^+ + HCO_3^-$

Normal value of blood is pH 7.4.

5. CARBOHYDRATES: MONOSACCHARIDES: CLASSIFICATION, STRUCTURE, CONFIGURATION, PROPERTIES AND DERIVATIVES, COMMON DISACCHARIDES, STRUCTURE AND FUNCTION, OCCURRENCE OF STORAGE AND STRUCTURAL POLYSACCHARIDES, GLUCOSAMINOGLYCANS, GLYCOPROTEINS: STRUCTURE AND FUNCTION.

Introduction

It is polyhydroxy aldehydes or polyhydroxy ketones or compounds that can be hydrolyzed to them. It makes up the majority of carbohydrates, H and O are present in the same ratio as in water hence also called as hydrates of carbon. Carbohydrates are polyhydroxyaldehydes or polyhydroxy-ketones of formula $(CH_2O)_n$, or compounds that can be hydrolyzed to them (aldehyde sugars or saccharides). It is classified into the following classes depending upon whether these undergo hydrolysis and if so on the number of products formed.

Monosaccharides

It consists of a single polyhydroxy aldehyde or ketone unit. It is simple sugars and they have the general formula $C_nH_{2n}O_n$. It is freely soluble in water but insoluble in non-polar solvents. Carbohydrates that cannot be hydrolyzed to simpler carbohydrates such as glucose or fructose.

Disaccharides

It is the simplest and most common oligosaccharides. A disaccharide consists of two monosaccharides joined by an O-glycosidic bond. Carbohydrates that can be hydrolyzed into two monosaccharide units such as sucrose, which is hydrolyzed into glucose and fructose.

Oligosaccharides

Oligosaccharides are containing three or more residues are relatively rare and occurring almost entirely in plants. Carbohydrates can be hydrolyzed into a few monosaccharide units.

Polysaccharides

It is ubiquitous in nature. It is also called glycans. They are two types such as structural and storage polysaccharide. Structural polysaccharide provides mechanical stability to cells, organs and organism.

Storage polysachharide serve as carbohydrate stores that release monosaccharide as required. Carbohydrates that are polymeric sugars ex Starch or cellulose. Triose, tetrose, pentose, hexose, etc. carbohydrates that contain three, four, five, six, etc. Hydrates of carbon is general formula $C_n(H_2O)_n$.

In plants: photosynthesis reaction has occured $6\ CO_2 + 6\ H_2O = C_6H_{12}O_6 + 6\ O_2$

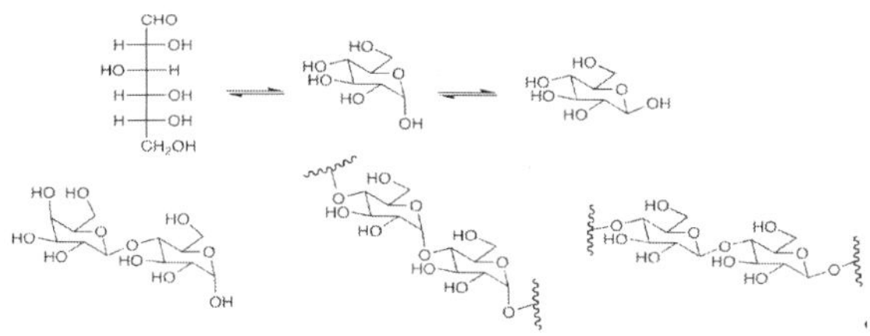

Structure of carbohydrate

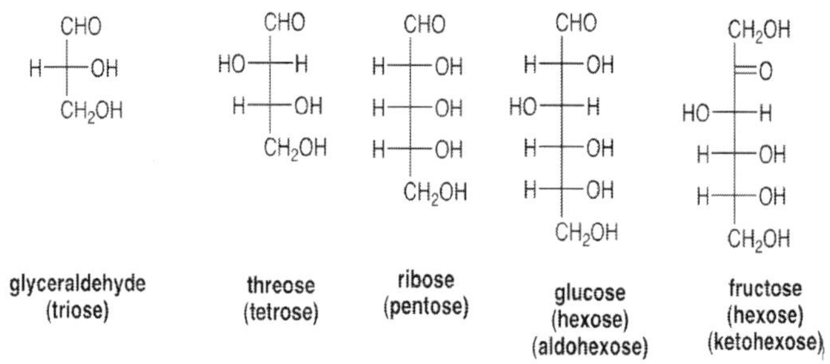

Structure of Polysaccharides

***Polymers*:** The large molecules are made up of repeating smaller units which is called monomer. Aldose: Polyhydroxyaldehyde example glucose. Ketose: Polyhydroxyketone example fructose. Five or six carbon eg. Aldohexose and ketopentose Reducing sugar: It is a carbohydrate that is oxidized by Tollen's, Fehling's or Benedict's solution.

Tollen's: $Ag^+ \rightarrow Ag$ (silver mirror)

Fehling's or Benedict's: Cu^{3+} (blue) $\rightarrow$ Cu^{2+} (red).

These are reactions of aldehydes and alpha-hydroxyketones. **All monosaccharides** (both aldoses and ketoses) and **most disaccharides** are reducing sugars. Sucrose (table sugar) is a disaccharide, is not a reducing sugar.

Glucose (a monosaccharide)

Plants:

photosynthesis

$$6\,CO_2 + 6\,H_2O \xrightarrow[\text{sunlight}]{\text{chlorophyll}} \underset{\text{(+)-glucose}}{C_6H_{12}O_6} + 6\,O_2$$

$$\text{(+)-glucose} \rightleftharpoons \text{starch or cellulose}$$

respiration

$$C_6H_{12}O_6 + 6\,O_2 \longrightarrow 6\,CO_2 + 6\,H_2O + \text{energy}$$

Glucose in plants

Animals

$$\text{plant starch} \longrightarrow \text{(+)-glucose}$$

$$\text{(+)-glucose} \longrightarrow \text{glycogen}$$

$$\text{glycogen} \longrightarrow \text{(+)-glucose}$$

$$\text{(+)-glucose} \longrightarrow \text{fats or aminoacids}$$

respiration

$$\text{(+)-glucose} + 6\,O_2 \longrightarrow 6\,CO_2 + 6\,H_2O + \text{energy}$$

Glucose in animals

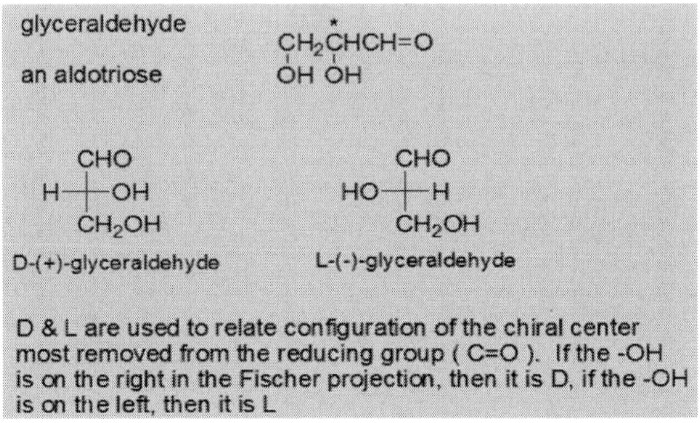

Fischer projections and the D, L Notation

Representation of a three-dimensional molecule as a flat structure. Tetrahedral carbon represented by two crossed lines.

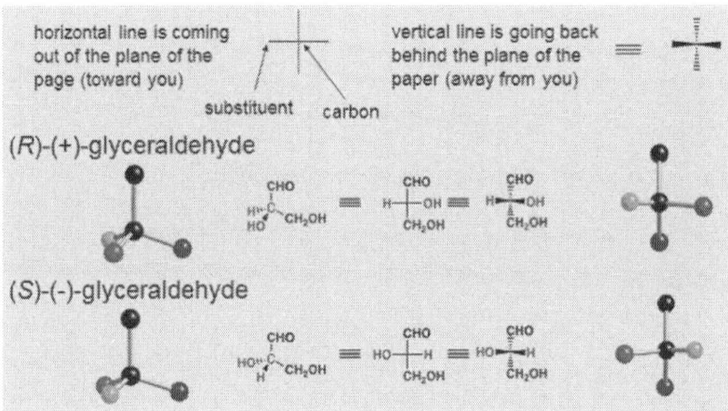

The absolute configuration of glyceraldehyde is specified by the RS system in Fisher projection formulas.

Before the *R*/*S* convention, stereochemistry was related to (+)-glyceraldehyde.

D-glyceraldehyde	**L-glyceraldehyde**
***R*-(+)-glyceraldhyde**	***S*-(-)-glyceraldhyde**
(+)-rotation = dextrorotatory = d	**(-)-rotation = levorotatory = l**

D-carbohydrates have the -OH group of the highest numbered chiral carbon pointing to the right in the Fischer projection as in R-(+)-glyceraldehyde. **For carbohydrates**, the convention is to arrange the Fischer projection with the carbonyl group at the top for aldoses and closest to the top for ketoses. The carbons are numbered from top to bottom. Carbohydrates are designated as D or L according to the stereochemistry of the highest numbered chiral carbon of the Fischer projection. If the hydroxyl group of the highest numbered chiral carbon is pointing to the right, the sugar is designated as D (*Dextro*: Latin for *on the right side*). If the hydroxyl group is pointing to the left, the sugar is designated as **L** (*Levo*: Latin for on *the left side*). Most naturally occurring carbohydrates are of the D-configuration. Chiral molecules such as glucose and arabinose can exit in two configurations that are non-superimposable mirror images of each other. These two configurations are called enantiomers.

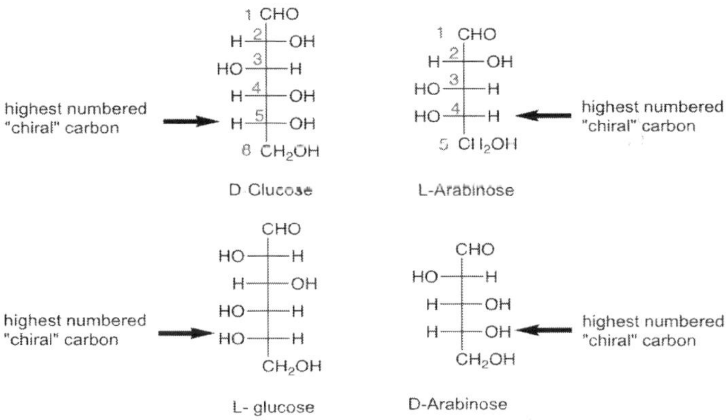

Enantiomers of Glucose and Arabinose

The Aldotetroses: Glyceraldehyde is the simplest carbohydrate (C_3, aldotriose, 2,3-dihydroxypropanal). The next carbohydrates are aldotetroses (C_4, 2,3,4-trihydroxybutanal).

aldotriose

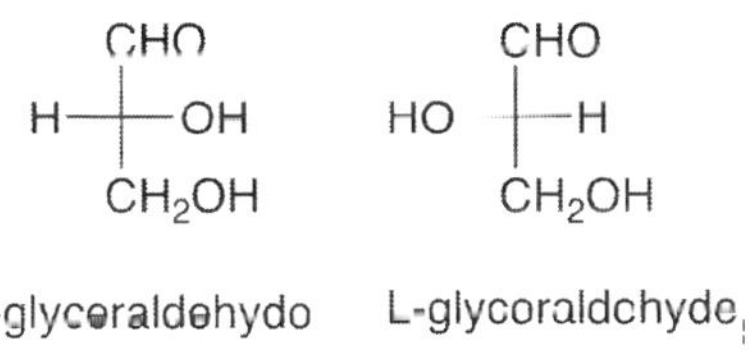

Structure of DL aldoses with three carbons

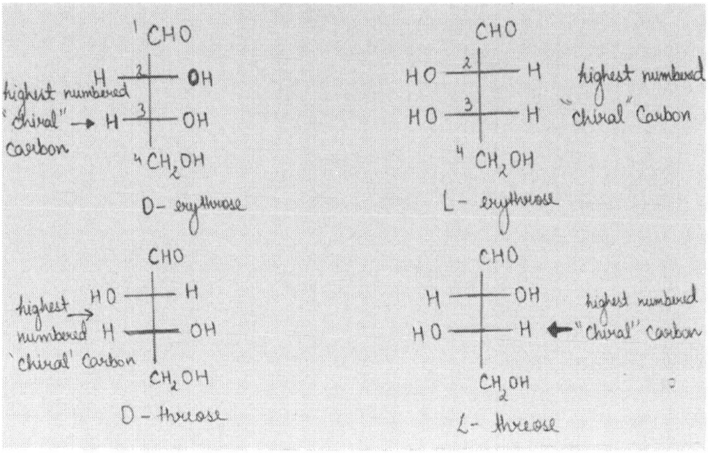

Aldopentoses and Aldohexoses

Aldopentoses: have C_5, three chiral carbons and eight stereoisomers

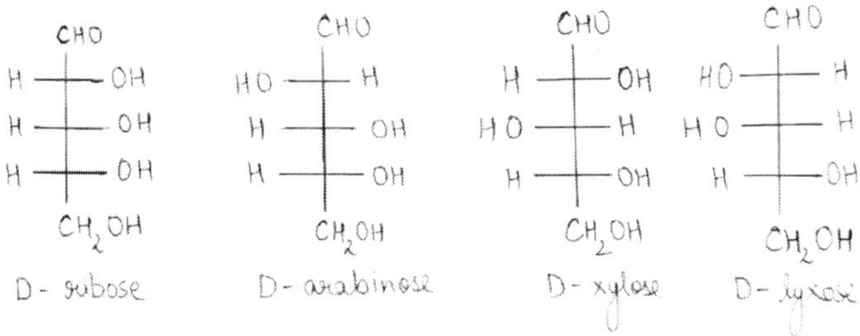

Aldohexoses: have C_6, four chiral carbons and sixteen stereoisomers

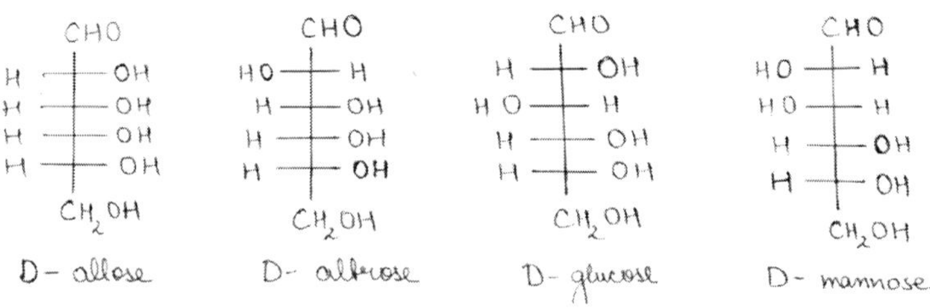

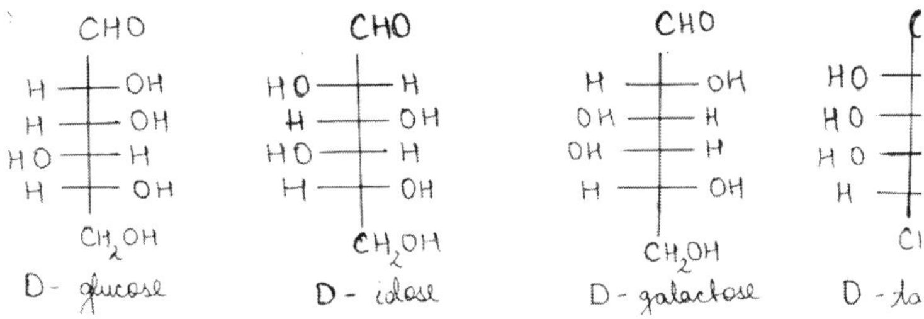

Manipulation of Fischer Projections

Fischer projections can be rotated by 180°.

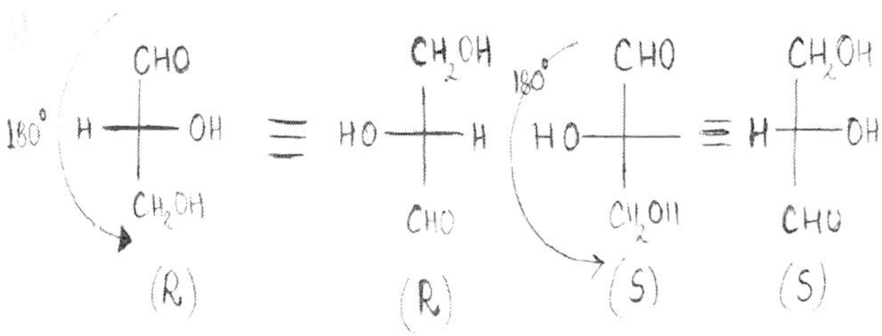

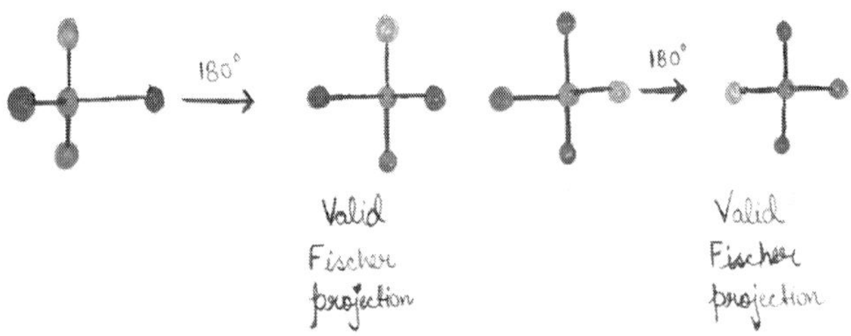

A 90° rotation inverts the stereochemistry and is not permissable.

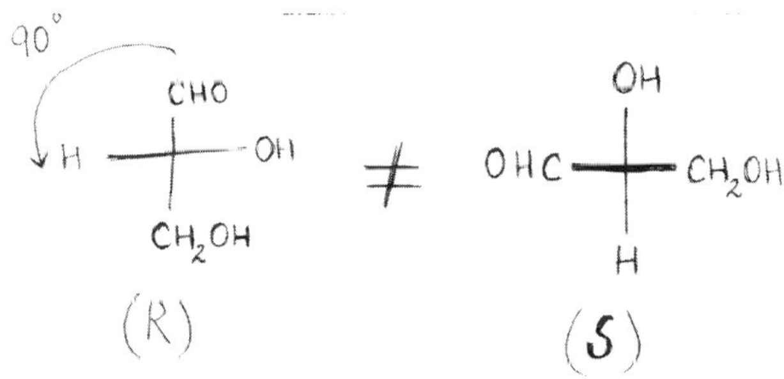

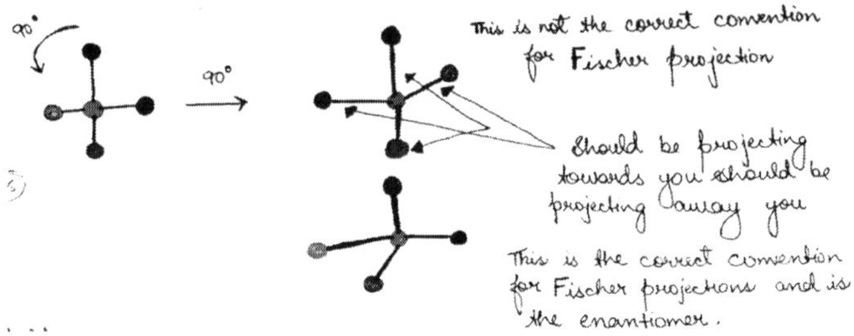

If one group of a Fischer projection is held steady, the other three groups can be rotated clockwise or counterclockwise.

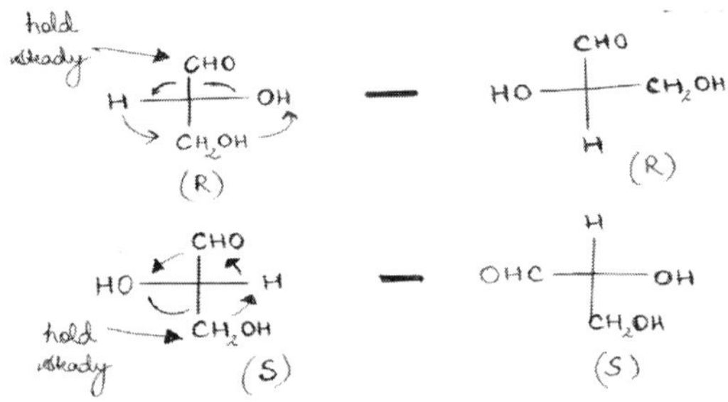

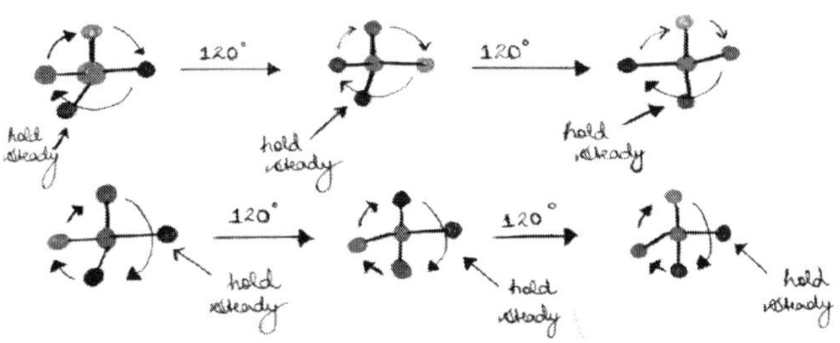

6. LIPIDS: FATTY ACIDS, TRIACYLGLYCEROL, PHOSPHOLIPIDS, SPHINGOLIPIDS: SPHINGOMYELINS, CEREBROSIDES GANGLIOSIDES, CHOLESTEROL, MICELLES, BILAYERS, LIPOSOMES, LIPOPROTEINS: STRUCTURE AND FUNCTION.

Introduction

A lipid is an organic substance found in living systems that is insoluble in water but is soluble in organic solvents. Lipids are organic molecules essential for life that are composed mostly of C, H, O. They serve as a storage form of metabolic fuel. They serve as a transport form of metabolic fuel. They provide the structural components of membranes. They have protective functions in bacteria, plants, insects, and vertebrates serving as a part of the outer coating between the body of the organism and the environment.

There are four types of lipids

- Fats (triglycerides)
- Phospholipids
- Steroids
- Waxes

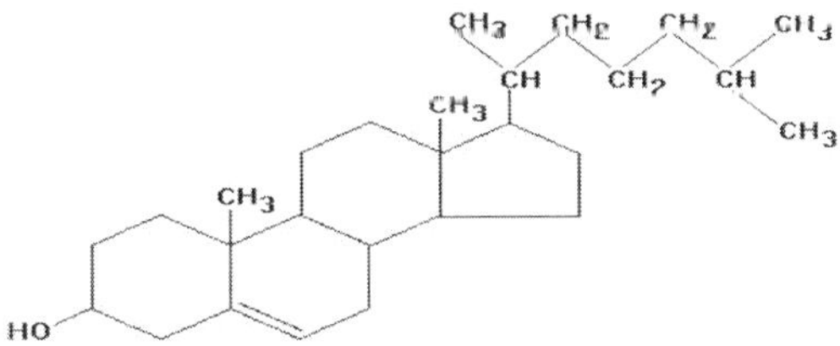

Structure of Lipid

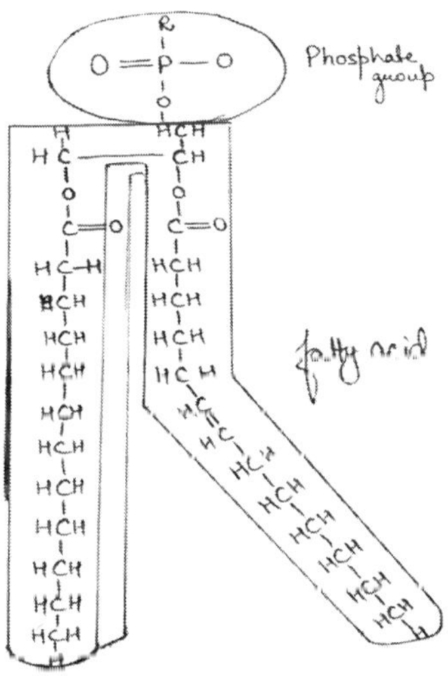

Fats & oils make up 95% of the nutritional lipids and the other 5% are steroids. Waxes are functional only. **Fats** are solid triglycerides and **oils** are liquid triglycerides.

Representative Fatty Acids

C12-C18 (also C20) Dietary Fats and Oils

$CH_3-(CH_2)_{12}-COOH$ (1)

$CH_3-(CH_2)_{14}-COOH$ (2)

$CH_3-(CH_2)_{16}-CO_2H$ (3)

$CH_3-(CH_2)_7-CH=CH-(CH_2)_7-CO_2H$ (4)

$CH_3-(CH_2)_4-CH=CH-CH_2-CH=CH-(CH_2)_7-CO_2H$ (5)

$CH_3-CH_2-CH=CH-CH_2-CH=CH-CH_2-CH=CH-(CH_2)_7-CO_2H$ (6)

Name	Mp (°C)	Class
(1) Myristic acid	58	Saturated C-14
(2) Palmitic acid	63	Saturated C-16
(3) Stearic acid	71	Saturated C-18
(4) Oleic acid	4	Monounsaturated C-18
(5) Linoleic acid	-5	Polyunsaturated C-18
(6) Linolenic acid	-11	Polyunsaturated C-18

Triglycerides

The length and the number of carbons in the side chains. The number of carbon-carbon double bonds in the side chains.

Fat molecules

It contain 3 fatty acid chains linked to glycerol. It has long term energy storage and insulation for animals. It can be saturated or unsaturated.

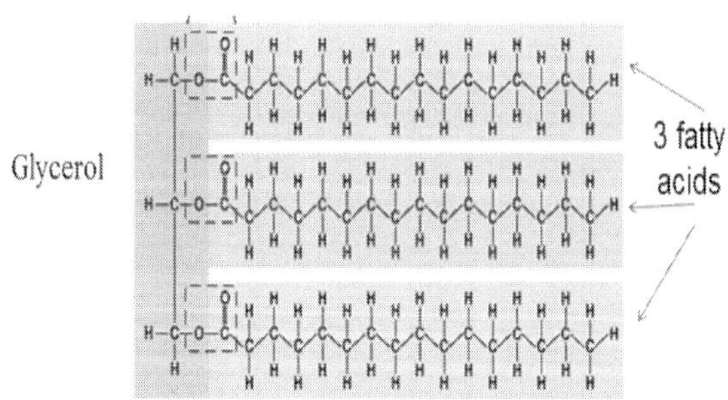

Body fat is stored

The body converts the unused carbohydrates, proteins and triglycerides that make up our macronutrients into small globules of fat that end up in the specialized cells of adipose tissue, the fatty tissue of the body. One pound of adipose tissue stores, and provides when needed, ~3500 Cal of energy. The high energy density of fat and its ability to store energy (9 Cal/gram) compactly in relatively little space and with relatively little weight, compared with carbohydrates and proteins (4 Cal/gram) which allows us to carry stores of energy with us. It gives humans and other animals the mobility and freedom necessary for survival.

Fat

Fat is essential energy. Most of our long-term energy supplies operate via the formation, storage and metabolism of body fat (triglycerides). It is short term energy storage, from one meal to another and occurs through a starch-like substance called glycogen (a carbohydrate).

Adipose tissue and fat

Adipose tissue forms cushioning shields around our major organs, protecting them against damage from physical shock and provides insulation to our bodies, guarding against a rapid loss of body heat to the external environment. Fats carry the flavors and vitamins of many of our foods although fats have no flavors of their own, eg. carrying vitamins A, D, E and K from our foods to our tissues. Fatty acids form not only the triglycerides but other compounds as well, including such vital classes as prostaglandins and phospholipids.

Saturated fats

In the saturated fats, all carbons are bonded to hydrogen. There are no C=C double bonds. It has a long, straight chain. It includes most animal fats and butter. It is solid at room temperature. It contributes to cardiovascular disease.

a) Saturated fat and fatty acid.

Unsaturated fats

At least one C=C double bond in the fatty acids such as plant & fish fats, vegetable oils and liquids at room temperature.

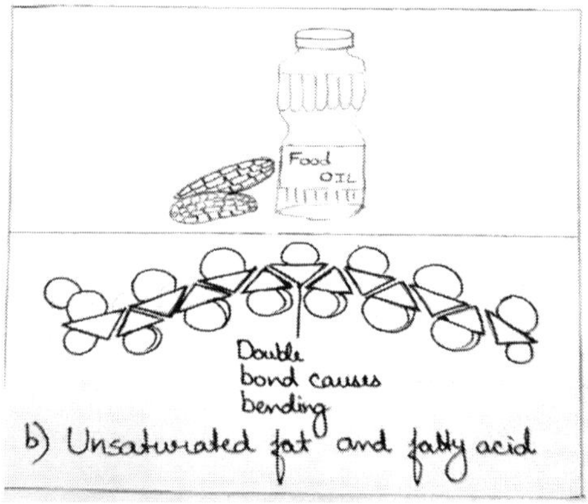

b) Unsaturated fat and fatty acid

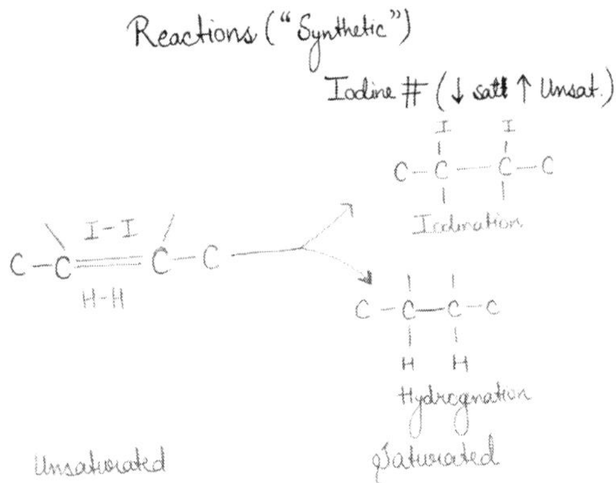

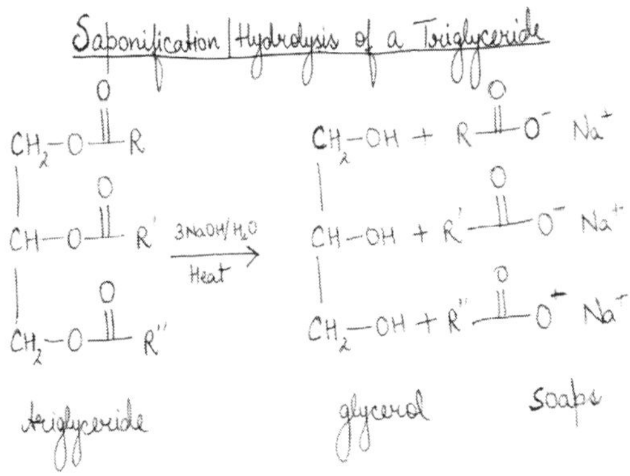

Omega (3 Fatty Acids, Especially in Fish Oils)

It contains mainly 20 & 22 C's, polyunsaturated and 1st 'ene' is 3 from Ω-end. It is eicosapentaenoic acid (EPA): 20 (5, 8, 11, 14, 17) mp -50°C. It is docosahexaenoic acid (DHA): 22 (5,8,11,14,17, 20). It contain α-linolenic: 18 (9,12,15).

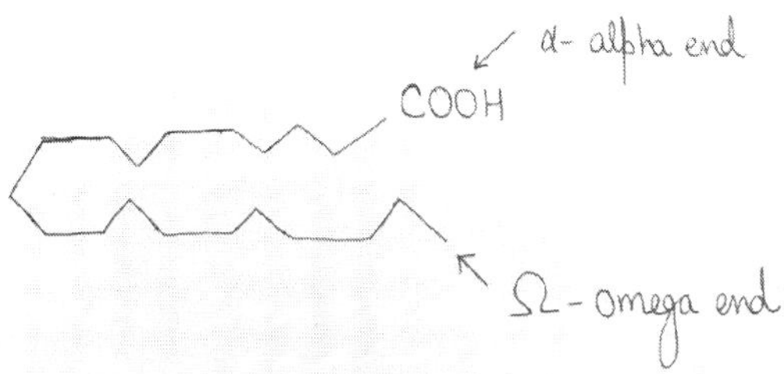

Phospholipids

Structure: It has two fatty acids chains + phosphate head and fatty acid chains are non-polar or Hydrophobic "water fearing". It also have phosphate head polar or Hydrophillic "water-loving". It is makes up cell membranes (phospholipid bilayer).

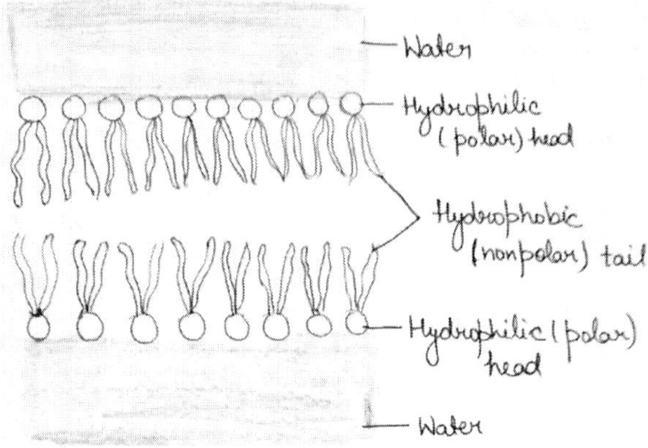

Steroids

It contains four fused carbon rings ex.

Cholesterol: It controls the fluidity of the cell membrane.

Hormones: It regulates processes in the body (such as pregnancy).

Vitamins (A, B, D): It supports metabolism and cell processes.

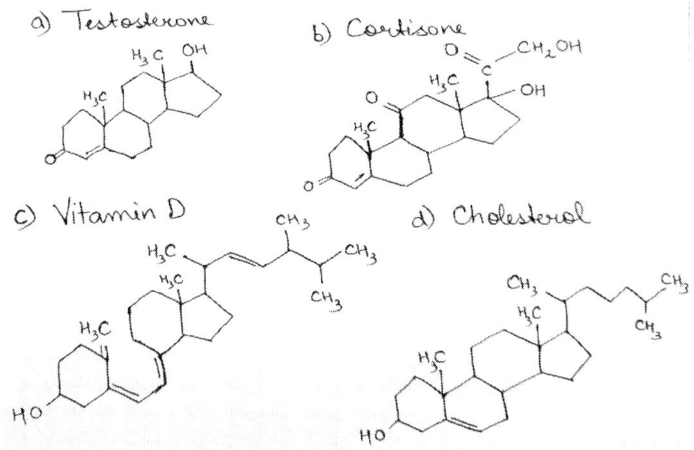

Waxes

It has solid at room temp and insoluble in water. It is used as a protective covering for animals and plants. *Examples*: Beeswax, earwax, cuticle of the leaf.

Lipids may be broadly defined as hydrophobic or amphiphilic small molecules that originate entirely or in part from two distinct types of biochemical subunits or "building blocks": ketoacyl and isoprene groups. Using this approach, lipids may be divided into eight categories: fatty acyls, glycerolipids, glycerophospholipids, sphingolipids, saccharolipids and polyketides (derived from the condensation of ketoacyl subunits) and sterol lipids and prenol lipids (derived from the condensation of isoprene subunits). (Fahy, E., et al, Journal of Lipid Research, Vol. 46, 839-862, May 2005).

Fundamental biosynthetic units of lipids

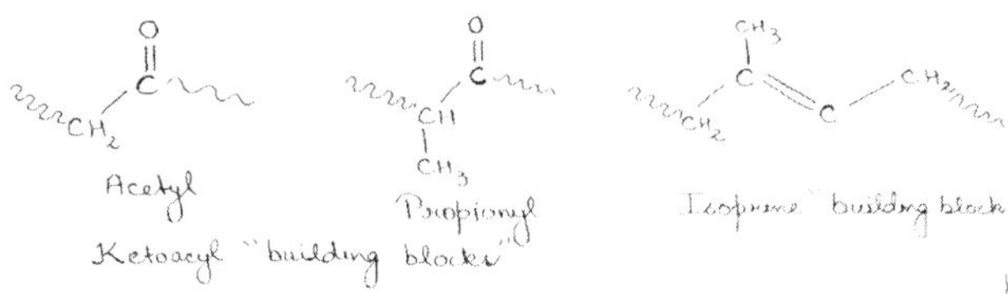

Table 4.1 List of lipids

Category	Abbreviation	Example
Fatty acyls	FA	Dodecanoic acid
Glycerolipids	GL	1-hexadecanoyl-2-(9Z-octadecenoyl)-sn-glycerol
Glycerophospholipids	GP	1-hexadecanoyl-2-(9Z-octadecenoyl)-sn-glycero-3-phosphocholine
Sphingolipids	SP	N-tetradecanoyl-sphing-4-enine
Sterol lipids	ST	Cholest-5-en-3ß-ol
Prenol lipids	PR	2E, 6E-farnesol
Saccharolipids	SL	UDP-3-O-3R-hydroxy-tetradecanyoyl-Ad-n-acetylglucosamine
Polyketides	PK	Aflatoxin B_1

Consistent structure representation across classes of lipids

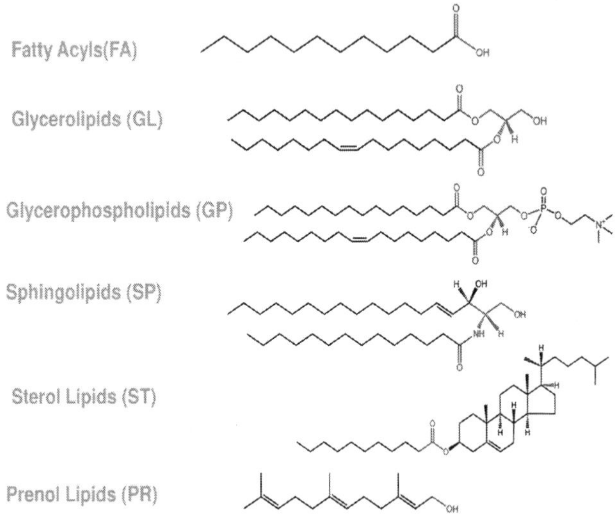

Structural comparison of Sphingomyelin and Phosphatidylcholine

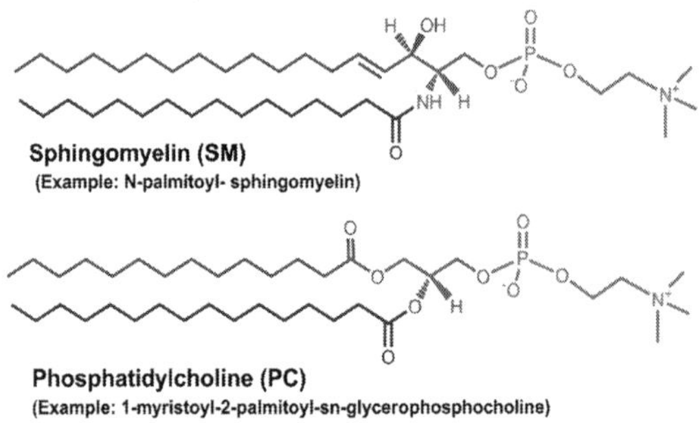

7. AMINO ACIDS AND PROTEINS: AMINO ACIDS: STRUCTURE, NOMENCLATURE AND GENERAL PROPERTIES. PEPTIDE BOND, PRIMARY STRUCTURE OF PROTEINS, END GROUP ANALYSIS, AMINO ACID COMPOSITION, SPECIFIC PEPTIDE CLEAVAGE AND SEQUENCE DETERMINATION. SECONDARY STRUCTURE: PEPTIDE GROUP, RAMACHANDRAN DIAGRAM, HELICAL STRUCTURES: ALPHA-HELIX AND OTHER POLYPEPTIDE HELICES, Β - PLEATED SHEETS. PROTEIN STABILITY: ELECTROSTATIC INTERACTIONS, HYDROGEN BOND AND HYDROPHOBIC FORCES, DISULFIDE BOND. TERTIARY AND QUATERNARY STRUCTURES OF PROTEINS.

Introduction

Amino acid is a compound containing carbon, hydrogen, oxygen and nitrogen and serves as monomers of proteins. It contains both α-Amino acid and carboxylic acids. The "building blocks" from which proteins are made. Note: except for glycine (R = H), all a-amino acids are chiral molecules. There are 20 naturally occuring R-groups.

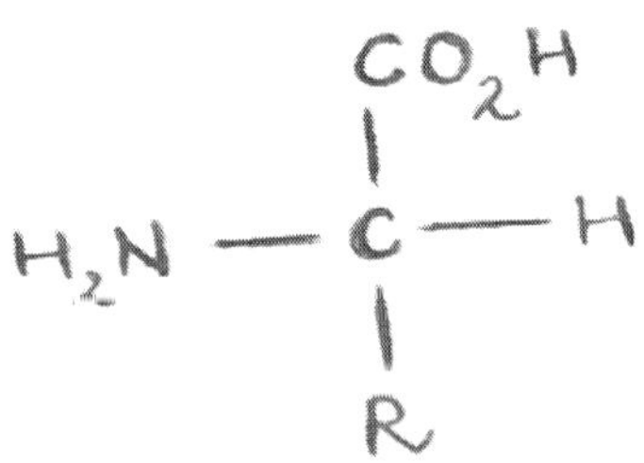

The use of D and L nomenclature with amino acids based on the configuration of glyceraldehyde. Naturally occurring amino acids generally have the same configuration as L-glyceraldehyde (*S* configuration at the a-carbon).

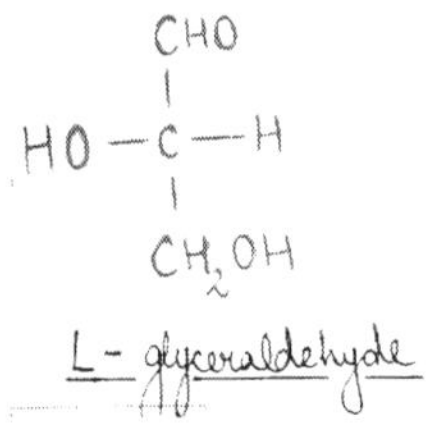

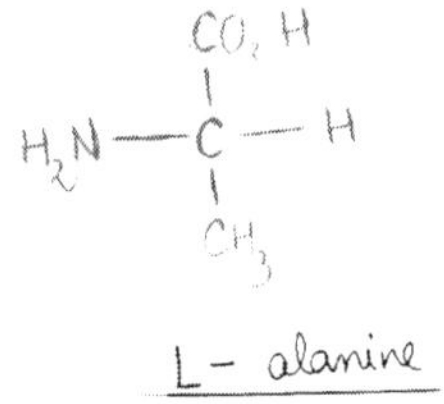

Acid–Base Properties

Since amino acids have both an acidic functionality and a basic functionality, we should expect the following equilibrium. In fact, the equilibrium lies to the right all amino acids are charged at any pH. Such species that are overall neutral molecules but contain charged ends are called **zwitterions.**

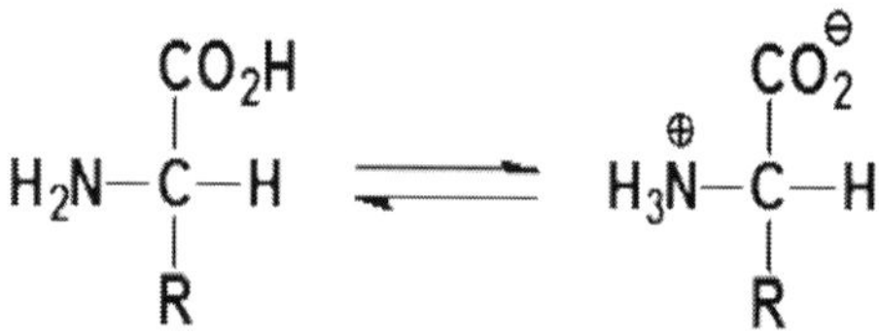

Amino acids can react as either acids or bases.

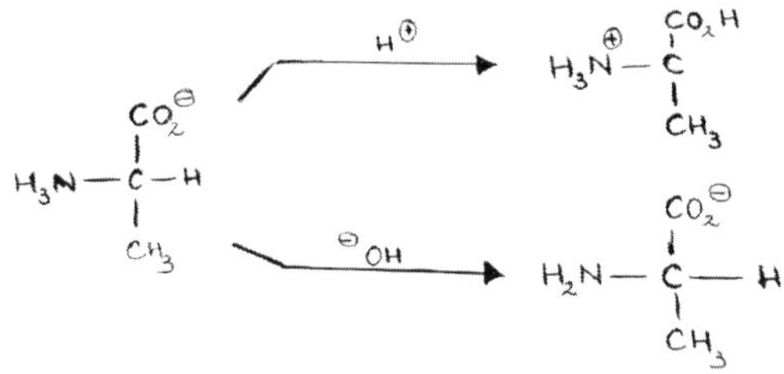

Amino acids can also have side chains containing acidic or basic groups.

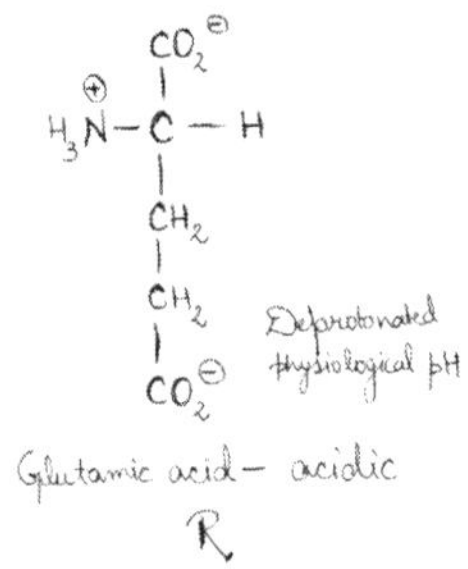

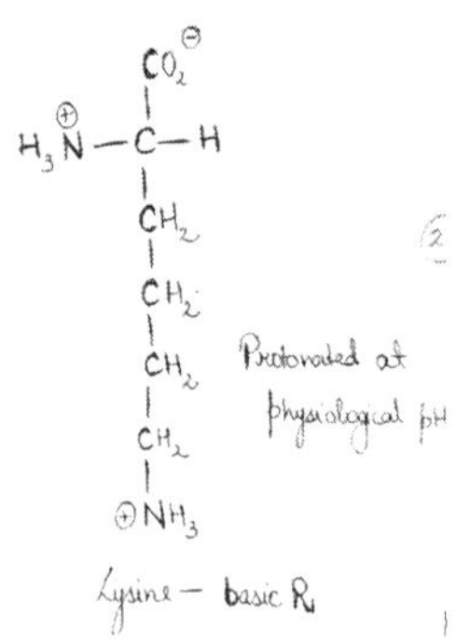

Classification

Amino acids are classified on the basis of the structure of R.

Aliphatic side chains: Hydrophobic

Polar side chains: Text classifies as HO-, S-, and amide containing – hydrophilic

Acidic – Hydrophilic

Basic – Hydrophilic

Heterocyclic/Aromatic: Hydrophilic or hydrophobic

Aliphatic side chains: Hydrophobic

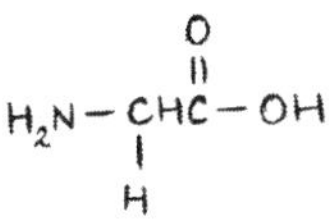

Gly G

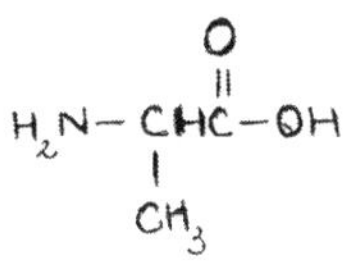

Ala A

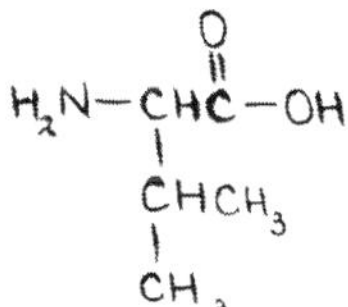

Val V

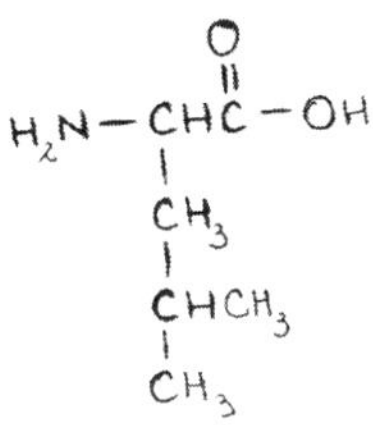

Leu, L

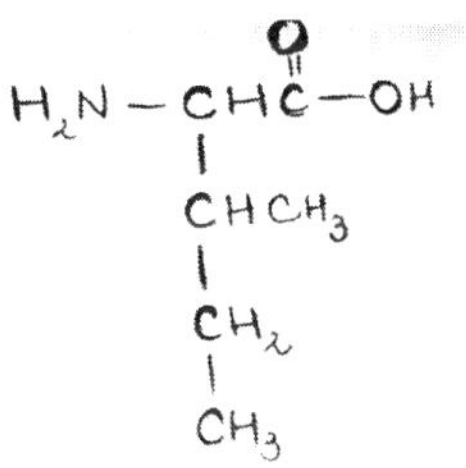

Ile, I

Polar side chains: Text classifies as HO-, S-, and amide containing hydrophilic

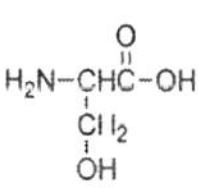

Ser, S

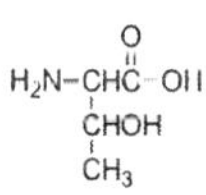

Thr, T

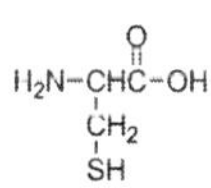

Cys, C

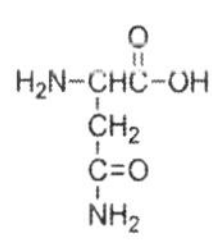

Met, M

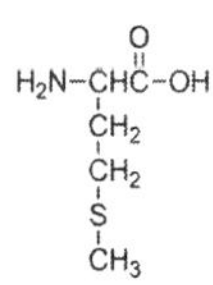

Asn, N

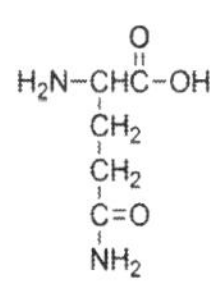

Gln, Q

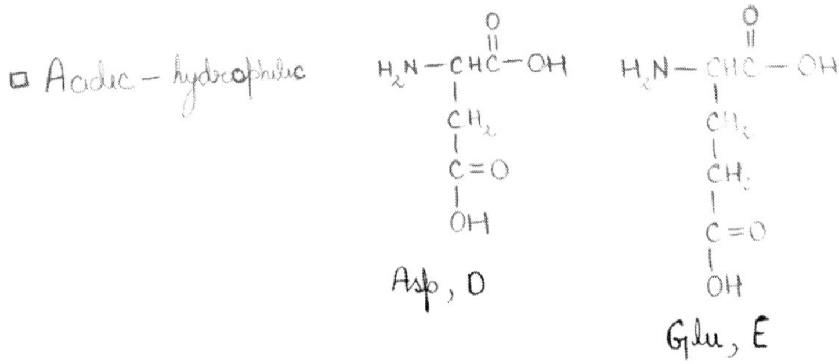

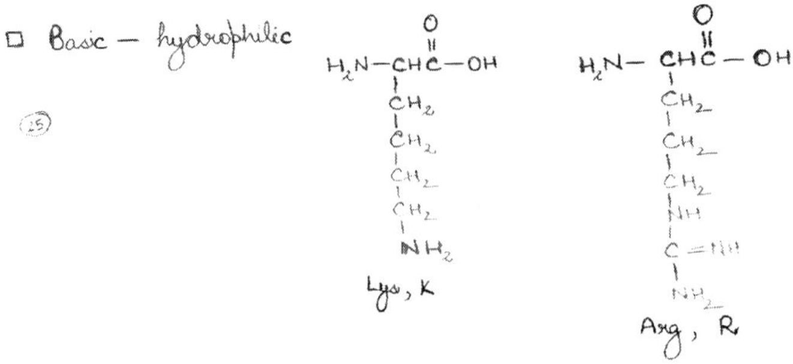

Heterocyclic/Aromatic: Hydrophilic or Hydrophobic

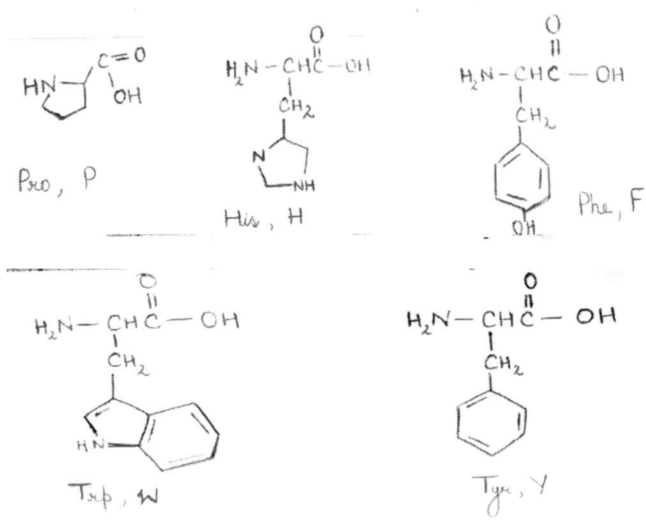

Acid-base properties

Nonpolar & polar side chains	pKa of α—COOH	pKa of α—NH_3^+
alanine	2.35	9.87
asparagine	2.02	8.80
glutamine	2.17	9.13
glycine	2.35	9.78
isoleucine	2.32	9.76
leucine	2.33	9.74
methionine	2.28	9.21
phenylalanine	2.58	9.24
proline	2.00	10.60
serine	2.21	9.15
threonine	2.09	9.10
tryptophan	2.38	9.39
valine	2.29	9.72

Acidic Side Chains	pKa of α-COOH	pKa of α-NH_3	pKa of side chain	Side Chain Group
aspartic acid	2.10	9.82	3.86	Carboxyl
glutamic acid	2.10	9.47	4.07	carboxyl
Cysteine	2.05	10.25	8.00	sulfhydril
tyrosine	2.20	9.11	10.07	phenolic

Basic Side Chains	pKa of α-COOH	pKa of α-NH_3^+	pKa of Side Chain	Side Chain group
arginine	2.01	9.04	12.48	guanidine
histidine	1.77	9.18	6.10	imidazole
lysine	2.18	8.95	10.53	1° amino

Acidity of a-COOH Groups

The average pK_a of an a-carboxyl group is 2.19, which makes them considerably stronger acids than acetic acid (pK_a 4.76). The greater acidity is accounted for by the electron-withdrawing inductive effect of the adjacent -NH_3^+ group.

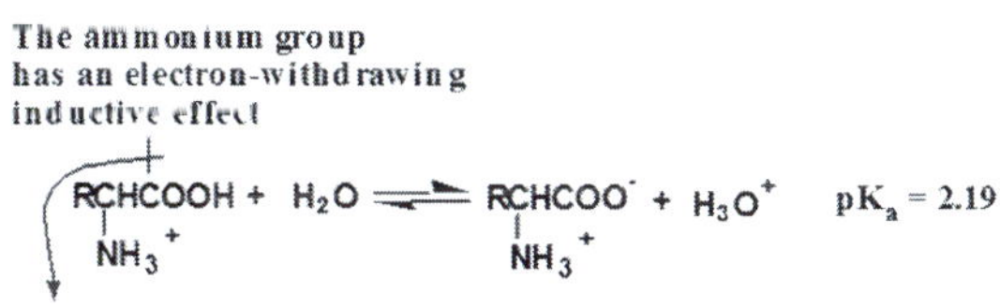

Side chain COOH groups

The a-NH_3^+ group and side-chain -COOH groups are also stronger than acetic acid due to the electron-with drawing inductive effect. The effect decreases with distance from the a-NH_3^+ group compare as

a-COOH group of alanine (pK_a 2.35)

b-COOH group of aspartic acid (pK_a 3.86)

g-COOH group of glutamic acid (pK_a 4.07)

Acidity of a-NH_3^+groups

The average value of pK_a for an a-NH_3^+ group is 9.47, compared with a value of 10.76 for 1° alkylammonium ions.

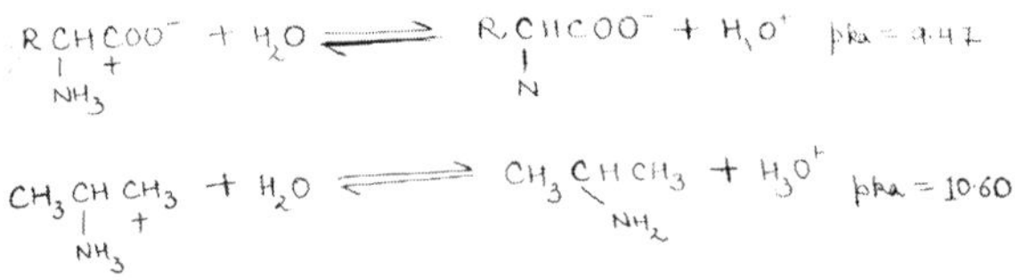

Basicity of Guanidine Group

The side chain of arginine contains a guanidine group.

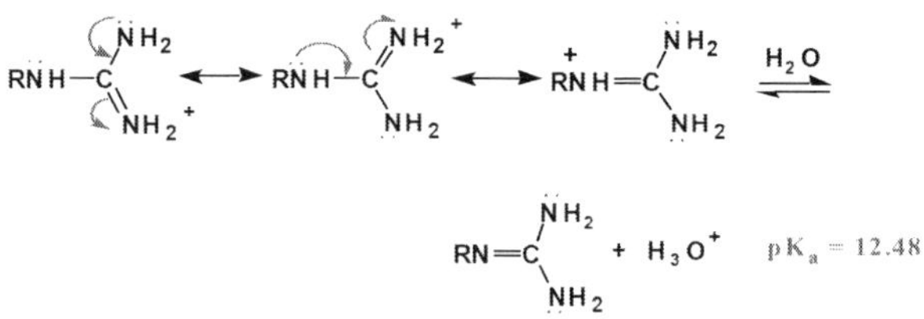

Imidazole groups of His

The imidazole group is a heterocyclic aromatic amine.

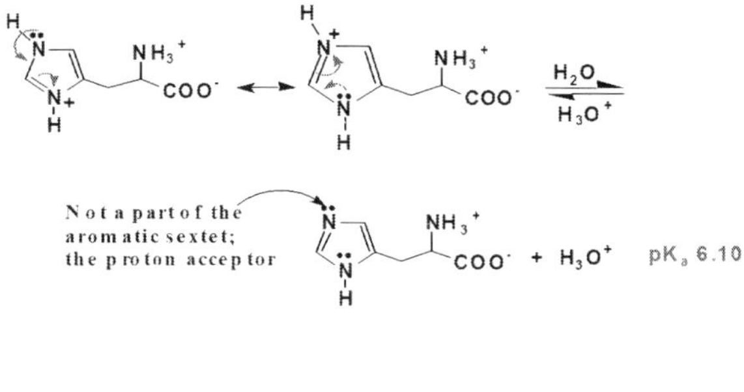

Titration of Amino Acids

Titration of glycine with NaOH

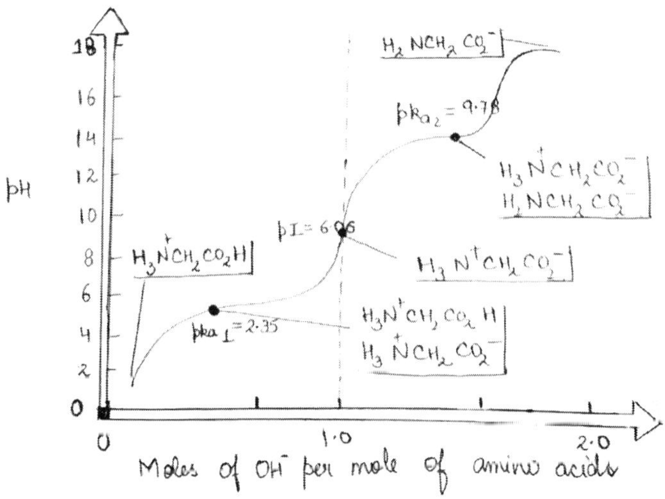

Isoelectric Point (pI)

The isoelectric point, pI: the pH at which the majority of amino acid molecules in solution have no net charge. The pI for glycine falls between the pK_a values for the carboxyl and amino groups. The following tables give isoelectric points for the 20 protein-derived amino acids.

$$pI = \frac{1}{2}\left(pka\ \alpha\text{-}COOH + pK_a\ \alpha\text{-}NH_3^+\right)$$

$$= \frac{1}{2}(2.35 + 9.78) = 6.06$$

Nonpolar & polar side Chains	pKa of α-COOH	pKa of α-NH_3^+	pKa of side Chain	pI
alanine	2.35	9.87		6.11
asparagine	2.02	8.80	–	5.41
glutamine	2.17	9.13	–	5.65
glycine	2.35	9.78	– –	6.06
isoleucine	2.32	9.76	– –	6.04
leucine	2.33	9.74	– –	6.04
methionine	2.28	9.21	– –	5.74
phenylalanine	2.58	9.24	–	5.91
proline	2.00	10.60		6.30
serine	2.21	9.15	– –	5.68
threonine	2.09	9.10	–	5.60
Valine	2.29	9.39	–	6.00
tryptophan	2.38	9.72	– –	5.88

Acidic Side chain	pKa of α-COOH	pKa of α-NH_3^+	pKa of Side chain	pI
aspartic acid	2.10	9.82	3.86	2.98
glutamic acid	2.10	9.47	4.07	3.08
cysteine	2.05	10.25	8.00	5.02
tyrosine	2.20	9.11	10.07	5.63

Basic Side chain	pKa of α-COOH	pKa of α-NH_3^+	pKa of Side chain	pI
arginine	2.01	9.04	12.48	10.76
histidine	1.77	9.18	6.10	7.64
lysine	2.18	8.95	10.53	9.74

A reagent commonly used to detect amino acid is ninhydrin.

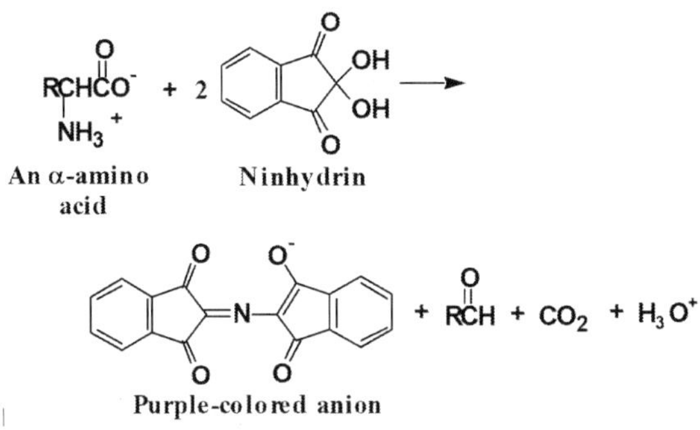

Polypeptides and Proteins

Emil Fischer proposed that proteins are long chains of amino acids joined by peptide bonds in 1902. Peptide bond: the special name given to the amide bond between the a-carboxyl group of one amino acid and the a-amino group of another.

Peptides

Peptide: the name given to a short polymer of amino acids joined by peptide bonds and they are classified by the number of amino acids in the chain.

Dipeptide: a molecule containing two amino acids joined by a peptide bond.

Tripeptide: a molecule containing three amino acids joined by peptide bonds.

Polypeptide: a macromolecule containing many amino acids joined by peptide bonds.

Protein: a biological macromolecule of molecular weight 5000 g/mol or greater, consisting of one or more polypeptide chains. Peptides are amino acid polymers containing 2–50 individual units. Peptides with >50 units are called proteins. Peptide structures are written with the *N*-terminal amino acid on the left and the *C*-terminal amino acid on the right.

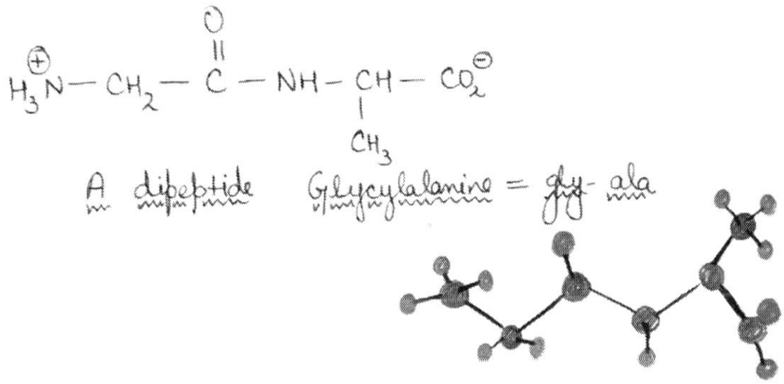

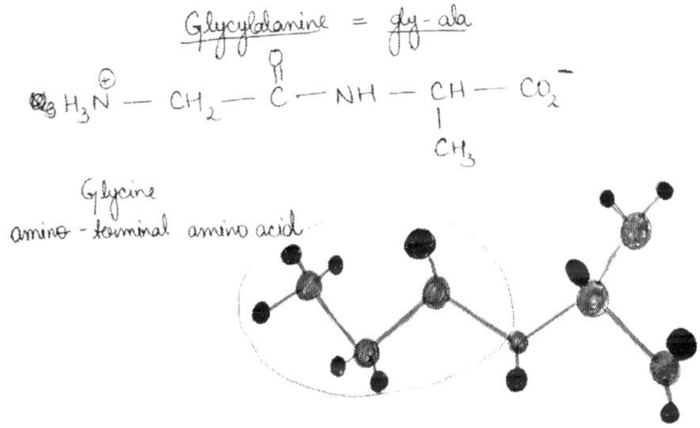

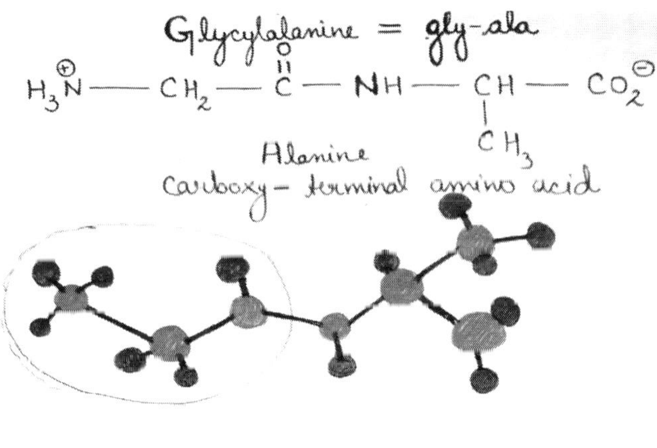

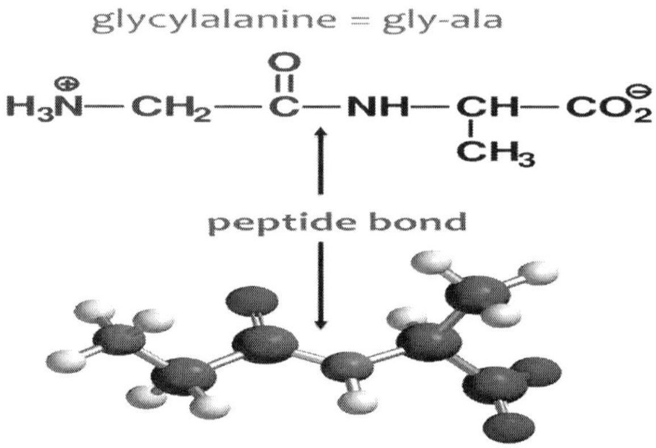

• A tetrapeptide:

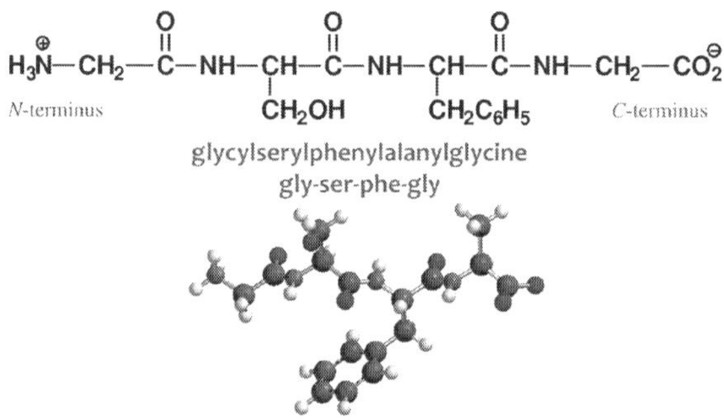

The Peptide (Amide) Bond

The amide nitrogen is sp^2 hybridized and the lone pair is conjugated with the carbonyl group. There is considerable C–N double-bond character. Rotation is about the C–N bond which is difficult.

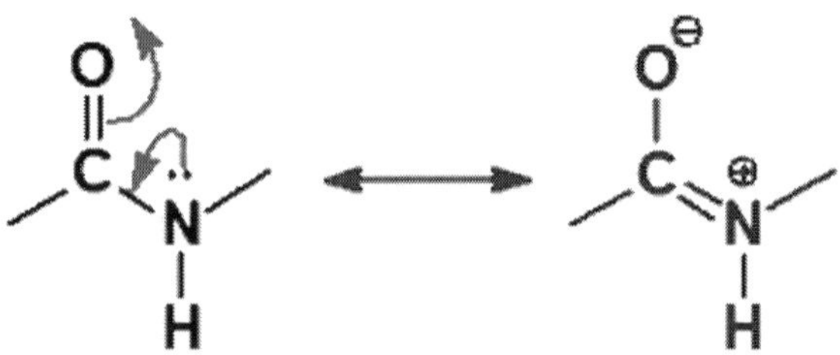

The amide groups are planar. The R-groups are on opposite sides of the plane.

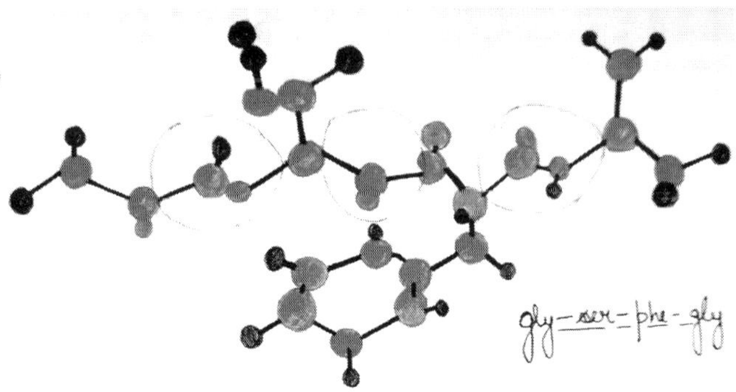

The amide groups are planar. The R-groups are on opposite sides of the plane.

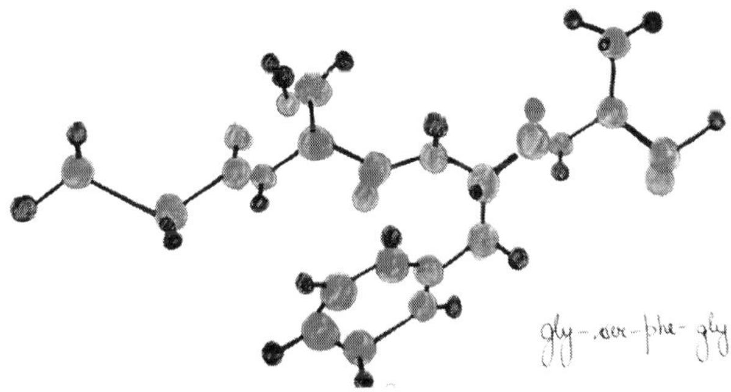

The covalent bond between amino acids in proteins and peptides is the disulfide linkage between two cysteine units.

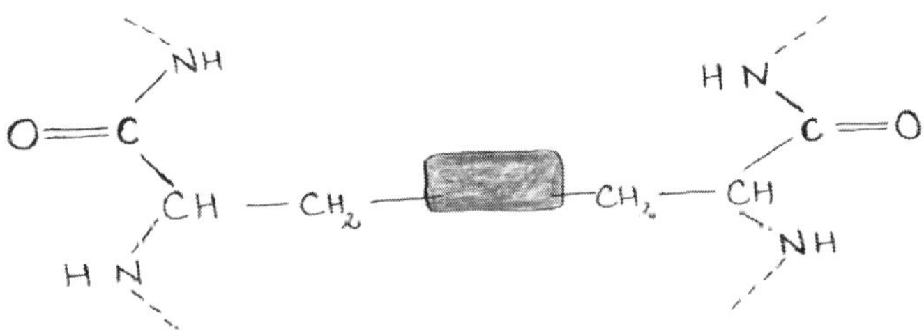

Thiols are readily oxidized to disulfides. Disulfides are readily reduced to thiols.

$$R{-}SH + HS{-}R \underset{[H]}{\overset{[O]}{\rightleftharpoons}} R{-}S{-}S{-}R$$

Disulfide links serve to connect polypeptide chains.

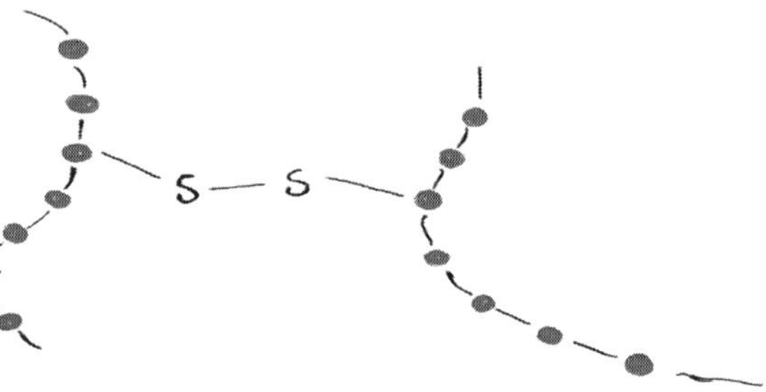

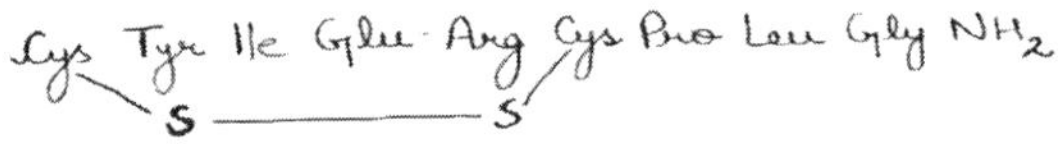

It can form a macrocycle.

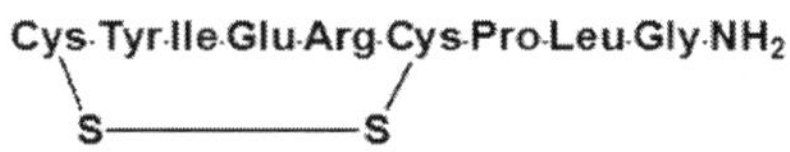

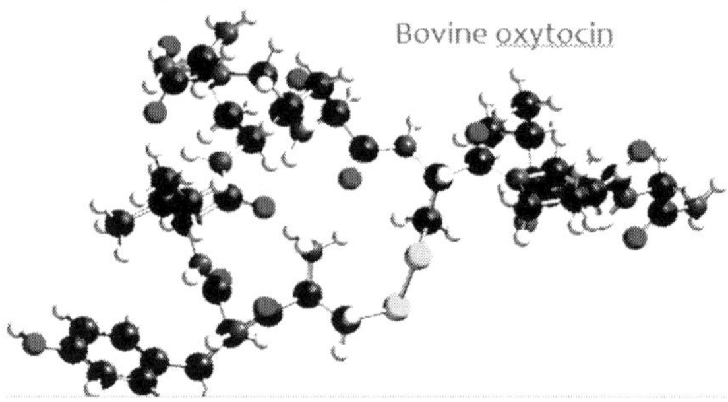

The structure determination of peptides which contain amino acid analysis, terminal residue analysis, N-terminal analysis (Edman degradation), C-terminal analysis (Carboxypeptidase), partial hydrolysis (enzymatic) and hydrolyze the peptide into smaller fragments.

- Trypsin - Cleaves at lys and arg.
- Chymotrypsin - Cleaves at phe, tyr, and trp.
- Pepsin - Cleaves at phe, tyr, trp, leu, asp, glu.
- Cyanogen bromide (not enzymatic) - Cleaves at met.
- Determine the sequence of the fragments.
- Successive edman degradations.

8. VITAMINS AND ENZYMES: WATER-SOLUBLE VITAMINS, THEIR COENZYME FORMS, SOURCES AND BIOCHEMICAL FUNCTIONS. FAT-SOLUBLE VITAMINS: SOURCES AND FUNCTIONS.

Introduction

It is an organic compound required by an organism in trace amounts to perform specific cellular functions. These compounds serve as the same roles in nearly all forms of life, but higher animals lost the capacity to synthesize them in the course of evolution. Vitamins are not synthesized by humans. Therefore must be supplied by the diet. It can be classified according to the solubility and their functions in metabolism. The requirement for any given vitamin depends on the organisms. Not all vitamins are required by all organisms. It may be water-soluble or fat-soluble. There are nine vitamins such as thiamine, riboflavin, niacin, biotin, pantothenic acid, folic acid, cobalamin, pyridoxine, and ascorbic acid are classified as water-soluble, whereas four vitamins (Vitamins A, D E and K) are termed fat-soluble. The water-soluble vitamins are all precursors' coenzymes, except vitamin C. There are different vitamins and minerals are needed for the body to function.

- 13 vitamins
- 22 minerals

Two types of vitamins

- Water-soluble
- Fat-soluble

Fat-soluble vitamins

The fat soluble vitamins are vitamin A, D, E and K. The vitamin is present in excess amount which is stored in the liver and in body fat. It is possible to build up to a toxic level.

Vitamin A (Retinol)

Beta-carotene is converted into vitamin A. It promotes good vision. It has promotes healthy skin. It helps with growth and maintenance of bones, teeth, and cell structure which require 900 micrograms for males; 700 micrograms for females.

High amount of vitamin A

It may turn your skin orange. It may cause fatigue, weakness, severe headache, blurred vision, hair loss and joint pain.

Toxicity

It may cause severe liver or brain damage and also cause birth defects.

Less amount of vitamin A

It may cause night blindness. It is severly effected the immune system.

Foods rich in vitamin A

The animal products such as Liver, Eggs and Milk, butter and cheese

Carotenoids

Carotenoids are present in Orange/Yellow fruits and vegetables such as Cantaloupes, carrots, sweet potatoes, winter squash and Leafy green vegetables such as Spinach, broccoli.

Vitamin D

It is sunshine vitamin. It is essential for building and maintaining bones and teeth. It has responsible for the absorption and utilization of calcium. It may boost the immune system. It may also help decrease certain cancers. RDA: 5 micrograms until age 50. 10 micrograms / day until 70; 15 mcg 70+

Less amount of vitamin D

Vitamin D deficiency has been in the news a lot lately. The deficiency may occur from the inadequate diet and also vegetarianism, lactose intolerance and milk allergy. Body unable to absorb needed vitamin D. It has also cause by limited exposure to sunlight. It may lead to osteomalacia and/or osteoporosis. Vitamin D gain from sun exposure for 10 minutes a day and also from food such as Fortified milk

- Tuna
- Salmon

Vitamin E

It is very important to red blood cells, muscles and other tissues. The deficiency is shown rare. Toxicity is rare seen. Vitamin E acts as a blood thinner. The vitamin E take from food such as vegetable oils, salad dressings, whole grain cereals, green leafy vegetables, nuts, seeds, peanut butter and wheat germ.

Vitamin K

It is most important for blood clotting. It also has a role in bone health. It is mostly made in the intestines. It also take from food such as turnip greens, cauliflower, spinach, liver, broccoli, kale and cabbage.

Water-soluble vitamins

Vitamins B and C

There are eight vitamins B such as Thiamin (B-1), Riboflavin (B-2), Niacin (B-3), Pyridoxine (B-4), Cobalamin (B-12), Folic acid, Pantothenic acid and Biotin.

Thiamin or B-1

It helps to convert carbohydrates to energy. **Deficiency** is oocur such as Fatigue, nausea, depression and nerve damage. It has take from foods such as Pork, beef, liver, peas, seeds, legumes, whole-grain products, and oatmeal

Riboflavin or B-2

It has key to metabolism and red blood cells. **Deficiency** is oocur such as Dry, scaly skin. It has take from food such as Milk, yogurt, cheese, whole-grain bread, green leafy vegetables, meat, and eggs.

Niacin or B-3

It is involved with energy production. It also helps with skin, nerves and digestive system. **Deficiency** is oocur such as Rare but causes: diarrhea, dermatitis, dementia and death. It also take from foods such as Meat, poultry, liver, eggs, brown rice, baked potatoes, fish, milk, and whole-grain foods.

Pyridoxine or B-6

It is involved in the chemical reactions of proteins and amino acids. **Deficiency** is oocur such as Skin changes, dementia, nervous system disorders and anemia. It has also take from food such as Lean meats, fish, legumes, green leafy vegetables, raisins, corn, bananas and mangos.

Cobalamin or B-12

It helps with the nervous system, red blood cells and DNA synthesis. **Deficiency** is oocur such as Nervous system disorders and pernicious anemia. It takes from foods only found in animal products such as Meat, fish, poultry, eggs, milk products and clams.

Folic acid (Folacin, Folate)

It has key role in red blood cell formation and cell division. **Deficiency** is oocur such as Anemia, digestive disorders. It has also take from foods such as Leafy, dark green vegetables and also found in liver, beans, peas, asparagus, oranges, avocados.

Pantothenic acid and biotin

They help with the metabolism and formation of some hormones. **Deficiencies** are rare. Foods: Almost any food, plant-based or animal-based.

Vitamin C

It has important to bone health, blood vessel health, cell structure and absorption of iron. **Deficiency** is rare. High amount of vitamin C is present in foods such as Melons, berries, tomatoes, potatoes, broccoli, fortified juices, kiwi, mangos, yellow peppers and citrus fruits.

Table 8.1 Roles of vitamins

Vitamin	Role of action
Water-soluble	
Thiamine (vit B1)	Cofactor in decarboxylations Deficiency disease: Beri-beri
Riboflavin (vit B2)	Constituent of flavoproteins
Niacin (vit B3)	Constituent of NAD^+ and $NADP^+$ Deficiency disease: Pellagra

Pantothenic acid (vit B5)	Constituent of CoA and acyl-group transfer
Pyridoxine (vit B6)	Group transfer to or from amino acids
Biotin (vit H)	ATP-dependent carboxylation and carboxyl group transfer
Folate (folic acid)	Coenzymes for 1-carbon transfer Deficiency disease: Megaloblastic anemia
Cyanocobalamin (vit B12)	Coenzyme in amino acid metabolism Deficiency disease: Pernicious anemia
Ascorbic acid (Vit C)	Antioxidant Deficiency disease: Scurvy
Fat-Soluble	
Vitamin A	Role in vision, growth, reproduction. Deficiency disease: Night blindness
Vitamin D	Regulation of calcium and phosphate metabolism Deficiency disease: Rickets (children), osteomalacia (adults).
Vitamin E	Antioxidants; cofactors in electron transport in cytochrome chain
Vitamin K	Blood coagulation, Post-translational modification of blood clotting factors.

Minerals

There are different minerals which are present in all human being and also have important role such as Calcium, Chromium, Folate, Iron, Magnesium, Selenium, Zinc, Selenium, Phosphorus, Sodium, Potassium, Chloride, Sulfur, Copper, Iodine, Cobalt and Fluoride.Our body is needed 22 minerals. There are two categories such as

- Major
  - Include calcium, chloride, magnesium, phosphorus, potassium, sodium, and sulfur
- Trace
  - Include iron, zinc, iodine, selenium, copper, manganese, fluoride, chromium, molybdenum, arsenic, nickel, silicon, boron and cobalt.

Sodium

It helps maintain fluid balance. It helps transmit nerve impulses. It influences the contraction and relaxation of muscles.

Sodium &health

High amount of sodium causes high blood pressure and it may lead to fluid retention.

Sodium savvy

The human body requires about 500 mg of sodium per day, while the average American usually ingests between 2,300-6,900 mg each day. It is recommended to stay in a range of 1,500 to 2,400 mg /day.

Where are you getting sodium?

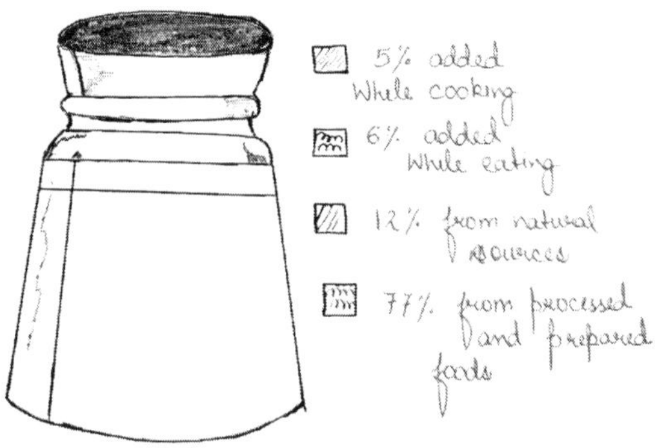

Sodium &food

- Monosodium glutamate (MSG)
- Baking soda
- Baking powder
- Disodium phosphate
- Sodium alginate
- Sodium nitrate or nitrite

Reducing sodium in your diet

Eat more fresh foods. Eat less processed foods. Look for low-sodium products. Limit the salt you add to foods. Experiment with other seasonings. Use salt substitutes with caution.

Calcium

The most abundant mineral in your body 99% is stored in the bones and known for bone health. Calcium needed in a day

- Males 19-50 years old: 1,000 mg / day
- Females 19-50 years old: 1,000 mg / day

Calcium get from Dairy products, fortified juices and sardines.

Yogurt, plain (low-fat) **Yogurt, flavored (low-fat)**	**1 cup - 415 mg** **1 cup – 345 mg**
Milk, skim **Milk, 1-2%**	**1 cup – 302 mg** **1 cup – 300 mg**
Ice cream	**½ cup – 88 mg**
Broccoli, cooked	**½ cup – 68 mg**
Salmon, canned	**3 oz – 165 mg**
Fortified orange juice	**8 oz – 300 mg**

Iron

Iron deficiency is the most widespread vitamin or mineral deficiency in the world.

- 70% of your body's iron is in your hemoglobin
- Too little iron = too little oxygen

Foods

It found in animal products such as Red meats, liver, poultry and eggs. Non-heme iron: It found in plant products such as Beans, nuts, seeds, dried fruits, fortified breads and cereals.

Potassium

Keeps the muscles and nervous system working properly. An electrolyte that conducts electricity in the body. Potassium-rich foods: bananas, broccoli, tomatoes, potatoes with skins, leafy green vegetables, citrus fruits, dried fruits, and legumes.

Zinc

Supports normal growth and development during pregnancy, childhood, and adolescence. Required for the proper sense of taste and smell. Involved in numerous aspects of cellular metabolism. Foods rich in zinc: beef, pork, lamb, & legumes.

Magnesium

Approximately 50% of total body magnesium is found in bone. The other half is found predominantly in cells of body tissue and organs. Helps maintain muscle and nerve function, keeps heart rhythm steady, supports a healthy immune system, keeps bones strong. Help regulate blood sugar levels, promotes normal blood pressure. Eating a wide variety of legumes, nuts, whole grains, and vegetables will help meet your daily need for magnesium.

Trace minerals

There are many trace minerals such as Iron, Manganese, Copper, Iodine, Zinc, Fluoride, and Selenium. Manganese is important for the production of enzymes and antioxidants. Copper is necessary for your body to make connective tissue and metabolism of iron. Iodine is used to synthesize thyroid hormones. Fluoride is used for strong bones and teeth. Reduces the number of cavities in your teeth. Selenium combines with proteins to make antioxidants and is also essential for normal thyroid function.

9. ENZYMES: HISTORICAL PERSPECTIVE, NAMING AND CLASSIFICATION, FACTORS AFFECTING ENZYME ACTIVITY, MECHANISM OF ENZYME ACTION, ENZYME UNITS.

Introduction

Enzyme is a biocatalyst that increases the rate of a chemical reaction without itself being changed in the overall process. All cellular reactions or processes are mediated by enzymes. Enzymes have several properties that make them unique. Enzyme is a globular protein thatfunctions as a biological catalyst, speeding up reaction rate by lowering activation energy without being affected by the reaction it catalyzes.

Enzymes are Protein In Nature

It is a globular protein. Ribozymes are RNA molecules with enzymatic activity. The catalytic behavior of any enzyme depends upon its primary, secondary, tertiary or quaternary structure. Enzymes of the digestive tract and those found in blood are present in an inactive form called zymogen or proezymes.

Active site

Enzymes are composed of long chains of amino acids that have folded into a very specific three-dimensional shape thatcontains an active site. An active site is a region on the surface of an enzyme to which substrates will bind and catalyzes a chemical reaction.

Enzymes are highly specific for the type of reaction they catalyze and for their substrate. Enzymes with the highest specificity and accuracy are involved in copying and expression of the genome. These enzymes have proofreading functions e.g DNA polymerase.

Enzymes make reactionseasier to occur at reasonable temperaturesbylowering the Activation energy$E_A$of the reaction.

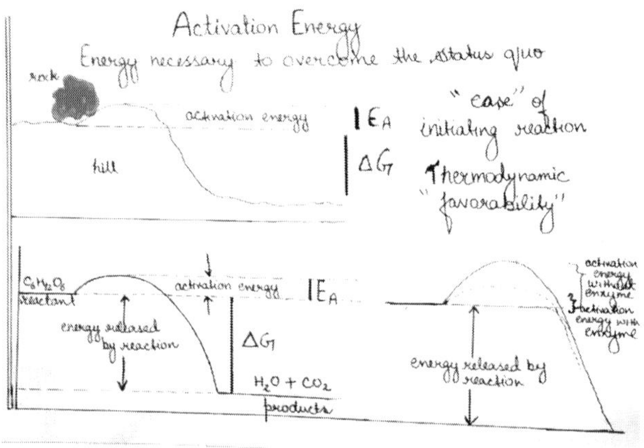

Some enzymes require a non-protein molecule for their activity. So enzymes are conjugated proteins known as holoenzymes. Holoenzyme = Apoenzyme +cofactor/coenzyme.

Enzyme without non-protein part is inactive and called apoenzyme. Cofactors can be either inorganic or organic compounds.

Mechanism of Enzyme Action

The enzymatic reactions take place by binding of the substrate with the active site of the enzyme molecule by several weak bonds.

E + S ‹--------› ES --------› E + P

Formation of the ES complex is the first step in the enzyme-catalyzed reaction then the ES complex is subsequently converted to product and free enzyme.

Lock and key model

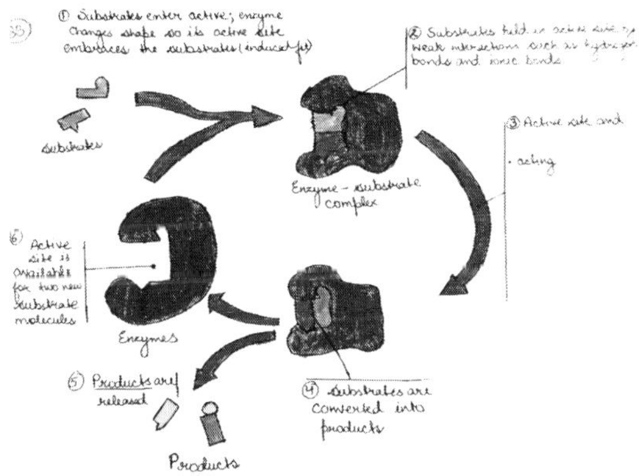

Induced-fit model

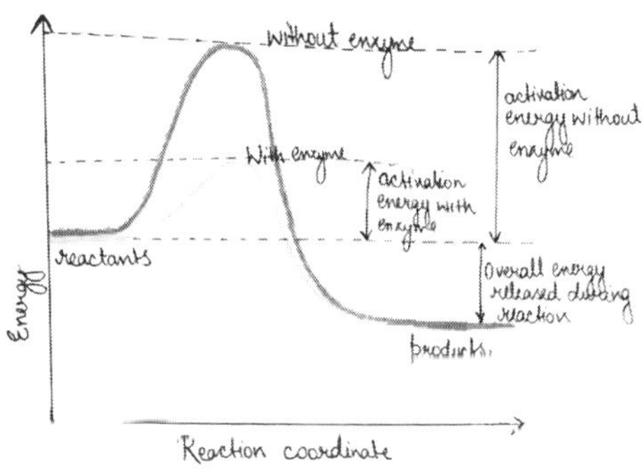

Nomenclature Enzyme Classification

IUBMB has recommended a system of nomenclature for enzymes & according to them, each enzyme is assigned with two names.

- Trivial name (common name, recommended name).
- Systemic name (official name).

Systemic name

Each enzyme is characterized by a code number called enzyme code no. or EC number and contain four figure (digit) separated by a dot.

- e.g. EC M.N.O.P
- The first digit represents the class.
- The second digit stands for sub-class.
- The third digit stands for the sub-sub-class or subgroup.
- The fourth digit gives the serial number of the particular enzyme in the list.
- ExampleEC 2.7.1.1 for hexokinase.

According to the IUBMB system of theenzyme nomenclature enzymes are grouped into 6 major classes.

1. EC 1 Oxidoreductases
2. EC 2 Transferases
3. EC 3 Hydrolases
4. EC 4 Lyases
5. EC 5 Isomerases
6. EC 6 Ligases

Factors affecting reaction velocity

- Temperature
- Hydrogen ion concentration (pH)
- Substrate concentration
- Enzyme concentration
- Products of the reaction
- Presence of activator/inhibitor
- Allosteric effects
- Time

Effect of Temperature

The rate of an enzymes catlysed reaction increases as a temperature increases, enzyme rate of reaction is decraeses at high temperature as due to denature enzyme.

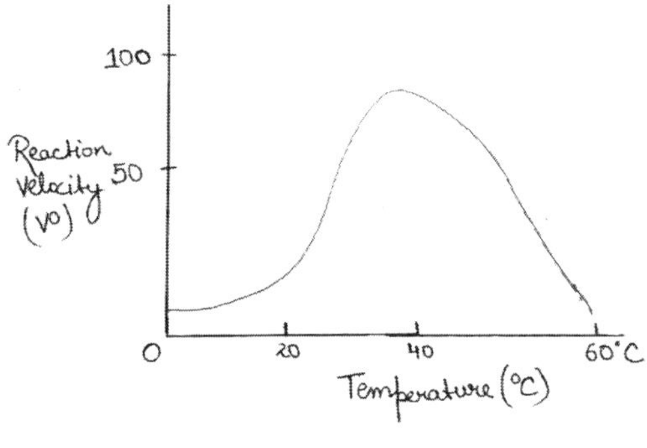

Effect of pH

Every enzyme has optimum range of pH, slow activity of rate of reaction due to changing the outside range of pH, the enzyme is denature at extreme high valuc of pH.

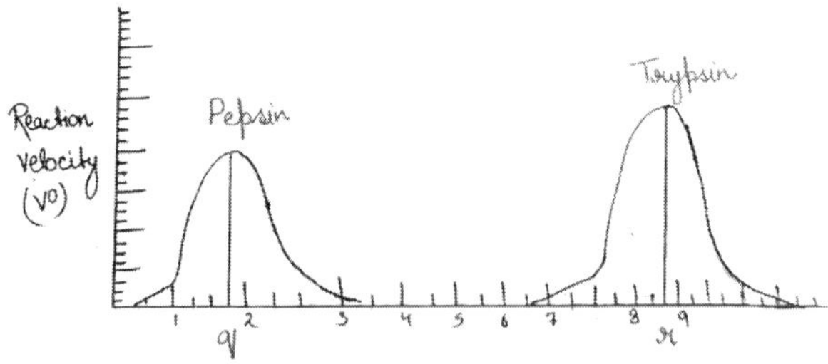

The rate of the reaction or velocity is directly proportional to the enzyme concentration when sufficient substrate is present.

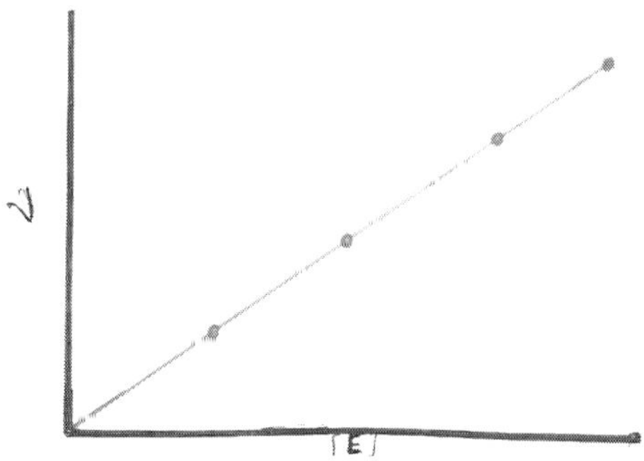

Accumulation of product in a reaction causes inhibition of enzyme activity.

Effect of Substrate Concentration

The rate of reaction is increases as enhance of substrate concentration at specific point. As the amount of substrate increases have no effect on rate of reaction once all the enzymes have bound. The available enzymes will be present on saturated state and it will work on maximum rate.

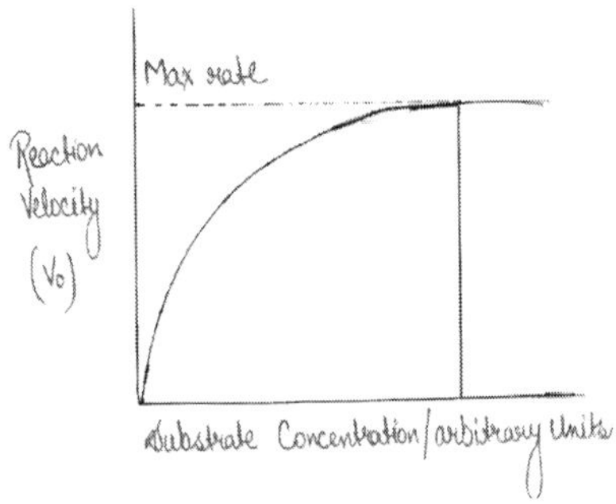

Enzyme kinetics

The study of reaction rate and how they changes in response to change in experimental parameter is known as kinetics. Amount of substrate present is one of the key factors affecting the rate of reaction catalyzed by an enzyme *in vitro*.

Effect of substrate concentration on reaction velocity

The initial velocity V_0 progressively increases with increasing concentration of the substrate during enzyme substrate rate of reation. The increase in V_0 will not depend on the susbstrate. The substrate concentration on the X axis and reaction velocity on Y axis have plot on graph.

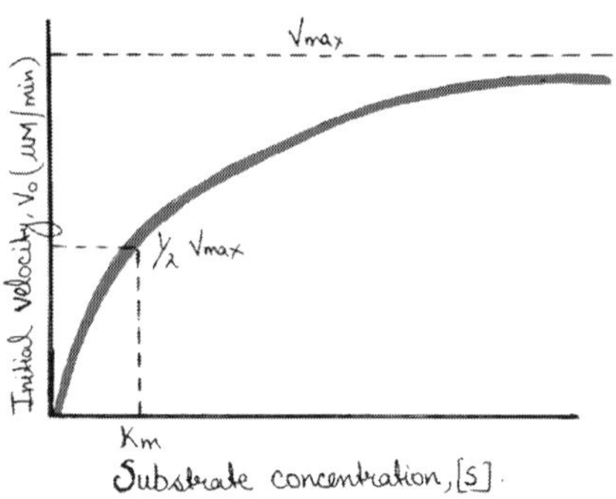

Michaelis- Menten kinetics

The model involves one substrate molecule.

k_1 k_2

$$E + S \;\langle\text{-------------}\rangle\; ES \;\text{------------}\rangle\; E + P$$

$\mathbf{k_{-1}}$

Where

- S is the substrate
- E is the enzyme
- K_1, k_{-1} and k_2 are the rate constants
- The mathematical equation that defines the quantitative relationship between the rate of an enzyme reaction and the substrate concentration is the *Michaelis-Menten equation*:

$$\mathbf{V_0 = \frac{V_{max}\,[S]}{K_m + [S]}}$$

- V_0 is the observed velocity at the given [S]
- K_m is the Michaelis-Menten constant
- $K_m = (K_{-1} + K_2) / K_1$
- V_{max} is the maximum velocity at saturating [S] conc.

Lineweaver-Burk (double reciprocal) plot

A linear representation is more accurate and convinient for determining V_{max} and K_m. This equation is obtained by taking reciprocal of both the side of Michelis-Menton equation.

$1/[S]$ vs. $1/V_0$

Lineweaver-Burk (Double Reciprocal) plot

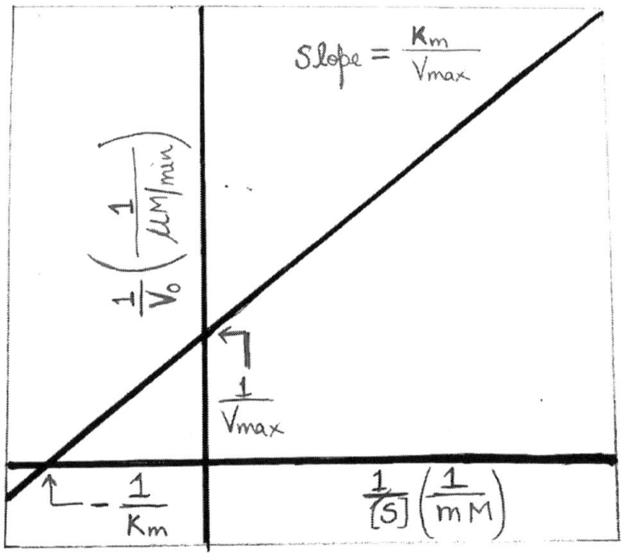

$$\frac{1}{v} = \frac{K_m}{V_{max}} \frac{1}{[S]} + \frac{1}{V_{max}}$$

Enzyme Inhibition

Any substance that can diminish the velocity of an enzyme-catalyzed reaction.These include drugs, antibiotics, poisons, and anti-metabolites. Useful in understanding the sequence of enzyme-catalyzed reactions, metabolic regulation, studying the mechanism of cell toxicity produced by toxicants. Forms the basis of drug designing.

Types of enzyme inhibition

- Reversible inhibitors
- Irreversible inhibitors

Reversible inhibitors can be classified as

- Competitive
- Non-competitive
- Un-competitive

Competitive inhibition

The structure of a competitive inhibitor closely resembles that of the enzyme's normal substrate. The competitive inhibitor binds reversibly to the enzyme's active site. The inhibitor forms an enzyme-inhibitor complex (EI) that is equivalent to the ES complex.

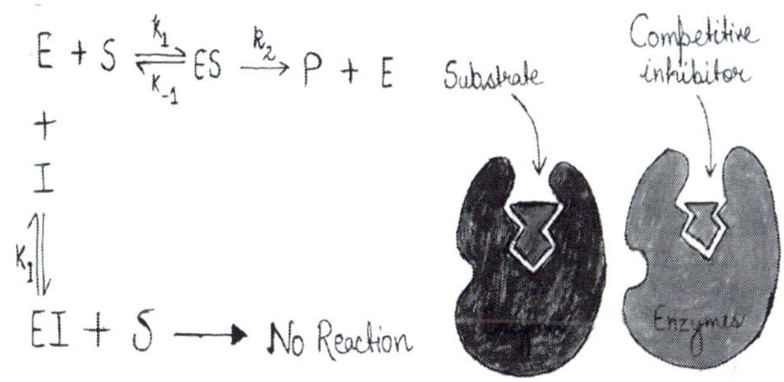

Non-competitive inhibition

The inhibitor binds to the enzyme at a site other than the active site. Inhibitor binding alters the enzyme's three-dimensional configuration and blocks the reaction.

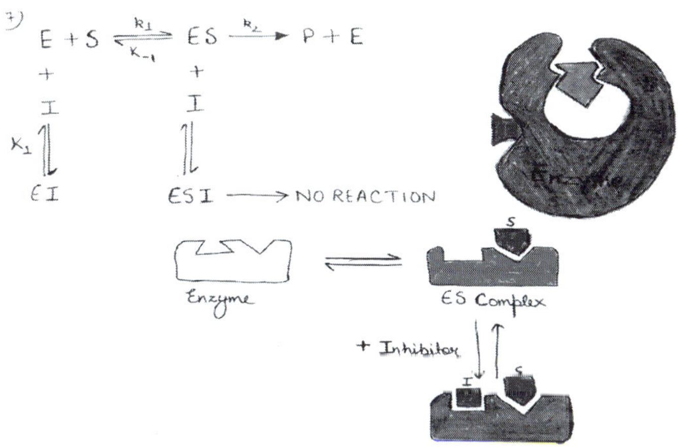

Un-competitive inhibition

Binds only to the enzyme-substrate complex. Does not have the capacity to bind to the free enzyme. Not overcome by increasing substrate concentration. Both the K_m and V_{max} are reduced.

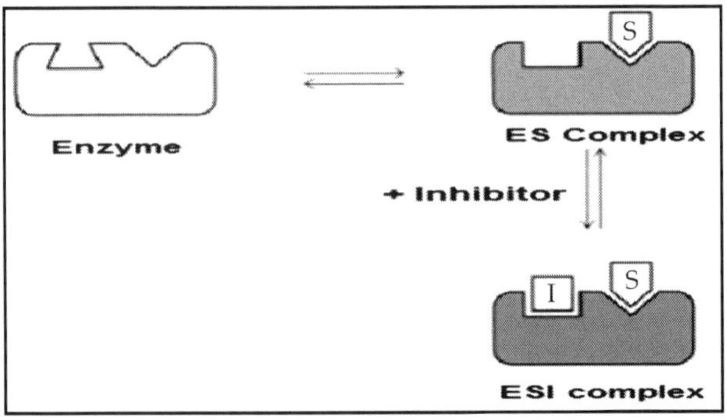

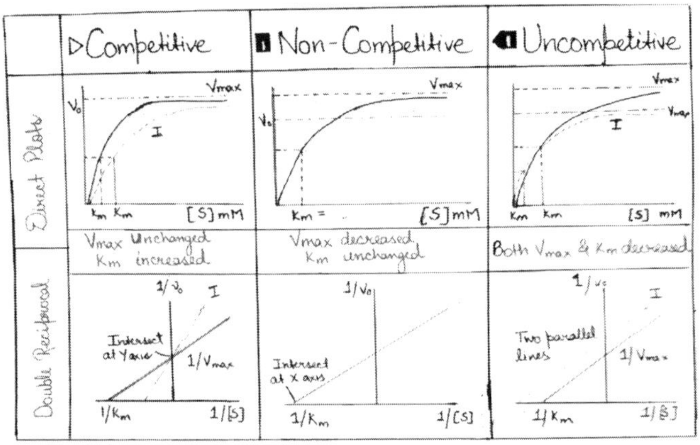

Isoenzymes

Isoenzymes (Isozymes): Some metabolic processes are regulated by enzymes that exist in different molecular forms. All forms demonstrate enzymatic enzyme activity and catalyze the same biochemical reaction, but they may differ in kinetics (different K, M and Vmax) regulatory properties (different effectors).

Ribozymes

Some forms of RNA that can serve as a biological catalyst wereacknowledged by the awarding of 1989 Nobel Prize in Chemistry Sidney Altman. Thomas Cech Ribozyme: RNA enzymes Ribonuclease P: Self-splicing RNA introns.

Significance of ribozymes

Their use as tools in specific cleavage of RNA has been designed to catalyze the *in vitro* cleavage of the RNA of HIV.

Abzymes: catalytic antibodies

It was reported in 1986 by Peter Schultz and Richard Lerner. It has catalyzed the hydrolysis of esters and carbonates.

10. NUCLEIC ACIDS AND CHROMOSOMES: NUCLEIC ACID STRUCTURE AND FUNCTION, PHYSICAL AND CHEMICAL PROPERTIES, SPECTROSCOPIC AND THERMAL PROPERTIES, IN VITRO DENATURATION AND RENATURATION OF DNA, DNA AS GENETIC MATERIAL.

Introduction

It was first discovered by Friedrich Miescher from the nuclei of the pus cells (Leukocytes) from discarded surgical bandages and called it nuclein. Nuclein was chemically a nucleoprotein. Later, Richard Altman in 1889 obtained the first protein-free material, to which he gave the name nucleic acid. It is the polymer of nucleotides. There are two types of nucleic acids: ribonucleic acid (RNA) and deoxyribonucleic acid (DNA).

The cell is the basic unit of life. All living organisms including bacteria, plants and animals are built from cells. Generally, cells are very small, with diameters much less than 1 mm, so they are invisible to the naked eye. In the simplest cells, bacteria, a cell wall surrounds a very thin fatty acid containing outer plasma membrane, which in turn surrounds a superficially unstructured inner region located in the bacterial DNA that carries the genomic information. These nuclei free bacteria and close relatives are known as prokaryotes. In prokaryotes, nuclear material is not separated from the cytoplasm by a discrete membrane.These also lack an extensive cellular architecture: membranous organelles are absent and the genetic material is not enclosed in a distinct structure. In all the cells other than bacteria, the inner cellular mass is partitioned into a membrane-bound spherical body called nucleus and an outer surrounding cytoplasm. In the nucleus is located the cellular DNA in the form of coiled rods known as chromosomes. Cells that contain a nucleus and with other distinctive membranous organelles such as mitochondria, vesicles and Golgi bodies are referred to as eukaryotes. In the cells of eukaryotes, the nuclear membrane separates the genetic material from the cytoplasm, which is then further subdivided by other distinct membranous structures.

Eukaryotes can be unicellular and multicellular and comprise all macroscopic forms of life such as plants, animals and fungi. Plant cells differ from animal cells in the presence of an outer cell wall surrounding the cell membrane and chloroplasts, whereas animal cells contain centrosomes, which are involved in cell division and lysosomes. The essence of cells is the ability to grow and divided to produce progeny of cells thatare capable ofgenerating new cellular molecules and replicating themselves. Thus cells are in effect tiny factories that grow by taking up simples molecular building blocks.

Considering the complexity of an organism, one is struck by the relatively small number of organic components used in the formation of substances; see**figure 10.1.** Cells contain molecules of great variety, but living cells arecomposed of restricted sets of elements six of which (C, H, N, O, P, and S) make up more than 99% of its weight. Certain simple combination of atoms such as methyl (-CH3), hydroxyl (OH), carboxyl (-COOH), and amino groups (-NH2) groups occur repeatedly in biological molecules. Each such group has distinct chemical and physical properties. Molecules can be broadly grouped into two categories.

Small Organic Molecules

These are compounds with a molecular weight in the 100-1000 range containing up to 30 or so carbon atoms. Molecules of this kind are usually found free in solution in the cytoplasm where they form a pool of intermediates from which large molecules are made. Some of the small organic molecules are sugars, amino acids and fatty acids. Atleast 750 different types of small molecules are found.

Macromolecules

These are invariably polymeric molecules that are formed by joining together specific types of small molecules. A large polymeric molecule or more than one polymer associated into large molecules can be referred to as macromolecules. Some of the macromolecules such as carbohydrates, proteins and nucleic acids are most important. Three simple molecules such as sugar, amino acids and nucleotides which form macromolecules have been briefly described.

Sugars

There are three types of sugar molecules that areinvolved different role in organisms.

- Monosaccharide $(CH_2O)n$
- Disaccharide $(C_{12}H_{22}O_{11})$
- Polysaccharide (C6H10O5)n

Nucleotides

Nucleotides have 3 basic parts:

1. A 5-carbon sugar.
2. A phosphate group.
3. A nitrogenous (nitrogen-containing) base; see **figure 10.2**.

Nucleic Acids

Nucleic acids contain instructions that cells need to carry out all the functions of life. It is a very large organic molecule made up of carbon, oxygen, hydrogen, nitrogen, and phosphorus.

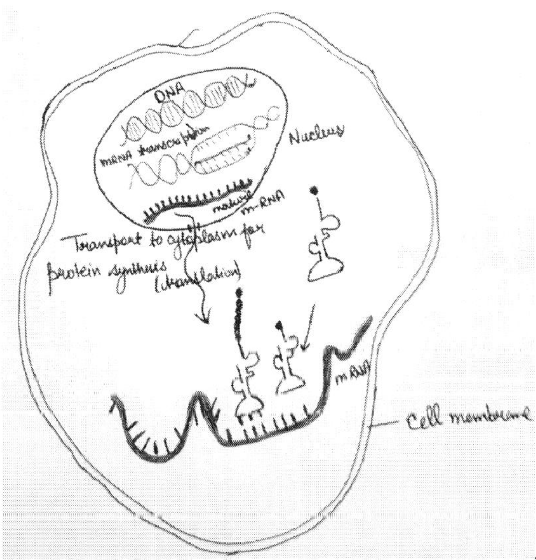

Figure 10.1. Cell structure

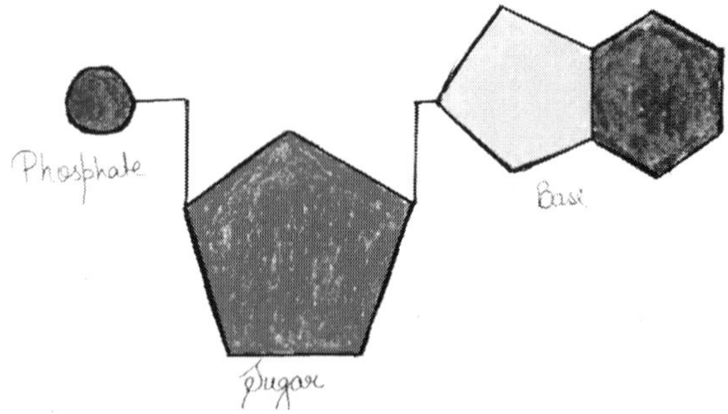

Figure 10.2. Nucleotides

Types of Nucleic Acid

DNA (deoxyribonucleic acid) - the genetic material that carries information about an organism. It is passed from parent to offspring. DNA directs all cellular activities and is only found in the nucleus of a cell. RNA (ribonucleic acid) – a nucleic acid found in the nucleus and cytoplasm of cells that plays an important role in the production of proteins; see figure 10.3.

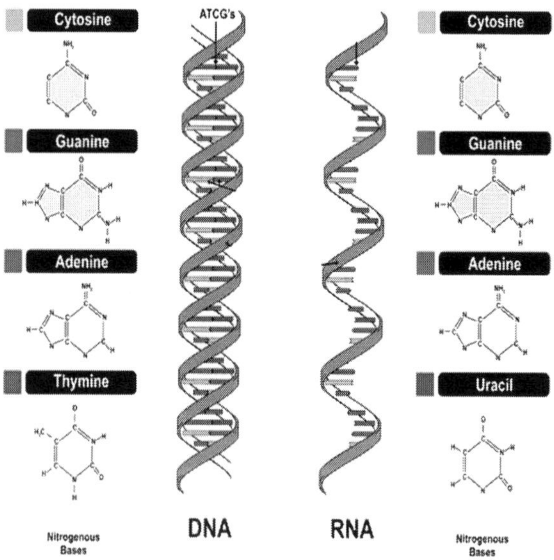

Figure 10.3. DNA and RNA sugar

Nucleotides &genetic organization

Nucleic acids are made up of nucleotides.

Nucleotides have 3 basic parts

1. A 5-carbon sugar.
2. A phosphate group.
3. A nitrogenous (nitrogen-containing) base.

Levels of genetic organization

Nucleotide	Gene	DNA	Chromosomes	Genome
Smallest building block of gene	Code for traits	Made of many genes	Tightly coiled strands of DNA	The entire genetic make-up of an organism

Composition of DNA

Major parts of a DNA nucleotide:A 5-carbon sugar called deoxyribose. Phosphates Four different nitrogenous bases (rungs of the ladder); see in **Figure 10.4.**

- Adenine (A)
- Thymine (T)
- Guanine (G)
- Cytosine (C)

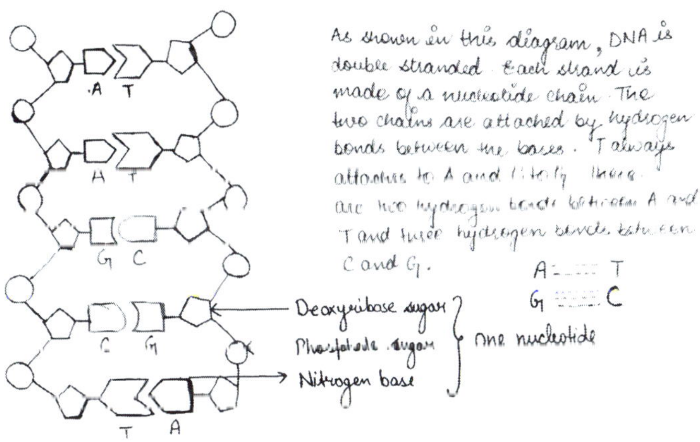

Figure 10.4. Nucleotide chain

Pairing the Nitrogenous Bases in DNA

Adenine always pairs (matches up) with thymine and guanine always pairs (is complementary to) with cytosine. These nitrogenous bases make up four different possible nucleotides in DNA. Many nucleotides (basic unit of nucleic acids) placed together to make one DNA molecule see in **figure10.5.**

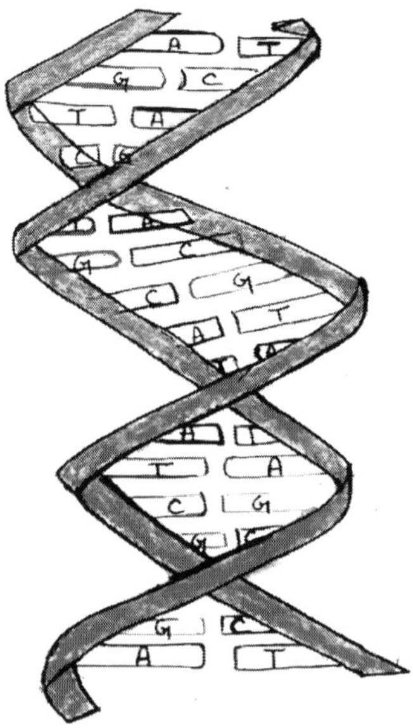

Figure 10.5. DNA structure

Structure of DNA

James Watson & Francis Crick – American biologist & British physicist who built the first accurate structural model of DNA (Nobel Prize in 1962). Watson & Crick's model of DNA was a double helix (twisted ladder) in which 2 strands were wound around each other. The sides of the ladder are made of deoxyribose sugar and phosphates. The stages of the ladder are the nitrogenous bases (connect to deoxyribose sugar in the outside/side of the ladder).

Identifying Base Pairs

DNA strands are complementary to one another (match up with each other). A & T and G & C. Given the following sequences of DNA, find the complementary strands.

AGT TCA	CAG GTC
CTT GAA	AGC TCG

DNA replication

They make copies of their DNA before cell divide. DNA replication ensures that each daughter cell have all of the genetic information it needs to carry out its activities. DNA replication is semiconservative, meaning that when DNA copies, each new strand has ½ of the original strand.

Major steps in DNA replication

DNA unwinds (untwists) exposing the nitrogenous bases. DNA unzips are weak hydrogen bonds between nitrogenous bases break. Hydrogen bonds are the weakest type of bond that occurs in molecules. New complementary strands form nitrogenous bases floating in the nucleus of the cell pair up with the bases on each half of the DNA molecule. Two new DNA molecules that are exactly alike are formed. Both new strands are identical to the original strand of DNA.

The function of genes in DNA

Proteins determine the size, shape, and many other traits of an organism, eg. eye color, hair color, type of hairline, etc. Each gene in a DNA molecule codes for the production of a particular protein. **Gene:** a small portion of the DNA molecule that codes for a particular trait.

Order of nitrogenous bases

The order of the nitrogenous bases along a gene forms a genetic code that specifies what type of protein will be produced. The order of the nitrogenous bases tells the cell which amino acids to put together to make a protein. If one amino acid is changed or out of place, a different protein is made.

Since DNA cannot leave the nucleus to make proteins at the ribosomes, RNA directs protein synthesis see in **figures10.6 and 10.7.**

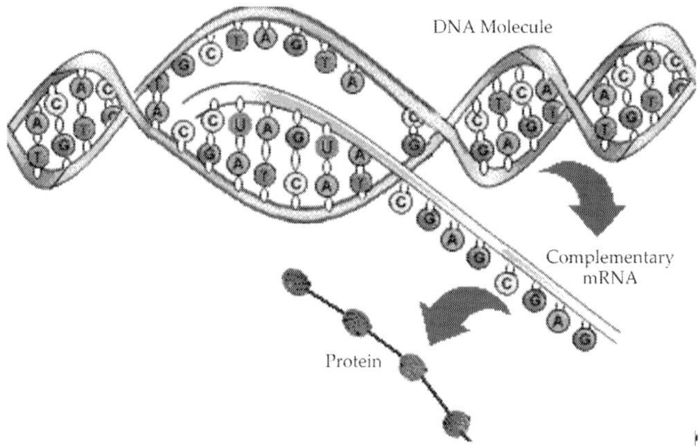

Figure 10.6. Nitrogenous bases involved to make protein.

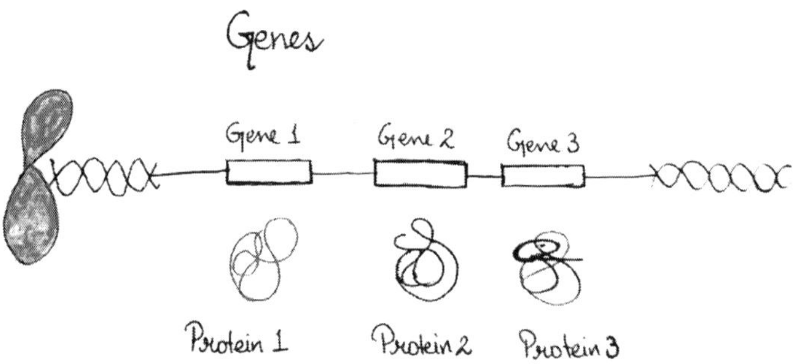

Figure 10.7. Sequence of genes

The beginning of our thinking of the possibility of genetic material begins with Mendel. He described the genetics of pea crosses, although he did not know what genetic material was he referred to "factors" as the things which gave his pea plants their physical characteristics. While Mendel's work gave a logical description of heredity it was largely ignored for many decades. His work was finally reproduced in the early 1900s and was thereafter regarded as fact read from genetics.

11. BIOENERGETICS: PRINCIPLES OF THERMODYNAMICS AND THEIR APPLICATIONS IN BIOCHEMISTRY -INTRODUCTION AND THERMODYNAMIC STATE FUNCTIONS, FIRST AND SECOND LAW OF THERMODYNAMICS, CONCEPT OF FREE ENERGY, STANDARD FREE ENERGY CHANGE, BIOLOGICAL OXIDATION-REDUCTION REACTIONS, REDOX POTENTIAL. HIGH ENERGY PHOSPHATE COMPOUNDS, FREE ENERGY OF HYDROLYSISOF ATP AND SUGAR PHOSPHATES.

INTRODUCTION

Living organisms must work to stay alive, to grow and to reproduce. All living organisms have the ability to produce energy and to channel it into biological work. Living organisms carry out energy transductions, conversions of one form of energy to another form. Modern organisms use the chemical energy in fuels (carbohydrates, lipids) to bring about the synthesis of complex macromolecules from simple precursors. They also convert the chemical energy into concentration gradients and electrical gradients, into motion and heat, and, in a few organisms into light (fireflies, some deep-sea fishes). Biological energy transductions obey the same physical laws that govern all other natural processes. Bioenergetics is the quantitative study of the energy transductions that occur in living cells and of the nature and function of the chemical process underlying these transductions.

Bioenergetics and Thermodynamics

Biological energy transductions obey the laws of thermodynamics

Laws of Thermodynamics

The total amount of energy in the universe remains constant for any physical or chemical change, energy may be changed from or it may be transported from one region to another, but it can not be created or destroyed. The universe always tends toward increasing disorder in all natural processes the entropy of the universe increases. The reacting system may be an organism, a cell or two reacting compounds. The reacting system and its surroundings together constitute the universe. In the laboratory, some chemical or physical processes can be carried out in closed systems and no material or energy is exchanged with the surroundings. However, living organisms are open systems. They exchange both material and energy with their surroundings. Living systems are never at equilibrium with their surroundings.

Gibbs free energy (G): G expresses the amount of energy capable of doing work during a reaction at constant temperature and pressure. When a reaction proceeds with the release of free energy (that is, when the system changes so as to possess less free energy) ΔG has a negative value and the reaction is said to be exergonic. In endergonic reactions, the system gains free energy and ΔG is positive. The unit of ΔG is joules/mole or calories/mole. **Enthalpy (H):** H is the heat content of the reacting system. H reflects the number and kinds of chemical bonds in the reactants and products. When a chemical reaction releases heat, it is said to be exothermic, the heat content of the products is less than that of the reactants and ΔH has a negative value. Reacting systems that take up heat from their surroundings are endothermic and have positive values of ΔH. The unit of ΔH is joules/mole or calories/mole. Entropy (S): S is a quantitative expression for the randomness or a disorder in a system. The unit of ΔS is joules/mole. **Kelvin**. Under the constant temperature and pressure changes in free energy, enthalpy and entropy in biological systems are related to each other by the equation

$$\Delta G = \Delta H - T\Delta S$$

ΔG= Change in Gibbs free energy of the reacting system ($G_{products} - G_{reactives}$)

ΔH= Change in enthalpy of the reacting system($H_{products} - H_{reactives}$)

T= Absolute temperature

ΔS= Change in entropy of the reacting system($S_{products} - S_{reactives}$)

Living organisms preserve their internal order by taking free energy from their surroundings in the form of nutrients or sunlight and returning to their surroundings an equal amount of energy as heat and entropy.

Oxidation of glucose

Aerobic organisms extract free energy from glucose obtained from their surroundings by oxidizing the glucose with oxygen (also obtained from surroundings). The end products of this oxidation reaction are CO_2 and H_2O and they are returned to the surroundings. At the end of this process, the surroundings undergo an increase in entropy, whereas the organism itself remains in a steady-state and no change occurs in its internal order.

$$C_6H_{12}O_6 + 6O_2 \longrightarrow 6CO_2 + 6H_2O$$

We can represent this schematically as

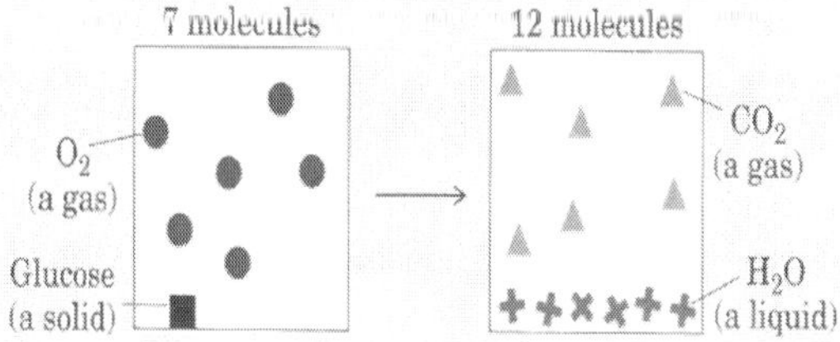

Whenever a chemical reaction results in an increase in the number of molecules the entropy of the surroundings increases. Whenever a solid substance is converted into liquid or gaseous forms the entropy of the surroundings increases because these transformations allow the molecule more freedom for movement. Living organisms acquire free energy from nutrient molecules. Cells transform this free energy into ATP and other energy-rich compounds.They are capable of providing energy for biological work at constant temperature. The composition of a reacting system tends to continue changing until equilibrium is reached. At the equilibrium the rates of the forward and reverse reactions are equal and no further change occurs in the system.

The K_{eq} is defined by the molar concentrations of products and reactants at equilibrium

aA+bB cC+ dD

$$K_{eq} = \frac{[C]^c[D]^d}{[A]^a[B]^b}$$

When a reacting system is not at equilibrium, the tendency to move toward the equilibrium represents a driving force. The magnitude of this driving force is expressed as free energy change (ΔG). Under standard conditions (25^0C), when reactants and products are initially at the 1 M concentrations the force driving the system toward equilibrium is defined as the standard free energy change (ΔG^0). By this defination, standard state for reactions involves $[H^+] = 1M$ or pH=0. However, most biochemical reactions occur in well-buffered aqueous solutions near pH=7. For the convenience of calculations, biochemists define a different standard state in which the concentration of $[H^+]$ is 10^{-7} M and for reactions that involve Mg^{2+} (available in most reactions involving ATP), its concentration in solution is commonly taken to be constant at 1mM. Physical constants based on this biochemical standard state are called standard transformed constants and written as $\Delta G'^0$ and K'eq to distinguish them from the untransformed constants which are used by chemists. $\Delta G'^0$ is the difference between the free energy content of the products and the free energy contents of the reactants under standard conditions.

$$\Delta G'^0 = \Delta G'^0_{products} - \Delta G'^0_{reactives})$$

When $\Delta G'^0$ is negative, the products contain less free energy than the reactants and the reaction will proceed spontaneously under standard conditions. When $\Delta G'^0$ is positive, the products contain more free energy than the reactants and the reaction will tend to go in the reverse direction under standard conditions. Each chemical reaction has a characteristic standard free energy change which may be positive, negative or zero depending on the equilibrium constant of the reaction. $\Delta G'^0$ tells us in which direction and how far a given reaction must go to reach equilibrium when the initial concentration of each component is 1M, the pH is 7, the temperature is 25^0C. Thus $\Delta G'^0$ is a constant; a characteristic for a given reaction.

Actual Free Energy Change

Actual free energy change (ΔG) is a function of reactant and product concentrations and of the temperature prevailing during the reaction which will not necessarily match the standard conditions as defined before. ΔG of any reaction proceeding spontaneously toward its equilibrium is always negative, becomes less negative as the reaction proceeds, and is zero at the point of equilibrium, indicating that no more work can be done by the reaction.

ΔG and $\Delta G›0$ for a reaction like that

- A+B C+D is written as

$$\Delta G = \Delta G›0 + RTln = \frac{[C][D]}{[A][B]}$$

The reaction is taking place at the standard temperature and pressure. But the concentrations of A,B,C and D are not equal and none of them at the 1M concentration. In order to determine actual ΔG under these non-standard concentrations as the reaction proceeds from left to right, we enter the actual concentrations of A,B,C and D in this equation.This principle of bioenergetics explains how a thermodynamically unfavorable (endergonic) reaction can be driven in the forward direction by coupling it to a highly exergonic reaction through a common intermediate. The main rule in biochemical reactions in living organisms. All endergonic reactions are coupled to an exergonic reaction. There is an energy cycle in cells that links anabolic and catabolic reactions. An example: The first step of glycolysis.

Glucose + P_iGlucose 6-phosphate+H_2O
$\Delta G'^0$ =13,8 kj/mol
$\Delta G'^0$>0 The reaction is not spontaneous

Another very exergonic cellular reaction. Hydrolysis of ATP (ATP + H_2O ------- ADP + P_i)

$\Delta G'^0$ = -30,5 kj/mol

These two reactions share the common intermediates H_2O and P_i and may be expressed as sequential reactions.

- Glucose + P_i Glucose 6-phosphate+H_2O
- ATP + H_2O ADP + P_i
- Sum: Glucose+ATP Glucose 6-phosphate+ADP
- The overall standard free energy changes:
- $\Delta G'^0$ =13,8 kj/mol + (-30,5 kj/mol)= -16,7 kj/mol
- The overall reaction is exergonic

Energy stored in ATP is used to drive the synthesis of glycose 6-phosphate, eventhough its formation from glucose and P_i is endergonic. This strategy works only if compounds such as ATP are continuously available.

Phosphoryl Group Transfer and ATP

Living cells obtain free energy in a chemical form by the catabolism of nutrient molecules. They use that energy to make ATP from ADP and P_i. ATP donates some of its chemical energy to

Endergonic processes such as the synthesis of metabolic intermediates and macromolecules from smaller precursors.

The transport of substances across membranes against concentration gradients. Mechanical motion (muscle contraction). This donation of energy from ATP can occur in the two form

A) ATP ADP+ P_i or

B) ATP AMP+ 2 P_i

The free energy change for ATP hydrolysis is large and negative. The hydrolytic cleavage of the terminal phosphoanhydride bond in ATP separates one of the three negatively charged phosphates and thus relieves some of the electrostatic repulsion in ATP. Released P_i is stabilized by the formation of several resonance forms not possible in ATP. ADP^{2-} is the other product of hydrolysis and it immediately ionizes, releasing H^+ into a medium of very low $[H^+]$ (~ 10^{-7} M). Because the concentrations of the products ofATP hydrolysis are far below the concentrations at equilibrium, mass action favors the hydrolysis in the cell. Although the hydrolysis of ATP is highly exergonic ($\Delta G'^0$ = -30,5 kJ/mol), the ATP is stable at pH 7, because the activation energy for ATP hydrolysis is relatively high. Rapid hydrolysis of ATP occurs only when catalyzed by an enzyme. The free energy change for ATP hydrolysis is -30,5 kJ/mol under standard conditions but the actual free energy change (ΔG) of ATP hydrolysis in living cells is very different.

The cellular concentrations of ATP, ADP and P_i are not the same and are much lower than the 1 M standard conditions. In addition, Mg^{2+} in the cytosol binds to ATP and ADP and for most enzymatic reactions that involve ATP as phosphorylgroup donor, the true substrate is $MgATP^{-2}$. The relevant $\Delta G'^0$ is therefore that for $MgATP^{-2}$ hydrolysis. Compounds have large free energy change such as Phosphorylated compounds and Thioesters (Acetyl-CoA)

Phosphorylated compounds

- Phosphoenolpyruvate
- 1,3-bisphosphoglycerate
- Phosphocreatine
- ADP
- ATP
- AMP
- PPi
- Glucose 1-phosphate
- Fructose 6-phosphate
- Glucose 6-phosphate

Thioesters

In thioesters, a sulfur atom is replaced the usual oxygen in the ester bond. Thioesters have large, negative standard free energy change of hydrolysis. Acetyl coenzyme A is one of many thioesters important in metabolism. The acyl group in these compounds is activated for trans-acylation, condensation or oxidation-reduction reactions. Hydrolysis of the ester bond generates a carboxylic acid which can ionize and assume several resonance forms.

$\Delta G'^0 = -31,4$ kj/mol for acetyl-CoA hydrolysis.

ATP is the chemical link between catabolism and anabolism. The exergonic conversion of ATP is coupled to many endergonic processes in living organisms. Cells also contain some high-energy compounds which have a high phosphorylation potential, like ATP. They are good donors of phosphoryl groups.

CHAPTER 2

ENZYMOLOGY

12. INTRODUCTION: REVIEW OF A BRIEF HISTORY, ENZYMES AS BIOLOGICAL CATALYSTS, CLASSIFICATION, NOMENCLATURE, PROXIMITY AND ORIENTATION, COVALENT CATALYSIS, ACID-BASE CATALYSIS.

Introduction

It is the most remarkable and highly specialized protein. Catalyze hundreds of stepwise reactions in biological systems. Regulate many different metabolic activities necessary to sustain life. Living systems use catalysts called enzymes to increase the rate of chemical reactions. The study of enzymes has immense practical importance. In medical science: to know the epidemiology, to diagnose, and to treat diseases (inheritable genetic disorders).

- In chemical industries
- In food processing
- In agriculture
- In everyday activities in the home (food preparation, cleaning, beauty care ,etc.)

Nomenclature of Enzymology

The International Enzyme Commission (EC) has recommended a systematic nomenclature for enzymes. This commission assigns names and numbers to enzymes according to the reaction they catalyze. An example of a systematic enzyme name is EC 3.5.1.5 urea aminohydrolases for the enzyme that catalyzes the hydrolysis of urea. The name of an enzyme frequently provides a clue to its function. In some cases, an enzyme is named by incorporating the suffix *-ase* into the name of its substrate, e.g., *pyruvate decarboxylases* catalyzes the removal of a CO_2 group from pyruvate. Certain protein cutting digestive enzymes are an exception to this general rule of enzyme nomenclature, e.g., *pepsin, trypsin, chymotrypsin and thrombin.* Isozyme:Within an organism, more than one enzyme may catalyze a given reaction. Multiple enzymes catalyzing the same reaction is called isozymes. Isozymes differ in their catalytic properties. Consequently, the various isozymes that are present in different tissues or at different developmental stages can carry out slightly different metabolic functions.

Pyruvate Decarboxylase

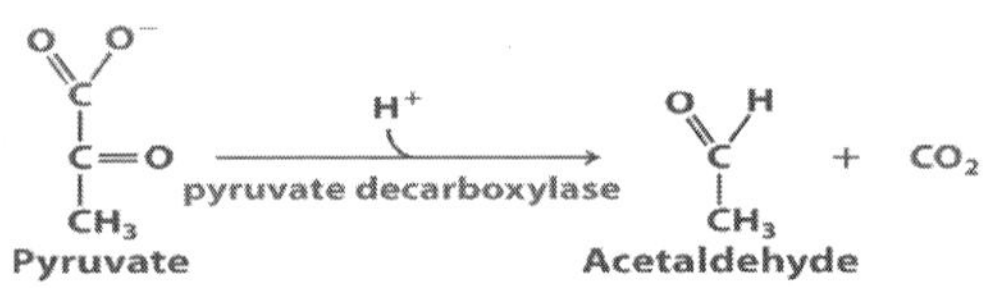

Alanine aminotransferase

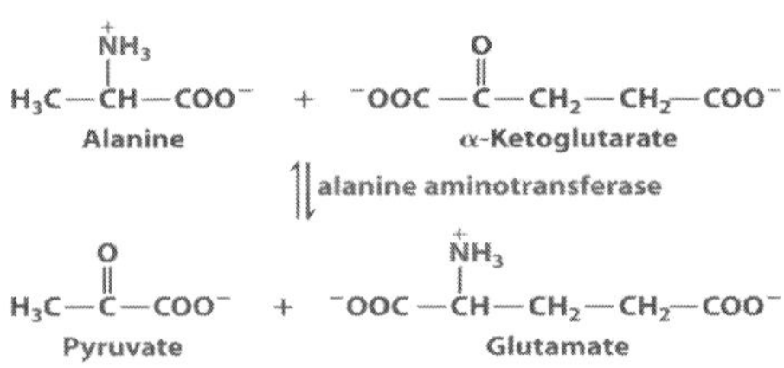

Enzyme classification

Class of Enzyme
1. Oxidoreductases: Oxidation-reduction reactions
2. Transferases: Transfer of functional groups
3. Hydrolases: Hydrolysis reactions
4. Lyases: Group elimination to form a double bond
5. Isomerases: Isomerization reactions
6. Ligases: Bond formation coupledwith ATP hydrolysis

Enzymes vs. Non-Biological Catalysts

Enzymes have their reaction specificity. The activities of many enzymes are regulated so that the organism can respond to the changing condition or follow genetically determined developmental programs. Enzyme activity is affected by the temperature and pH of the medium.

Properties of Enzymes

All enzymes are proteins or protein in nature (except a small group of catalytic molecules). Enzymes have a molecular weight ranging from 20,000 - 1 million. Some enzymes require no chemical group other than their amino acid residue for activity. Some enzymes may require an additional chemical compound called cofactors (Zn^{2+}, Mg^{2+}, Fe^{2+}, Mn^{2+}etc.). Some enzymes may require a complex organic or metallorganic molecule called coenzymes. Sometimes they need both cofactors and coenzymes. A coenzyme or metal ion that is covalently bound to the enzyme protein is called a prosthetic group. A complete catalytically active enzyme together with its coenzymes and or metal ions is called a holoenzyme. The protein part of such an enzyme is called the apoenzyme or apoprotein. Enzymes are classified by the reactions they catalyze. Many enzymes have been named by adding the suffix "-ase" to the name of their substrate or to a word or phrase describing their activities. Thus urease catalyzes the hydrolysis of urea. An enzyme circumvents many problems providing a specific environment within which a given reaction is energetically more favorable. An enzyme-catalyzed reaction occurs within the confinement of a pocket on the enzymes called the **active site**. The molecule that is bound by the active site and acted upon by the enzyme is called the **substrate.** The enzyme-substrate complex is central to the action of enzymes.

Enzyme Specificity

Enzymes are specific and efficient catalysts. In nonenzymatic reactions, substrates are transformed into a product by bond-making and bond-breaking processes. In an enzymatic reaction, these processes occur through interactions of the substrates with appropriate side chains of the amino acid residues of the enzyme or with the coenzyme of the enzyme. One property of many enzymes is specificity: catalyzing one chemical reaction with only one substrate. An example iscarbonic anhydrase. This enzyme catalyzes only one reversible reaction, the breakdown of carbonic acid (H_2CO_3) to water and carbon dioxide. At ideal conditions, a single molecule of Carbonic anhydrase is capable of catalyzing the breakdown of about 36 million molecules of carbonic acid in one minute.

In a biological reaction, the reacting species must come together and undergo electronic rearrangement that results in the formation of a product. Let us consider an idealized transfer reaction in which compound A—B reacts with compound C to form two new products, A and B—C. In order for the first two compounds to react, they must approach closely enough for their constituent atoms to interact. Normally, atoms that approach too closely repel each other; but if the groups have sufficient free energy, they can pass this point and react with each other to form products. The energy-requiring step of the reaction is known as an energy barrier, called free energy of activation or activation energy ($G^{\ddagger}$).

Idealized transfer reaction

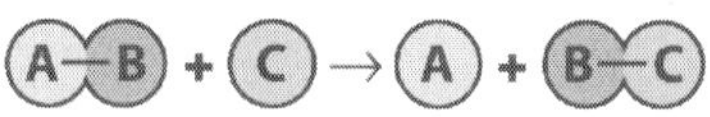

Reaction coordinate diagram for the reaction A—B + C   A + B—C

The progress of the reaction (reaction coordinate) is shown on the horizontal axis, and the free energy(*G*) is shown on the vertical axis. The transition state of the reaction, represented as a..b..c, is the point of highest free energy. The free energy difference between the reactant and the transition state is the free energy of activation *($G^{\ddagger}$)*.

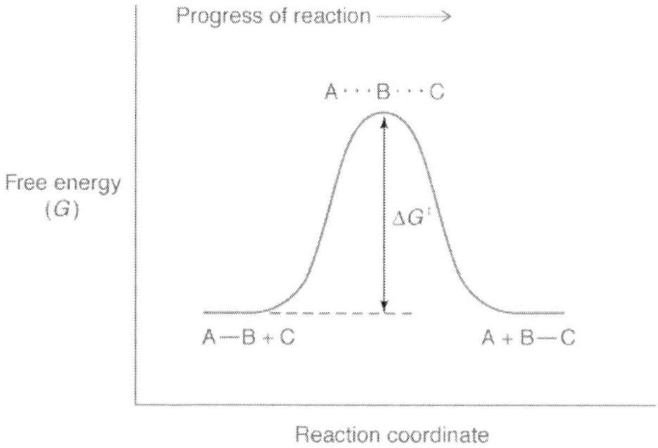

Reaction Coordinate of A Reaction: Reactants And Products Have Different Free Energies

When the free energy of the reactant is greater than that of the product, the free energy changes for the reaction is negative, so the reaction proceeds spontaneously (a). When the free energy of the product is greater than that of the reactant, the free energy change of the reaction is positive, so the reaction does not proceed spontaneously (however the reverse reaction does proceed).

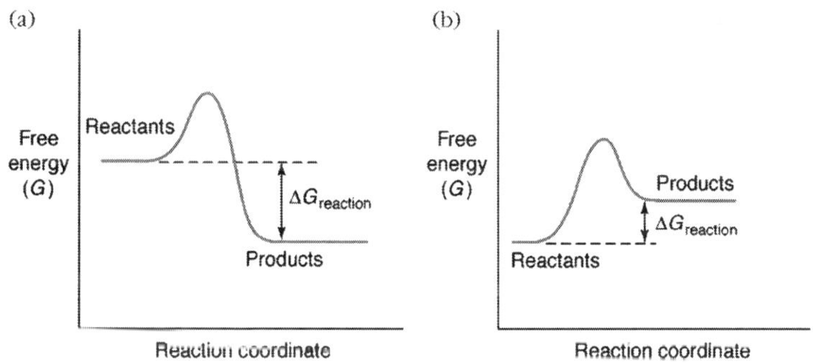

Effect of a catalyst on a chemical reaction

Here the reactants proceed through a transition state ($X^{\ddagger}$) during their conversion to products. A catalyst lowers the free energy of activation ($\Delta G^{\ddagger}$) for the reaction so that $\Delta G^{\ddagger}_{cat} < \Delta G^{\ddagger}_{uncat}$. Lowering thefree energy of the transition state ($X^{\ddagger}$) accelerates the reaction because more reactants are able to achieve the transition state per unit time.

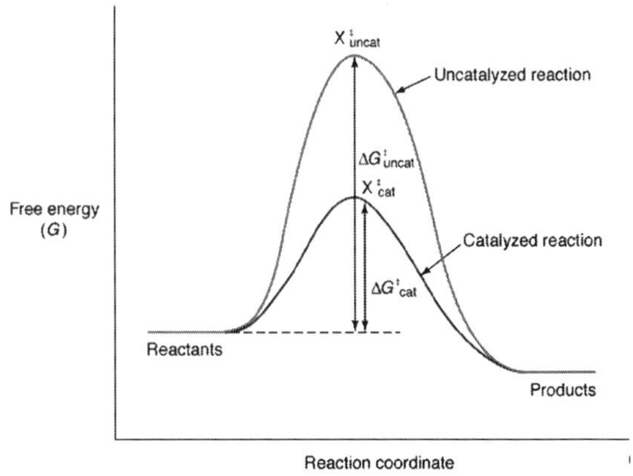

Enzymes use a chemical catalytic mechanism

There are three basic kinds of chemical catalytic mechanisms used by enzymes:

- Acid-base catalysis
- Covalent catalysis
- Metal ion catalysis

Acid-Base Catalysis

In **acid-base** catalysis, a proton is transferred between the enzyme and the substrate. It may be either **acid** catalysis or **base** catalysis or both. Many enzymes use both acid and base catalysis.

Tautomerization of a ketone

Tautomer's are interconvertible isomers that differ in the placement of hydrogen and a double bond. The dotted lines represent bonds in the process of breaking or forming.

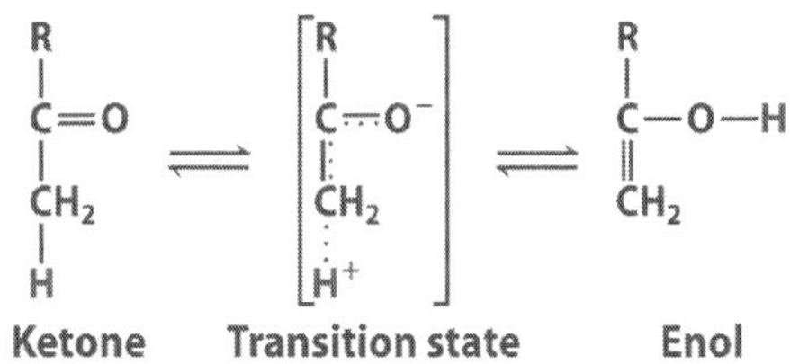

Acid catalysis

If a catalyst (H—A) donates a proton to the ketone's oxygen atom, it reduces the unfavorable character of the transition state, thereby lowers its energy. Hence, it also lowers the activation energy barriers for the reaction. This is an example of an acid catalyst, since the catalyst acts as an acid by donating a proton.

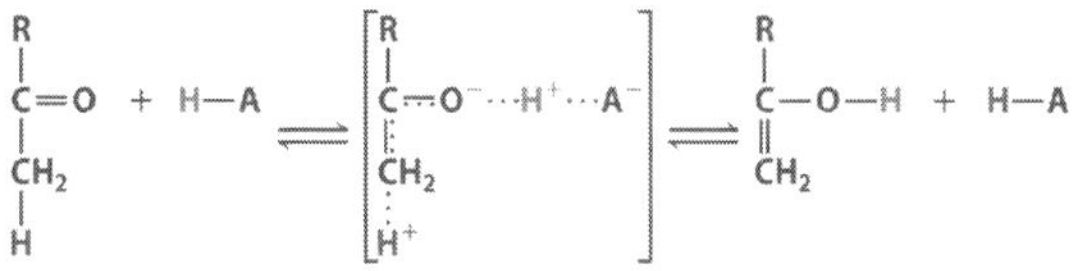

Ketoenol Tautomerization (base catalysis)

The same keto-enol tautomerization reaction shown in the previous slide can be accelerated by a catalyst that can accept a proton. In this case, it is a base catalyst (B), where the dots represent unpaired electrons.

Base catalysis lowers the energy of the transition state and thereby accelerates the reaction.

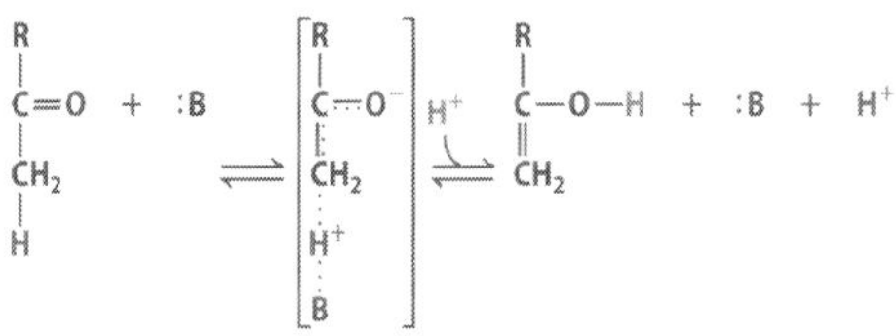

Amino acid side chains that can act as acid-base catalysts

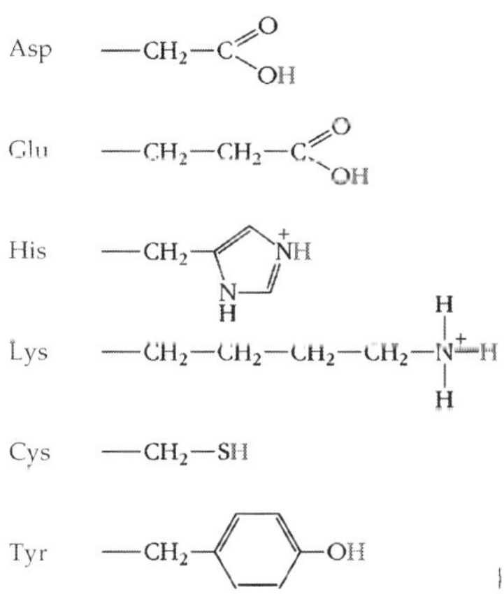

Covalent Catalysis

In covalent catalysis, a covalent bond is formed between the catalyst and the substrate during the formation of the transition state. In covalent catalysis, an electron-rich group in the enzyme active site forms a covalent adduct with a substrate. This covalent complex is much more stable than a transition state and can be isolated. Covalent catalysis is also called nucleophilic catalysis because the catalyst is a nucleophile (an electron-rich group in search of an electron -poor center (a compound with an electron deficiency is known as an electrophile). Enzymes that use covalent catalysis undergo a two-part reaction process. So the reaction coordinate diagram contains two energy barriers with the reaction intermediate between them.

Reaction Coordinate Diagram For A Reaction Accelerated By Covalent Catalysis

Two transition states flank the covalent intermediate. The relative heights of the activation energy barriers (the energies of the two transition states, $X^{\ddagger}_1$ and $X^{\ddagger}_2$) vary depending on the reactions.

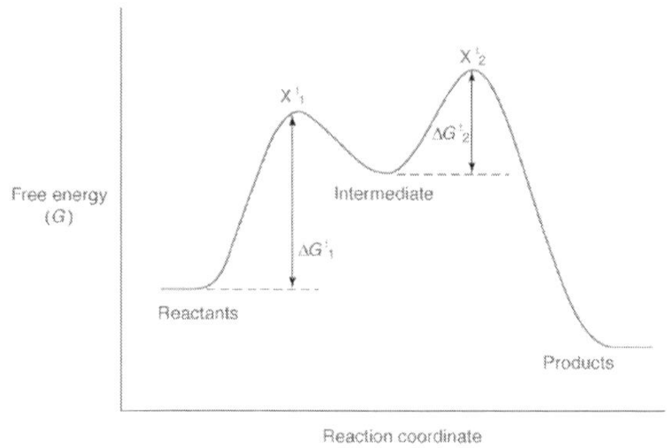

Metal ion catalysis

Metal ions participate in enzymatic reactions by mediating oxidation-reduction reactions, or by promoting the reactivity of other groups in the enzyme's active site through electrostatic effects. A protein-bound metal ion can interact directly with the reacting substrate **ex.** The conversion of acetaldehyde to ethanol catalyzed by the liver enzyme alcohol dehydrogenase. Here a zinc ion stabilizes the developing negative charge on the oxygen atom during the formation of the transition state.

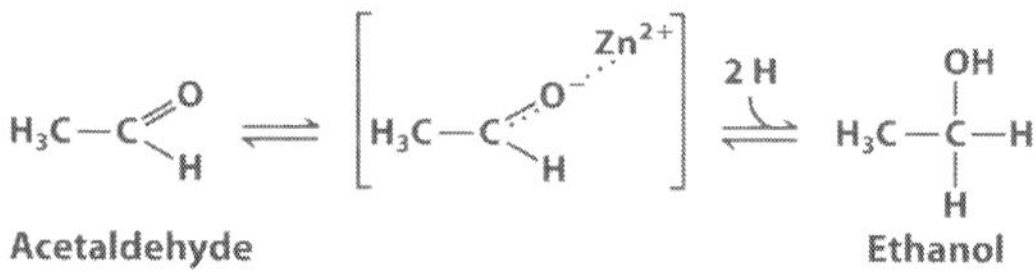

Enzyme–substrate complexes

Enzymes are specific and efficient catalysts. Lock and key fit of enzymes and substrates help produce enzyme specificity.

Lock-and-key model of enzyme action

In this early model, the substrate (key) was believed to bind tightly to a site on the enzyme (lock) that was exactly complementary to it in its size, shape and charge.

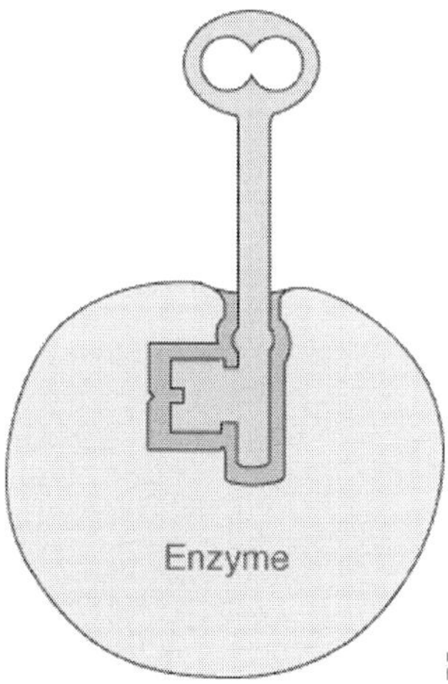

Enzyme-Substrate Complexes

Lock and key fit of enzymes and substrates helps produce enzyme specificity: A substrate molecule must contact an enzyme molecule before it can be transformed into products. Once the substrate has made contact, it must bind to the enzyme at the active site: a place where the conversion of substrate to a product can take effect.

The active site is usually a dimple, pocket or crevice formed by fold in the chain conformation of the protein. When the substrate key fits the enzyme lock, an enzyme-substrate complex is formed. The fit between enzyme and substrate in binding helps to produce enzyme specificity. Complementarity between enzyme and the substrate is governed by factors besides shape– e.g., the electrical charge. The substrate may be positively charged and the enzyme's active site negatively charged and vice versa. Hydrogen bonds and other weak forces may also hold the enzyme- substrate complex together. Hydrophobic regions of the substrate associate with similar regions on the active site of the enzyme. For most enzymes contributions from shape, charge, and a large number of weak forces are responsible for the formation and strength of the complex.

Transition State Stabilization

Linus Pauling (1946) was formulated an enzyme increases the reaction rate not by binding tightly to the substrates, but by binding tightly to the reaction's transition state i.e., a substrate that hasbeen strained toward the structure of the product. In other words, the tightly bound key of the lock and key model is the 'transition state', not the substrate. Pauling's theory of enzyme's actions is consistent with the chemical catalytic mechanisms, in which the catalyst stabilizes the transition state of the reaction. In an enzyme, tight binding(stabilization) of the transition state occurs in addition to acid-base, covalent, or metal ion catalysis.

Effect of very tight substrate binding on enzyme catalysis

The red line indicates the free energy changes as the reactants pass through a transition state to products. The blue line indicates free energy changes for the same reaction when the reactants (substrates) are very tightly bound to the enzyme (they are at a very low free energy level) before they react. The activation energy barrier ($G^{\ddagger}$) is therefore increased because the tight-binding must be overcome before the substrates can reach the transition state.

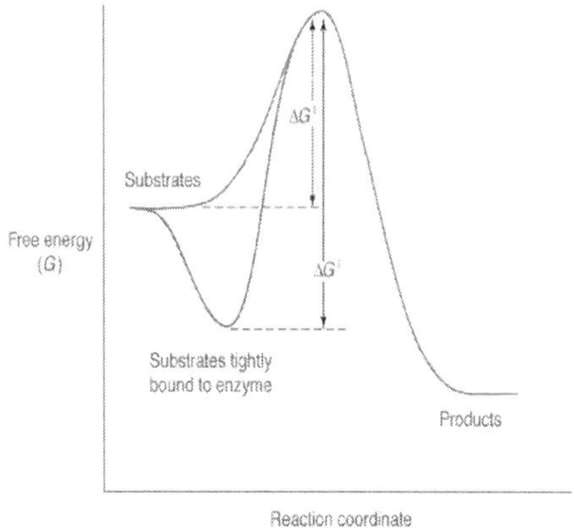

Enzyme assay

Enzyme activity is expressed as the rate at which the enzyme catalyzes the conversion of a substrate to products. Activity depends on the concentration of the enzyme and substrate, and reaction conditions (pH, temperature). Experiments in which the enzyme activity is measured are called enzyme assays. Many diagnostic procedures consist of assays for blood serum enzymes. The concept of this diagnostic procedure is simple: slight leakage of enzymes from tissues and organs to the circulatory system is normal; additional leakage is abnormal and indicates damage or disease. Damage to the heart muscle in heart attacks results in the release of the enzyme creatine phosphokinase (CPK) into the bloodstream. Over 40 other blood serum enzymes can be subjected to clinical testing.

13. ISOLATION AND PURIFICATION OF ENZYMES: OBJECTIVES, STRATEGY, CHOICE OF SOURCE, METHODS OF HOMOGENIZATION AND SEPARATION.

Introduction

Steps involved in isolation of a protein or enzyme

- Cell disruption (thiscan be done via a number of different processes of choice e.g.: Detergents lysis, Osmolysis, freeze-thaw cycles, enzymatic lysis, ultrasonication, Homogenisation).

- Centrifugation (at a specific speed depending on the organ, tissue, organelle or fluid).
- Removal of supernatant (Decantation to obtain supernatant).
- Huge number of enzymes (> than the # of genes encoded in a genome). Most cell types in multicellular organisms express tens of thousands of different proteins. Relative abundances of various proteins vary widely.

Goals of Protein Purification

It obtains a particular protein free of others and other cell components. It obtains a good yield (absolute amount and proportion of starting amount). They maintain the activity of the protein.

Problems during purification

- Denaturation
- Proteolysis
- *In vitro* mixing

A measurement of how well the process worked. Some characterization of the purified protein.

Make an extract from the source

Almost always want to keep it very cold (just above freezing). Gentle breaking. In some cases can obtain big purification in one step by separating cellular compartments (e.g., purify nuclei from cytoplasm before extracting nuclear proteins).

Begin separating components

Remove nucleic acids, polysaccharides, cell membrane debris. Ammonium sulfate precipitations, other crude fractionations (pH or other salt precipitations, antibody clearing, "autolysis"). These crude steps are often needed to avoid ruining or overloading chromatography agents.

Fractionate by chromatography

Several steps are almost always needed. Need to assay for amount and purity at each step. Need a way to decide when you are finished.

Separation principles

Sizing

The principle is based on the exclusion of larger molecules from pores in resin; smaller molecules require longer times of transit. Sizing resins = "gel filtration". Different matrices have different size ranges.

Examples: Sephadex, Sephacryl, Ultragel, some HPLC resins.

Ion exchange

Separate on the basis of net charge.Wash and elute with higher salt concentrations. Determine highest salt that your protein binds to the resin, and use this to load or to pre-wash before eluting your protein.

Examples: DEAE cellulose, phosphocellulose, some HPLC resins.

Affinity chromatography

Separate based on binding to residues specific or semi-specific to the protein you are purifying. Easy to overload with non-specific proteins **ex.** sequence-specific DNA sites and antibody resin.

Enzyme Characterization

- Enzyme characterization simply refers to the determination of the various chemical and physical properties (characteristics) of an enzyme. It involves the use of a series of laboratory procedures. Examples of assays for commonly characterized enzyme properties in biochemical researches include:
- Effect of changes in temperature on the enzyme's activity and optimum temperature.
- Enzyme's thermal stability.
- Effect of changes in pH on the enzyme's activity and optimum pH.
- pH stability.
- Effect of changes in substrate concentration on enzyme's activityand kinetic constants e.g Vmax, Km, Kcat, Km/Kcat, etc.
- Substrate specificity
- Molecular weight (Mw) of enzyme

- Effect of metal ions, chelating agent or denaturing agents.
- Enzyme's isoelectric point (pI)
- Effect of duration of incubation.
- Active site fractional saturation (V/Vmax) at a particular substrate concentration.
- Enzyme's turnover number (Vmax/ET)
- Activation energy (Ea)
- Salt tolerance

Within the cell, enzymes are generally found along with other proteins, nucleic acids, polysaccharides and lipids. The activity of the enzyme in relation to the total protein present (i.e. the specific activity) can be determined and used as a measure of enzyme purity. A variety of methods can be used to remove contaminating material in order to purify the enzyme and increase its specific activity. Enzymes that are used as diagnostic reagents and in clinical therapeutics are normally prepared to a high degree of purity because great emphasis is placed on the specificity of the reaction that is being catalyzed. Clearly the higher the level of purification, the greater the cost of enzyme production. In the case of many bulk industrial enzymes, the degree of purification is less important, and such enzymes may often be sold as very crude preparations of culture broth containing the growth medium, organisms (whole or fragmented) and enzymes of interest. However, even when the cheapest bulk enzymes are utilized (e.g. proteases for use in washing powders), the enzyme cost can contribute around 5–10% of the final product value.

Pre-treatment

At the end of a fermentation in which a microorganism rich in the required enzyme has been cultured, the broth may be cooled rapidly to 5°C to prevent further microbial growth and stabilize the enzyme product. The pH may also be adjusted to optimize enzyme stability. If the enzyme-producing organism is a fungus, this may be removed by centrifugation at low speed. If the enzyme source is bacterial, the bacteria are often flocculated with aluminum sulfate or calcium chloride, which negate the charge on the bacterial membranes, causing them to clump and thus come out of suspension.

Treatment

Extracellular enzymes are found in the liquid component of the pretreatment process. However, intracellular enzymes require more extensive treatment. The biomass may be concentrated by centrifugation and washed to remove medium components. The cellular component must then be ruptured to release the enzyme content. This can be done using one or more of the following processes:

- Ball milling (using glass beads)
- Enzymic removal of the cell wall
- Freeze–thaw cycles
- Liquid shearing through a small orifice at high pressure (e.g. within a French press)
- Osmotic shock
- Sonication.

Separation of enzymes from the resulting solution may then involve a variety of separation processes, which are often employed in a sequential fashion. The first step in an enzyme purification procedure commonly involves the separation of the proteins from the non-protein components by a process of salting out. Proteins remain in an aqueous solution because of interactions between the hydrophilic (water-loving) amino acids and the surrounding water molecules (the solvent). If the ionic strength of the solvent is increased by adding an agent such as ammonium sulfate, some of the water molecules will interact with the salt ions, thereby decreasing the number of water molecules available to interact with the protein. Under such conditions, when protein molecules cannot interact with the solvent, they interact with each other, coagulating and coming out of the solution in the form of a precipitate. This precipitate (containing the enzyme of interest and other proteins) can then be filtered or centrifuged and separated from the supernatant. Since different proteins vary in the extent to which they interact with water, it is possible to perform this process using a series of additions of ammonium sulfate, increasing the ionic strength in a stepwise fashion and removing the precipitate at each stage. Thus such fractional precipitation is not only capable of separating protein from non-protein components, but can also enable separation of the enzyme of interest from some of the other protein components. Subsequently, a wide variety of techniques may be used for further purification, and steps involving chromatography are standard practice.

Ion-Exchange Chromatography

It is often effective during the early stages of the purification process. The protein solution is added to a column containing an insoluble polymer (e.g. cellulose) that has been modified so that its ionic characteristics will determine the type of mobile ion (i.e. cation or anion) it attracts. Proteins whose net charge is opposite to that of the ion-exchange material will bind to it, whereas all other proteins will pass through the column. A subsequent change in pH or the introduction of a salt solution will alter the electrostatic forces, allowing the retained protein to be released into the solution again.

Gel filtration

It can be utilized in the later stages of a purification protocol to separate molecules on the basis of molecular size. Columns containing a bed of cross-linked gel particles such as Sephadex are used. These gel particles exclude large protein molecules while allowing the entry of smaller molecules. Separation occurs because the larger protein molecules follow a path down the column between the Sephadex particles (occupying a smaller fraction of the column volume). Larger molecules ,therefore, have a shorter elution time and are recovered first from the gel filtration column.

Affinity chromatography

The procedures can often enable purification protocols to be substantially simplified. Typically, with respect to enzyme purification, a column would be packed with a particulate stationary phase to which a ligand molecule such as a substrate analog, inhibitor or cofactor of the enzyme of interest would be firmly bound. As the sample mixture is passed through the column, the enzyme interacts with, and binds, to the immobilized ligand, being retained within the column as all of the other components of the mixture pass through the column unrewarded. Subsequently, a solution of the ligand is introduced to the column to release (elute) and thereby recover the bound enzyme from the column in a highly purified form. Nowadays numerous alternative affinity chromatography procedures exist that are able to separate enzymes by binding to areas of the molecule away from their active site. Advances in molecular biology enable us to purify recombinant proteins, including enzymes, through affinity tagging. In a typical approach, the gene for the enzyme of interest would be modified to code for a further short amino acid sequence at either the N- or C- terminal.

For example, a range of polyhistidine tagging procedures are available to yield protein products with six or more consecutive histidine residues at their N- or C- terminal end. When a mixture containing the tagged protein of interest is subsequently passed through a column containing a nickel-nitrilotriacetic acid (Ni-NTA) agarose resin, the histidine residues on the recombinant protein bind to the nickel ions attached to the support resin, retaining the protein, whilst other proteins and non-protein components pass through the column. Elution of the bound protein can then be accomplished by adding imidazole to the column, or by reducing the pH to 5-6 to displace the His-tagged protein from the nickel ions. Such techniques are therefore capable of rapidly and highly effectively isolating an enzyme from a complex mixture in only one step, and typically provide protein purities of up to 95%. If more highly purified enzyme products are required, other supplemental options are also available, including various forms of preparative electrophoresis e.g. disc-gel electrophoresis and isoelectric focusing.

Finishing of Enzymes

Enzymes are antigenic, and since problems occurred in the late 1960s when manufacturing workers exhibited severe allergic responses after breathing enzyme dust, procedures have now been implemented to reduce dust formation. These involve supplying enzymes as liquids wherever possible or increasing the particle size of dry powders from 10 μm to 200 500 μm by either prilling (mixing the enzyme with polyethylene glycol and preparing small spheres by atomization) or marumerizing (mixing the enzyme with a binder and water, extruding long filaments, converting them into spheres in a marumerizer, drying them and covering them with a waxy coat).

14. ENZYME KINETICS: MONOSUBSTRATE REACTIONS, MICHAELIS-MENTEN EQUATION AND ITS LINEAR TRANSFORMATIONS, KM AND VMAX: DEFINITION, DETERMINATION AND SIGNIFICANCE.

Introduction

Kinetics is the study of the rates of reactions. Enzymes provides cells with the remarkable capacity to exert kinetic control over thermodynamic potentiality. Enzymes are the agents of metabolic function which increases rate of reaction.

It is able to determine:

- Maximum velocity
- Substrate affinity
- Inhibitor affinity

It can tell us:

- Flow-through metabolic pathways
- Utilization of substrates

It can we do with the information:

- Control and manipulate metabolic events

The Michaelis-Menten equation

The ES complex is the key to understanding this kinetic behavior. In 1903, this kinetic pattern led Victor Henry to propose that an enzyme combines with its substrate molecule to form the ES complex as a necessary step in enzyme catalysis. In 1913, Leonor Michaelis and Maud Mentenexpanded the idea of Victor Henry into a general theory of enzyme actions.

Michaelis and Menten postulated that:

The enzyme first combines reversibly with its substrate to form an ES complex in a relatively fast reversible step. The ES complex then breaks down in a slower second step (rate-limiting step) to yield the free enzyme and the product. Michaelis-Menten equation is the rate equation for an enzyme-catalyzed reaction and is the mathematical description of the hyperbolic curve we have discussed earlier. The formula is:

$$V_o = \frac{V_{max}\,[S]}{K_m + [S]}$$

Where,

- V_0 is the initial velocity
- V_{max} is the maximum velocity
- [S] is the substrate concentration
- Km (Michaelis-Menten constant) is the substrate concentration at which the reaction velocity is the half of the maximum velocity.

$$E + S \underset{k_{-1}}{\overset{k_1}{\rightleftharpoons}} ES \xrightarrow{k_2} P + E$$

Conclusions of Michaels - Menten kinetics

- When $[S] = K_M$, the equation reduces to

$$V = \frac{V_{max}\,[S]}{K_M + [S]} = \frac{V_{max}\,[S]}{[S] + [S]} = \frac{V_{max}}{2}$$

- when $[S] >> K_M$, the equation reduces to

$$V = \frac{V_{max}\,[S]}{K_M + [S]} = \frac{V_{max}\,[S]}{[S]} = V_{max}$$

- when $[S] << K_M$, the equation reduces to

$$V = \frac{V_{max}\,[S]}{K_M + [S]} = \frac{V_{max}\,[S]}{K_M} = \frac{V_{max}}{K_M}\,[S]$$

Lineweaver – Burk Double Reciprocal Plots

- It is difficult to determine Vmax experimentally
- The equation for a hyperbola can be transformed into the equation for a straight line by taking the reciprocal of each side
- The formula for a straight line is $y = mx + b$
- A plot of 1/V versus 1/[S] will give a straight line with the slope of KM/Vmax and y-intercept of 1/Vmax

- Such a plot is known as a Lineweaver-Burk double reciprocal plot

$$\frac{1}{V} = \frac{K_M}{V_{max}} \cdot \frac{1}{[S]} + \frac{1}{V_{max}}$$

$$y = m \cdot x + b$$

Lineweaver – Burk Double Reciprocal Plots

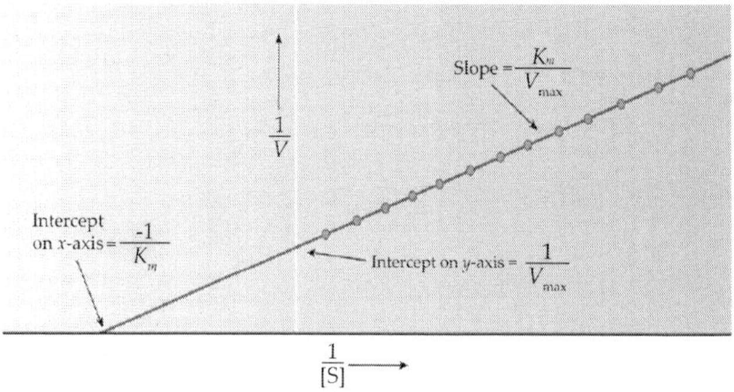

Significance of K_m

- Km is a constant
- Small Km means tight binding; high Km means weak binding
- Useful to compare Km for different substrates for one enzyme

  Hexokinase: D-fructose – 1.5 mM

  D-glucose – 0.15 mM
- Useful to compare Km for a common substrate used by several enzymes

  Hexokinase: D-glucose – 0.15 mM

  Glucokinase: D-glucose – 20 mM

Kinetic vs Chemical Mechanism:

- An enzyme kinetic mechanism is the order of substrate addition and product release in an enzyme-catalyzed reaction
- A chemical mechanism is the chemical pathway of conversion of S → P, including the structures of any intermediates

Bi-substrate Reactions:

- The Michaelis –Menten model of enzyme kinetics was derived for single substrate reactions

- The majority of enzymatic reactions have multiple substrates and products
- Bi-substrate reactions account for ~ 60% of the known enzymatic reactions.

$$A + B \overset{E}{\rightleftharpoons} P + Q$$

Substrate Addition / Product Release

- The order of substrate addition and product release in most enzymatic reactions follow two reaction mechanism

  – Sequential reaction - all substrates must bind to the enzyme before the reaction occurs and products are released

    • Ordered sequential

    • Random sequential

  – Ping-pong reaction - one or more products are released before all substrates have been added and an alternate stable enzyme form, F, is produced in the half reaction

Sequential reaction

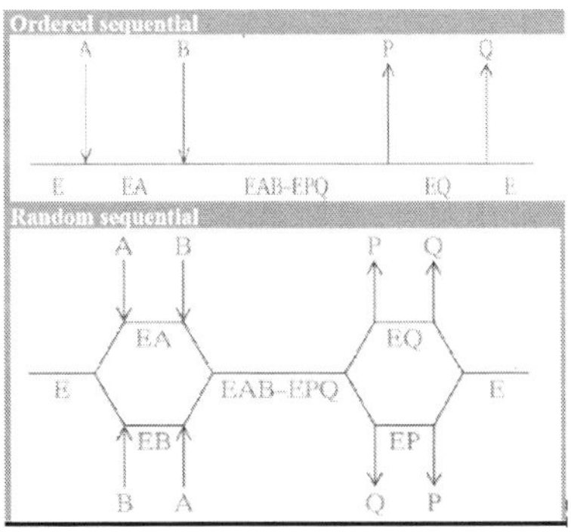

Ping-pong reaction

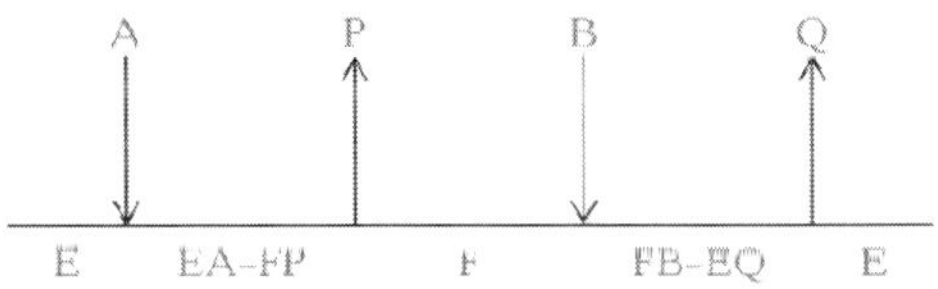

Initial Velocity Plots

Sequential reaction exhibits an intersecting pattern of lines. Order and random substrate additions cannot be distinguished in this type of plot.

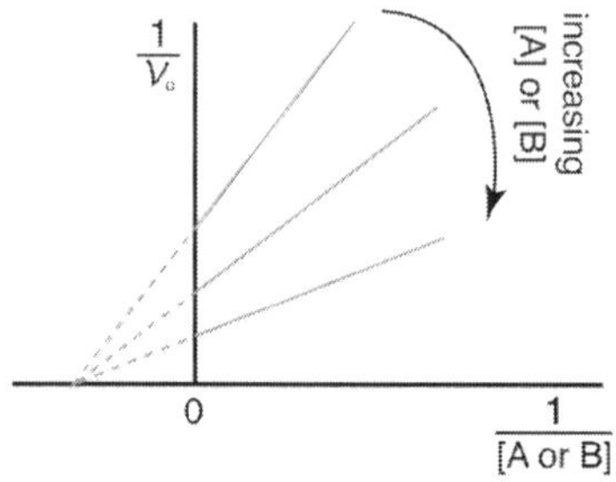

Ping-pong reaction shows parallel or non-intersecting lines

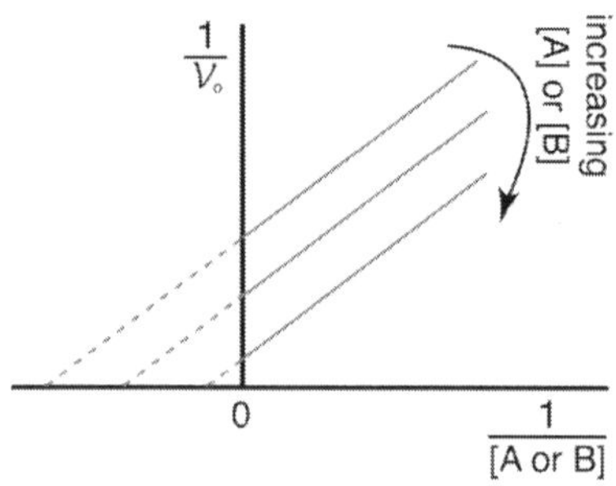

Influence of Enzyme Concentration

$$V = \frac{K_3[E][S]}{K_m + [S]}$$

$v = k_3$ [E], as [S]>>[E]

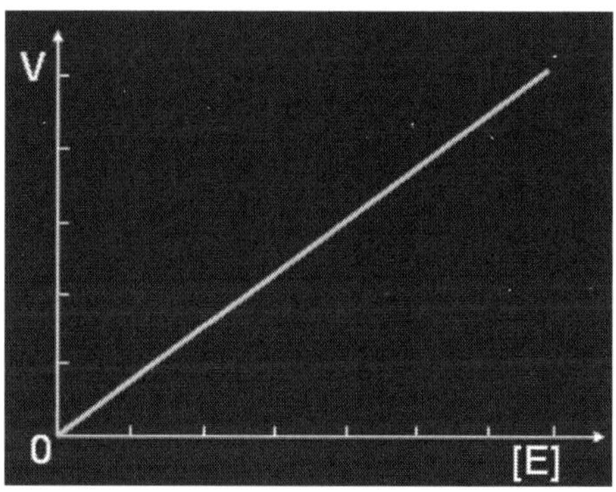

Influence of Temperature

Optimum temperature, most of them are in the range from 35 to 45°C .

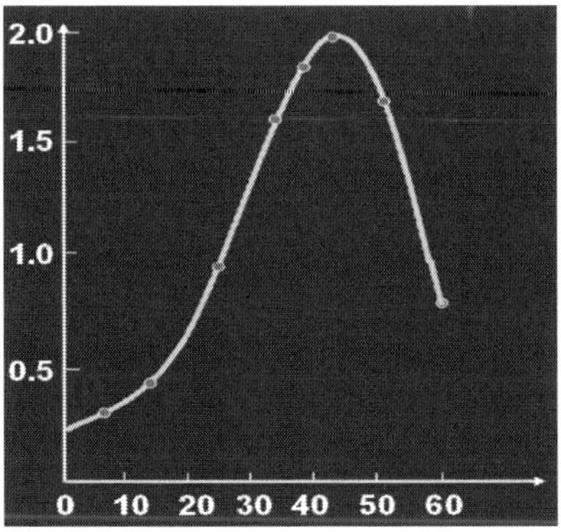

Influence of pH

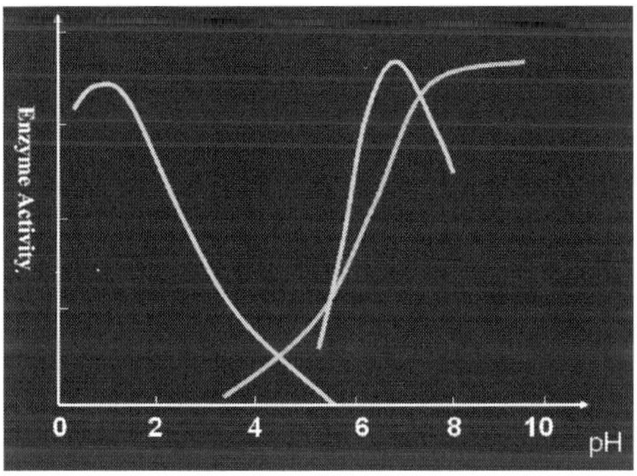

15. ENZYME INHIBITION: REVERSIBLE INHIBITION, COMPETITIVE, NON- COMPETITIVE AND UNCOMPETITIVE INHIBITIONS. IRREVERSIBLE INHIBITION, SPECIFIC EXAMPLES

Enzyme inhibitors are important:

- They can be used to gain information about the shape ofthe enzyme active site and the amino acid residues in the active site.
- They can be used to gain information about the chemical mechanism.
- They can be used to gain information about the regulation or control of a metabolic pathway.
- They can be very important in drug design.

Reversible Inhibitor

A substance that binds to an enzyme to inhibit it, but can be released.It usually involves the formation of non-covalent bonds.

Irreversible inhibitor

A substance that causes inhibition that cannot be reversed. It is usually involves the formation or breaking of covalent bonds to or on the enzyme.

Inhibitors

- Reversible
- Irreversible inhibition

Irreversible inhibition

The inhibitor combines with an essential group of enzyme active center by a covalent bond, resulting in enzymatic activity loss.

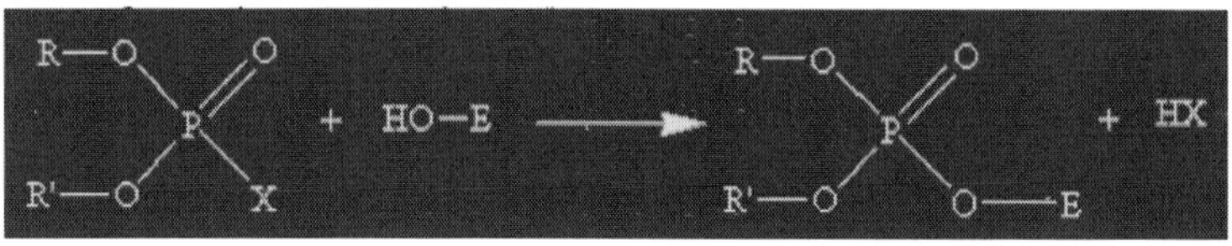

Inhibition Patterns

An inhibitor may bind at the same site as one of the substrates. These inhibitors structurally look like the substrate. An inhibitor may bind at an alternate site affecting catalytic activity without affecting substrate binding. Many inhibitors are do both

- Competitive
- Mixed or Non-competitive
- Uncompetitive

Competitive inhibition

Competitive inhibitor is competes with a substrate for the enzyme - substrate binding site. Malonate is a competitive inhibitor of succinate for succinate dehydrogenase.

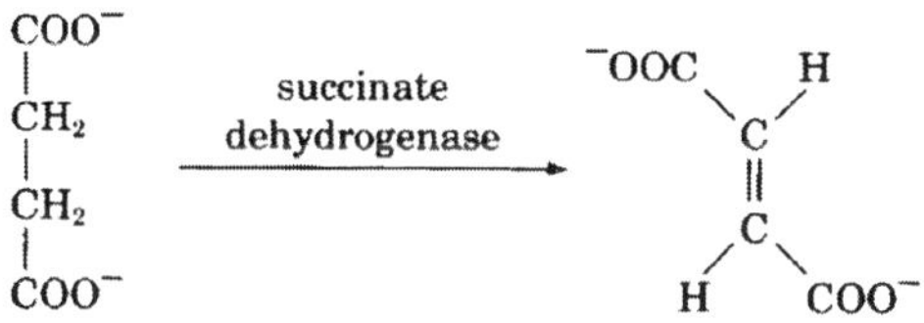

Succinate **Fumarate**

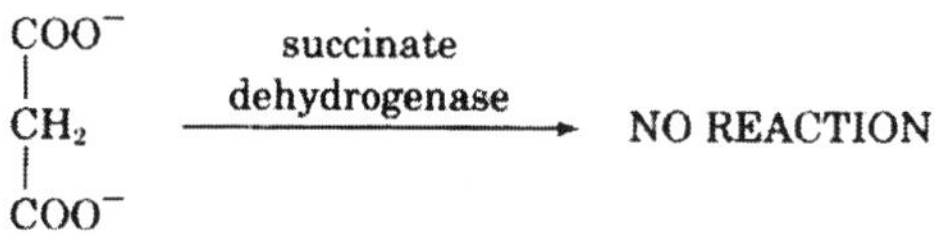

Malonate

Competitive inhibition

A competitive inhibitor reduces the amount of free enzyme available for substrate binding thus increasing the Km for the substrate. The effect of a competitive inhibitor can be overcome with high concentrations of the substrate.

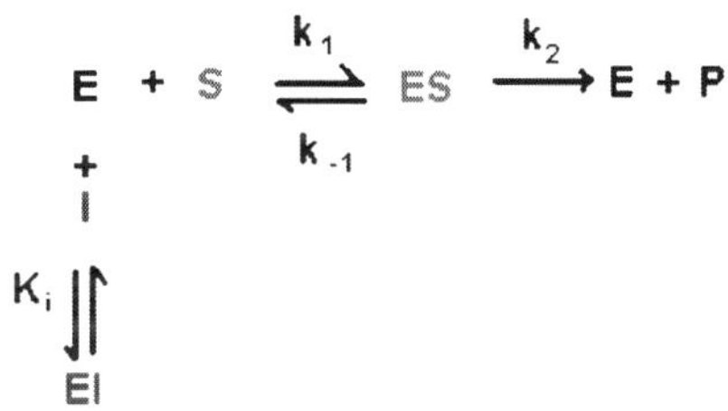

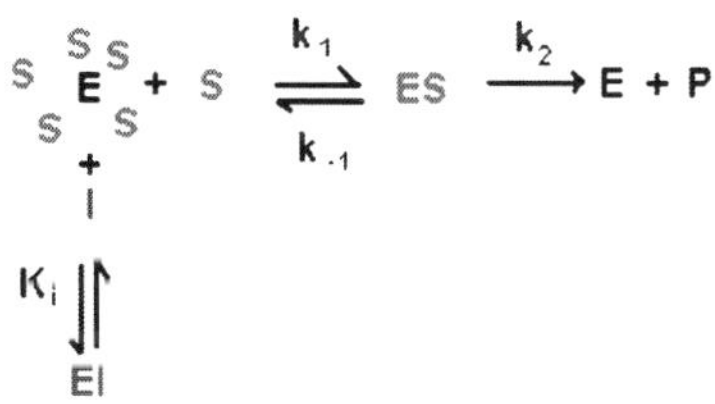

Uncompetitive inhibition

An uncompetitive inhibitor is binds to the enzyme-substrate complex but not to free enzyme. The result is a decrease in Vmax and Km. The effect of an uncompetitive inhibitor cannot be overcome by high concentrations of the substrate.

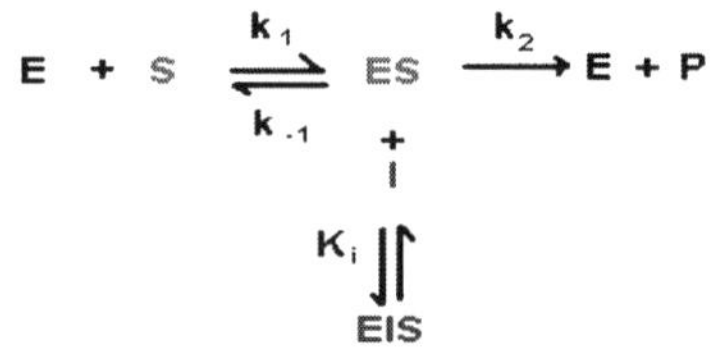

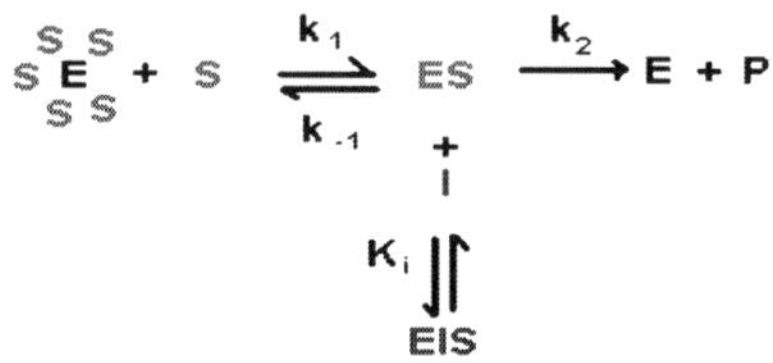

Mixed or non-nompetitive inhibition

- The inhibitor can bind to both free enzyme and the ES complex
- The affinity of the inhibitor to the two complexes might be different
  – If the binding of inhibitor changes the affinity for the substrate, Km will be changed and called mixed inhibition
  – If only Vmax affected called Non-competitive inhibitor

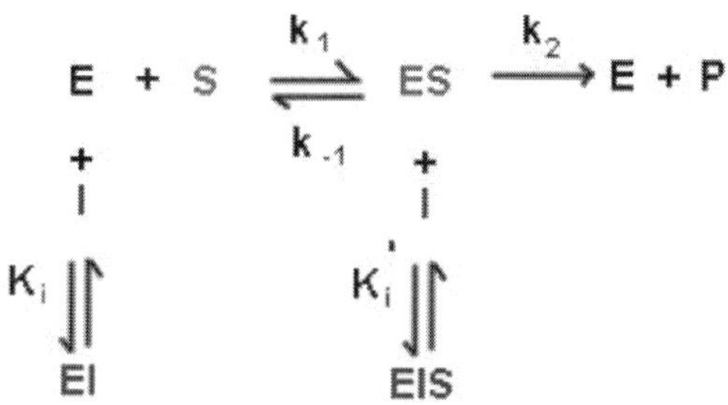

Allosteric Inhibition

Those with multiple active sites in one multi-subunit protein are commonly subjected to allosteric inhibition. Catalytic proteins containing both an active site and a modulator site are called allosteric enzymes. In allosteric inhibition, the inhibitor binding to one subunit of an enzyme may decrease the catalytic activity of all the subunits

Feedback Inhibition

Cells have other ways of controlling the activity of enzymes. A fine slowdown control of enzyme activity built into some enzymes relies on feedback inhibition i.e., the concentration of a product at the end of a long series of steps builds up and then "feedback" to inhibit the enzyme in a preceding step.

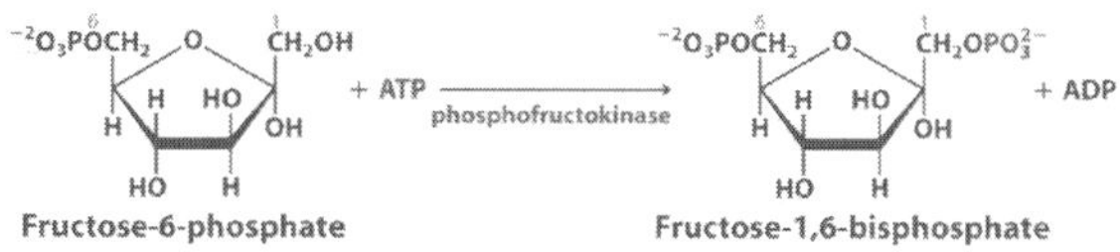

Denaturation of Enzymes

Enzymes are sensitive to pH and temperature. Enzymeactivity varies with changes in pH. Most enzymes exhibit maximum activity at a pH near 7, the pH of most body fluids. The pH where maximum activity occurs is called the optimum pH of the enzyme. Enzyme activity usually varies with temperature. At a temperature near 20 ^{0}C, most enzymes are active and relatively stable. An enzyme's activity decreases beyond the optimum temperature. The decrease in activity is usually due to heat denaturation of the enzyme. In the case of a non-biological catalyst, an increase in temperature increases the reaction rate.

16. IDENTIFICATION OF FUNCTIONAL GROUPS ESSENTIAL FOR CATALYSIS, RIBONUCLEASE AND CHYMOTRYPSIN AS SPECIFIC EXAMPLES.

Introduction

Enzymes provide functional groups that act as proton donors or acceptors.

General acid-base catalysis

- Enzymes provide functional groups that act as proton donors or acceptors.
- Amino groups
- Carboxyl groups
- Sulfhydryl groups

Example – Lysozyme (donates proton)

Metal ion catalysis

Metal ion acts as an electrophilic catalyst. Stabilizes are negative charge on intermediate. The metal ions may generate a nucleophile by increasing the acidity of nearby molecules.

Example – Carbonic anhydrase

Metal ions may bind to a substrate and increase binding energy.

Example – NMP kinases

Binding to enzyme brings two substrates together.

Example – NMP kinases

Binding to enzyme orients substrate.

Susceptible bond is close to catalytic groups of the active site

Orbital steering (Koshland and Storm)

Proteases

Protein degrading enzymes

- Recycling of amino acids
- Digestion of proteins in the diet
- Enzyme regulation

Chymotrypsin

Covalent catalysis

A nucleophilic group in enzyme attacks unreactive carbonyl group of the substrate forming a covalent bond.

- Serine 195

Three-dimensional structure of enzymes has helped to reveal its catalytic mechanism. Three amino acid residues are involved in catalysis: serine 195, histidine 57, aspartate 102. Unstable tetrahedral intermediate is stabilized by interactions with NH groups from protein at oxyanion hole.

Serine proteases

Trypsin

- Cleaves at peptide bond after residue with long, positively charged side chain

Elastase

- Cleaves at the peptide bond after residue with small side chains

Specificity is depends upon residues in a pocket. Subtilisin from *Bacillus amyloliquefaciens* also uses catalytic triad and oxyanion hole. Carboxypeptidase II which has a very different structure also uses the catalytic triad and oxyanion hole.

Cysteine proteases – cysteine residue acts as a nucleophile.

- Papain

Aspartyl proteases – pair of aspartate residues act to enable water molecule to attack peptide bond.

- Renin

Metalloproteases – active site contains a metal ion that activates water to act as a nucleophile.

- Metal is usually zinc
- Carboxypeptidase A

Protease Inhibitors

Several useful drugs

- Catopril – ACE inhibitor
- Regulator of blood pressure
- Crixivan – HIV protease
- AIDS treatment

17. ENZYME REGULATION

Introduction

Cells of our body are accomplished of making different enzymes which can be regulated by other molecules that increase and decreases molecules activity. The molecule is increases the activity of reaction is called activators although molecules that decrease the activity of an enzyme are called inhibitors.

Regulation by modification

- Proteolytic cleavage
  - Covalent modification
  - Protein-protein interaction

Allosteric regulation

- Properties of allosteric enzymes (important).
- Sigmoid kinetics (what does Km mean in this case) (important).
- Positive and negative modulators (where do they act and how do they modify activity at constant substrate concentration) (most important).

Models of Allosteric Transitions (important)

It has more than one site, where effector binding at one site induces a conformational change in the enzyme, altering its affinity for a substrate. An allosteric activator increases the enzyme rate of activity, an allosteric inhibitor decreases its activity.

Regulation Mechanism

Reversible, non-covalent binding of allosteric effectors. Covalent modification (phosphorylation, adenylation, etc.). Binding by separate regulatory proteins. Proteolytic activation (irreversible). In most cases, the first enzyme of the multi-reaction pathway (catabolism and anabolism) is a regulatory enzyme to avoid an unneeded accumulation of the intermediates.

Regulatory enzymes are generally more complex than other enzymes

- ✓ Aspartate transcarbamoylase – the first step in CTP synthesis, converts Asp to *N*-carbamoyl Asp
  $CO_2 + Gln + ATP \rightarrow H_2N\text{-}(C{=}O)\text{-}OPO_3^{2-}$
  (carbamoyl phosphate)
- ✓ Asp transcarbamoylase catalyzes the following reaction:
- ✓ Carbamoyl phosphate + Asp →*N*-carbamoylAspartate →→→ CTP (building block of DNA) CTP, the end product of the reaction, decreases the rate of enzyme activity – allosteric inhibitor.
- ✓ ATP increases the rate of enzyme activity – an allosteric activator. Normal metabolic control may be exerted in a variety of ways. Examples are:

1. Proteolytic Cleavage of inactive Proenzymes to active enzymes
   Pepsinogen pepsin + small peptide
   in the gastrointestinal tract for protein digestion
1. Coagulation cascade—a series of proenzymes are converted to active enzymes. The last step is
   Fibrinogen Fibrin

18. ENZYME AS TOOLS IN BIOTECHNOLOGY: METHODS OF ENZYME IMMOBILIZATION AND ITS INDUSTRIAL APPLICATIONS.

Introduction

Enzymes are biological catalysts (also known as biocatalysts) that speed up biochemical reactions in living organisms, and which can be extracted from cells and then used to catalyze a wide range of commercially important processes. **For example,** they have important roles in the production of sweetening agents and the modification of antibiotics, they are used in washing powders and various cleaning products, and they play a key role in analytical devices and assays that have clinical, forensic and environmental applications. The word 'enzyme' was first used by the German physiologist **Wilhelm Kuhne** in 1878 when he was describing the ability of yeast to produce alcohol from sugars, and it is derived from the Greek words *en* (meaning 'within') and *zume* (meaning 'yeast').

Immobilization of enzymes

Enzyme immobilization can be defined as restricting the enzyme molecule in a distinct phase from the one in which substrates and products are present. The materials used for immobilization of enzymes are called carrier matrices, these are usually inert polymers or inorganic materials.

Properties

- Low cost
- Inert
- Stable
- Physical strength
- Enhancement of enzyme specificity

- Regenerability
- Reduction in product inhibition
- Reduction in microbial contamination
- Non-specific adsorption

Methods for immobilization

- Adsorption
- Covalent binding
- Entrapment
- Membrane confinement

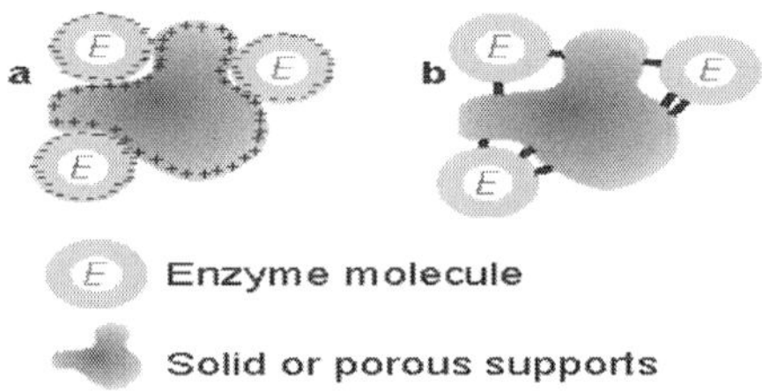

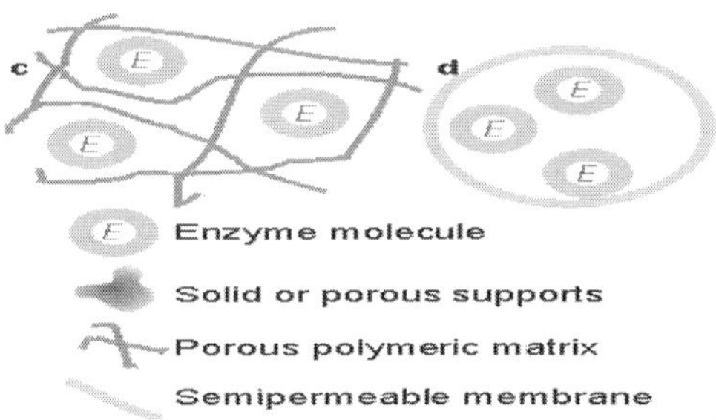

Adsorption

The enzyme is adhering to carrier matrix due to hydrophobic effects or by forming salt links. Binding is strong but may be weakened by substrate and pH etc. The commonly used matrices are ion exchange matrices, porous carbon,clays, hydrous metal oxides, polymeric aromatic resins etc.

Covalent binding

The enzyme is attached to certain matrix by forming covalent bonds. The covalent binding is very strong. It generally occurs with the side chain of the enzyme. Lysine is very useful, as they are very reactive, usually exposed to the surface and very rarely occur at the active site of an enzyme and commonly used matrices:-Agarose, cellulose, polyacrylamides, etc.

Entrapment

The enzyme is entrapped within suitable gels or fibers, with or without covalent bonds and commonly used matrices: Cellulose acetate fibers, calcium alginate ,etc.

Membrane confinement

The enzyme molecules are confined to semipermeable membrane. Thus in an aqueous phase of the membrane is free to move without allowing an enzyme to come out.

Advantages

Enzymes are costly and immobilization permits their repeated use. The product can be easily separated without any additional cost. Immobilized enzymes can be used in a non-aqueous system. It continuous production systems can be used. Thermo stability of some enzyme can be increased by immobilization. An enzyme can be used at a much higher concentration than the free enzyme.

Disadvantages

Some enzymes are very expensive. It sometimes affects the stability and activity of an enzyme. This approach can not be used when one of the substrates is insoluble. Some immobilization methods are restrict the diffusion of substrate.

Applications of immobilized enzymes

Enzyme	Substrate	Product
Glucose isomerase	D-glucose	HFCS
Invertase	Sucrose	Invert sugar
Lipase	Vegetable oil	Cocoa butter substitute
Lactase	Milk and whey	Lactose-free milk &whey

Nitrile hydratase	Acrylonitrile	Acrylamide
Raffinase	Raffinose	Raffinose free solution
Glucoamylase	Dextrins	D-glucose

CHAPTER 3

CELL BIOLOGY

19. CELLULAR BASIS OF LIFE AND APPLICATION OF MICROSCOPE.

Introduction

Microbes to mammals are composed of chemical substances from both the inorganic and organic world that appear in roughly the same proportions and perform the same general tasks. The elements are hydrogen, oxygen, nitrogen, carbon, phosphorus, and sulfur normally make up more than 99% of the mass of living cells and when combined in various ways, form virtually all known organic biomolecules. They are initially utilized in the synthesis of a small number of building blocks that are, in turn, used in the construction of a vast array of vital macromolecules. There are four general classes of macromolecules within living cells such as nucleic acids, proteins, polysaccharides and lipids. These compounds, which have molecular weights ranging from 1 x 10^3 to 1 x 10^6, are created through polymerization of building blocks that have molecular weights in the range of 50 to 150 MW. Although many differences do exist between cells (such as erythrocyte, liver, muscle or fat cell), they all generally contain a greater variety of proteins than any other type of macromolecules, with about 50% of the solid matter of the cell being protein.

Cells generally contain many more protein molecules than DNA molecules, yet DNA is typically the largest biomolecule in the cell. About 99% of cellular molecules are water molecules, with water normally accounting for approximately 70% of the total wetweight of the cell. Although water is obviously important to the vitality of all living cells, the bulk of our attention is usually focused on the other 1% of biomolecules. Data regarding the chemical composition of the unicellular *Escherichia coli* (*E. coli*), are not greatly different for multicellular organisms which including mammals. Each *E. coli,* and similar bacterium, contains a single chromosome; therefore, it has only one unique DNA molecule. Mammals contain more chromosomes and thus have different DNA molecules in the nucleus. The elements such as hydrogen, oxygen, nitrogen, carbon, sulfur, and phosphorus normally makeup more than 99% of the mass of living cells. Homogenous polymers are non-informational. All non-essential lipids can be generated from acetyl-CoA. Different inorganic elements are dietary "essential" Like certain amino acids and unsaturated fatty acids. Most diseases in animals are manifestations of abnormalities in biomolecules, chemical reactions, or biochemical pathways.

Major elements

- Carbon (C)
- Oxygen (O)
- Hydrogen (H)
- Nitrogen (N)

Trace elements

- Calcium (Ca)- blood clotting
- Iron (Fe)- hemoglobin
- Iodine (I)- controls metabolism
- Electrolytes- carry an electrical charge

Water in the cell

- Cells are 60% water
- Interstitial fluid
  - Dilute, saltwater around cells
  - Site of all exchanges between cells and blood

Cells- shape

Red blood cells are disk-shaped. Nerve cells are spindle-shaped. Smooth muscle Cells are Threadlike. Epithelial cells are Cubes. Over 200 cell types found in the body. A general cell has three parts

- Nucleus
- Plasmamembrane
- Cytoplasm

Microorganisms are very small to be seen with the unaided eye. The study of microorganisms is greatly enhanced by the use of microscopes. The word microscope, first coined by Giovanni in 1625, derived from the Latin word micro (small) and the greek word scopos (to look at), is an important tool for microbiologists and the technology of making very small things visible to the human eye is called microscopy (Figure 19.1). The simple microscope used by Anton van Leeuwenhoek in the 17th century in 1673 had only one lens and was similar to a magnifying glass that could magnify a microbe 300X.

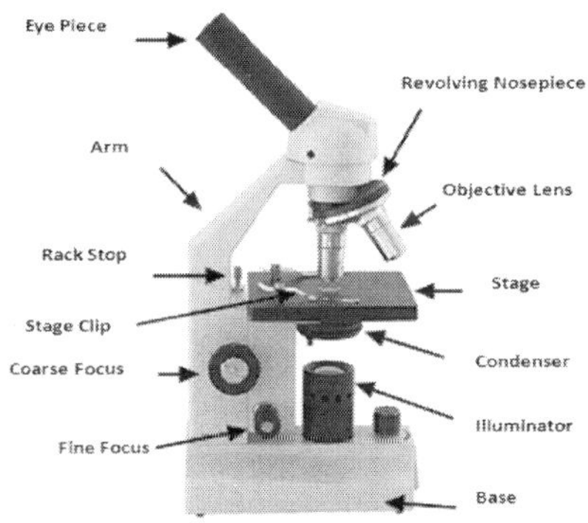

Figure 19.1. Microscope

The simple microscopes enabled to see bacteria and protozoa. Robert Hooke, a contemporary of van Leeuwenhoek built compound microscopes thathave multiple lenses. The microscopes used today are dividing into two categories. light and electron. Light microscopes are sub-divide into bright-field and fluorescent microscopes, while electron microscope is subdivided into scanning and transmission electron microscopes. The bright-field microscope or compound light microscope that allows light rays to pass directly through to the eye without being deflected by an intervening opaque plate in the condenser is routinely used in all the microbiology laboratory to study the size, shape, arrangement of microbial cells and sporulating structures. Dark-field microscopy is a kind of light microscopy that uses a special dark-field condenser that contains an opaque disk that causes the light to reflect off the specimen at an angle rather than pass directly through it. This microscopy is used to examine live microorganisms. Fluorescent microscopy uses ultraviolet light instead of white light to excite molecules within the specimen or fluorescent dye. Fluorescence microscopy must be intrinsically fluorescent or stained with a dye such as auramine O, acridine orange, fluorescein and DAPI (diamidino-2-phenylinodole).

Bright-Field Light Microscopy

A microscope is a compound optical system, a compound magnifying glass. The real meaning of the functioning of a microscope is a test object positioned between the single and double focus points of the objective. The light coming from the object and passing through the lens creates a magnified, inverted and real image of the subject on the other side of the objective, behind the double focus point. The eyepiece is at a distance from the objective lens so that the image formed by the objective is generated within the focus of the eyepiece. Thus, looking throughout the eyepieces, one can see a further enlarged, direct but virtual image of this real, inverted and magnified image. The magnification power of a bright-field microscope can be calculated by multiplying the magnification of the objective and of the eyepiece, respectively. The objective lens system consists of multiple lenses. The first member is the front lens facing the object. This determines the magnification and resolution of the microscope. Other elements are responsible for the exclusion of lens errors. The features of an image formed by the objective depend on the optical characteristics of the objective lens.

The quality of a lens represents its property of how sharp the object image is drawn. The image defect, i.e. blurring is caused by lens errors (aberrations). The spherical aberration (spherical divergence) is caused when the lens away from the optical axis increasingly breaks the light rays passing through it, therefore a point-like object in the image will be blurred at the edge. The reason for chromatic aberration is that the focal points of different wavelengths of light do not coincide on the optical axis. The shorter wavelength rays unite closer, while longer wavelength rays unite farther. Thus, a sharp image cannot be obtained using white light, and the image has a colored (rainbow) border. A concave flint glass lens with lead content and high refractive index is fitted to the convex lens to correct for chromatic aberration (the achromatic lens has two-colour correction, while the apochromatic lens has three-colour correction). The resolution of the objective lens is the ability of how detailed the image of a subject can be drawn. The resolving power is quantified with the minimum distance between two points that are just distinguishable.

The resolution (d) depends on the illumination wavelength of light used (λ), the half-angular aperture of the objective lens (α) and the refractive index of the material between the front lens and the coverslip (n).

$d = 1.22\ \lambda / 2\ n \times \sin\alpha$

where $n \times \sin\alpha$ = numerical aperture (NA).

The greater the resolution and the smaller the value of d are. This can be achieved by reducing the wavelength of light used, increasing the angular aperture of the objective, or increasing the refractive index of the material between the front lens and the coverslip. For this purpose, cedar oil (oil immersion) is the most suitable material, because it has almost the same refractive index as that of glass, so the light passes through a virtually homogeneous medium.

The denominator of the above formula and $2\ nX \sin\alpha$ is the value of numerical aperture (NA), which may vary between 0.20 and 1.4 (always indicated on the lens). The eyepiece draws a direct image of the test object. The fine structure of an image observed in the microscope depends on the details of the real picture, which in turn is determined by the resolution of the objective. This image, however, is not visible to the naked eye; it can only be visualized in the magnification of the eyepiece.

Microorganisms Examination From Natural Waters By Bright-Field Light Microscopy

Principle

A microscope is a series of finely ground lenses and visible light as a source of illumination isused. The microscope field or area observed is brightly lighted and the objects being studied appear dark because they absorb some of the light. However, microbes absorb light after staining them with a dye greatly increases their light-absorbing ability resulting in greater contrast and color differentiation.

Requirements

- Environmental water sample (e.g. from a lake, a stream, a creek or an aquarium)
- Glass slides
- Cover slips
- Glass dropper dispenser alcohol (for sterilization)
- Bunsen burner
- Light microscope

Procedure

1. Dry the surface of a glass slide with alcohol over a Bunsen burner and then label the slide.
2. Place one drop of the sample to the slide with a glass dropper.
3. Put the edge of a cover slip on the slide so that it touches the edge of the water drop.
4. Slowly lower the cover slip to prevent the formation and trapping of air bubbles.
5. Put the sample under the microscope, and locate the focal plane.
6. Apply first an objective lens of 16X, and then 40X magnification.
7. Observe the shape and movement of microbes, in the case of protists and eukaryotic algae, try to identify the different cell organelles (e.g. chloroplasts and contractile vacuoles)
8. Make drawings about the observations.

Fluorescence Microscopy

Specimens that absorb light of color and subsequently emit light of another color (fluoresce) can be visualized by using the fluorescence microscope. The basis of fluorescence microscopy is the principle of the removal of incident illumination by selective absorption, where light absorbed by the specimen and reemitted at another wavelength is transmitted. The light source must produce light of appropriate wavelength and ineffective wavelengths that are unable to excite the fluorochrome used are removed by an excitation filter. The phenomenon when a sample (e.g. cell organelle of a microbe) shows fluorescence without prior staining is called auto-fluorescence. Otherwise, bacterial cells can be stained with a fluorescent dye. Using a fluorescent dye, e.g. DAPI (4',6-diamidino-2-phenylindole), cells are stained bright blue because the dye forms a complex with the cell's DNA. Fluorescent dyes are therefore widely applied to visualize and enumerate bacteria in different natural habitats or clinical samples

Microbes Examined with DAPI Staining (Microbes from Water Sample)

DAPI molecules are able to penetrate cell membranes and bind to the double helix of the DNA. Cells can be easily counted, if a known volume of fixed water sample is filtered through a membrane and, after staining the filter, the surface is investigated by a fluorescence microscope (Figure 14.2). A microscope is a series of finely ground lenses and visible light as a source of illumination are used. The microscope field or area observed is brightly lighted and the objects being studied appear dark because they absorb some of the light. However, microbes do not absorb much light but staining them with a dye greatly increases their light-absorbing ability resulting in greater contrast and color differentiation. The effective magnification of a microscope depends upon its resolving power.

- Typical magnification values are:
- Scanning 10X × 4X= 40X magnification
- Low power 10X × 10X = 100X magnification
- High power 10X × 40X = 400X magnification
- Oil immersion 10X ×100X = 1000X magnification
- Total magnification = Magnifying power of the objective lens* magnifying power of the ocular lens.

Requirements

- Disposable gloves
- Laboratory scales
- paraformaldehyde (PFA)
- Beaker
- Magnetic stirrer
- Phosphate buffered saline (PBS) NaOH solution
- Pasteur pipette
- Membrane filtration apparatus
- Polycarbonate and cellulose nitrate membrane filters (0.22 or 0.45 µm pore size)

- 50 ml Falcon tube
- Plastic Petri dishes scalpel
- Pipette with pipette tips DAPI solution (1μg/ml)
- 80% ethanol
- Double distilled water
- Glass slide
- Cover slip
- Vectashield Mounting Medium (H-1000, Vector Laboratories Ltd) immersion oil (non-fluorescent)
- Fluorescence microscope with a mercury lamp digital camera computer with software

Procedure

1. For the preparation of the fixative solution, dissolve 1 gm PFA in 50 ml PBS. (PFA causes irritation when inhaled; therefore the use of a fume hood is recommended).
2. Dissolution can be aided with heating (60°C), permanent stirring and adding some drops of NaOH solution.
3. Adjust the pH at 7.0.
4. Filter the solution through a 0.22 μm pore size membrane filter. (The prepared 2% PFA solution can be stored in the fridge for one week).
5. Filter the water sample (2-50 ml, depending on the type of sample) using a polycarbonate membrane filter (slowly with occasional stirring).
6. To help uniform cell distribution, place a 0.45 μm pore size cellulose nitrate membrane filter between the sieve of the filtration unit and the polycarbonate filter.
7. Fill up the Falcon tube with fixative (PFA solution) and immerse the filter in it with sterile forceps (PFA solution must cover the entire membrane filter).
8. Incubate the filter overnight at 4°C.
9. Fill PBS into an empty Petri dish, then transfer the filter into the PBS solution for 1-2 minutes (liquid must cover the entire membrane filter).
10. Transfer the filter to another empty Petri dish and let it dry.

11. Cut a 0.5 by 0.5 cm piece from the filter with a scalpel or scissors, and pipette 30 µl DAPI solutions onto its surface. From this step onwards, work in a dark place.
12. The filter piece can be marked with a soft pencil.
13. After 2 minutes, transfer the filter into 80% ethanol for a few seconds.
14. Dip the filter paper into double distilled water for a few seconds.
15. Dry the filter.
16. Place the filter onto the surface of a glass slide, put a drop of Vectashield Mounting Medium onto the filter and then cover with a cover slip.
17. Cover with a paper towel and gently press the cover slip to remove any excess of mounting medium.
18. Examine the slide with an epifluorescence microscope using a 100X objective and immersion oil under UV excitation (the absorption maximum of DAPI is at 358 nm, and the emission maximum is at 461 nm).
19. Record images from at least 20 different microscopic fields with a digital camera.
20. Count the cells on each picture and determine the mean values.
21. Determine the cell count for one ml of water sample based on the amount of filtered water and the size of the membrane filter field.

Phase-Contrast Light Microscope

Most of the microorganisms are colorless and also their organelles, which often too difficult to see by dark field microscopy and common bright-field microscopy. The observations are clear by a phase-contrast microscope such as the structure of endospores, unstained bacteria and living organisms. The species are often arranged in pairs or chains, rod-shaped and rounded or square ends known as *Bacillus* species and their endospores are mostly oval shape or sometimes cylindrical. *Clostridium* species are repeatedly set in short chains or pairs, mostly with rounded or sometimes pointed ends. The endospores regularly swell up the cell. Consequently, by means of ready slides of *Clostridium* and *Bacillus* the student gains knowledge about the phase-contrast microscope and observing precise bacterial structures such as diverse endospores are present in bacteria. Pond water is generally crowded with protists and bacteria, which observe the internal structure by using phase-contrast microscope. Apart from this we can also see the internal structure of paramecium (protist) and slowing down the different microorganisms present in pond water.

Principle

The colorless and transparent living microorganisms having organelles are generally impossible to see by bright-field microscope due to their absorption, refract and reflect limits. Another reason is a sufficient amount of light diffract with the nearby environment for the rest of the microbes. The microbes are visible when they absorb much more light or to refract and reflect from the environment. Thoroughly, we can see microbe organelles by this process. The observations of living and invisible unstained microbes are by using only phase-contrast microscope. The condenser has an annular diaphragm and hollow cone which produces light, glass disk (phase plat) has an objective with a thin film transparent material deposit on it for phase changes produced in the specimen. The phase change of images is observed due to the difference inintensity in light. The positive phase plate is diffracted due to undiffracted light in dark phase-contrast light microscopy or negative phase plate same as above given light relative to the directed light-producing by phase-contrast light microscopy.

Requirements

- Pond waste water
- Phase-contrast light microscope
- New slides and coverslips
- Pipette
- Microorganisms
- Methyl cellulose "Protoslo, Carolina Biological Supply"
- Tweezers
- Lens paper
- Lens cleaner
- Prepared slides of "*Bacillus* or *Clostridium*"

Procedure

1. Create waste pond water.
2. Add one drop of methyl cellulose to slow down the swimming of the microorganisms.
3. Prepared slides of selected microbes which are given above in the introduction part.
4. The slide is placed on the stage of phase contrast microscope so that specimen is under the light hole.
5. The lenses are set into place and rotate the position (annular diaphragm) of the specimen if it is required.
6. It is necessary that the cone of light produced by the annular diaphragm below the condenser be centered.
7. Center with the 10X objective and observe the specimen which is under over light.
8. Rotate the annular diaphragm into the correct position for the observation with 40X objective if the image is not clear.
9. Do the same with the oil immersion lens.
10. In the report for exercise 4, sketch several of the microorganisms that you have observed.
11. If you examine pond water (collection of microbes), use the pictorial guides provided by your teacher to help you in identifying microorganisms that are present in pond water.

Scanning Electron Microscope

A scanning electron microscope (SEM) is a very specific type of microscope; it uses scanned electron beams to produce an image of a specimen. A complete labeled picture of the SEM instrument is shown in figure 19.2. The scanned electrons when falling on a specimen it interacts with the atoms of it and starts producing various signals that are captured by SEM software to produce an image. SEM is used to know the structure of a specimen with their composition as well as topography. By using SEM we can achieve a much better resolution (better than 1 nm). Mostly the image formation and detection of the specimenare done with the help of a secondary electron beam of SEM. Different states of the sample can be imaged in their dry, wet, hot or cool state

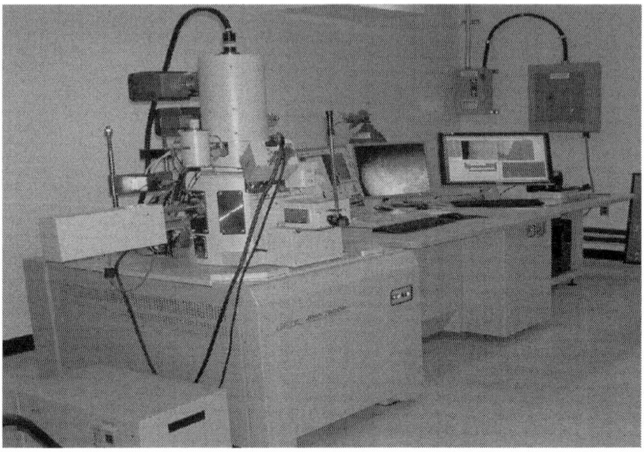

Figure 19.2. A complete abelled picture of the SEM instrument.

In 1935, a well-known scientist named M. Knoll first describes the concept of a scanning electron microscope, he was working in Germany. One other scientist named M. Von Ardenne constructed a scanning transmission electron microscope (STEM) by adding scan coils to a transmission electron microscope (TEM) after 3 years in 1938. ZnO crystal imaged first time by using STEM, the operating voltage was 23 Kv and magnification 8000 times with spatial resolution between 50-100 nm. But this instrument has no system to store the image. The first SEM was used to examine the surface of a solid specimen, which was described by Zworykin in 1942, After STEM. A scientist of Cambridge University, England named C. W. Oatley start reinvestigating the SEM with Dennis McMullan and they built their first SEM in 1948. In 1952, SEM had achieved various advancements like the resolution of 50 nm and three-dimensional imaging capacity. A research student named O.C. Wells in 1953 under Oatley's supervision was started building a second SEM, this instrument had the gun at the bottom of the column a configuration considered better for experimental work. In 1955, Oatley's one other research student named start working on SEM and he improved the secondary electron (SE) detector by using a scintillator to convert electrons to photons, which were then transmitted by a light pipe directly interfaced to the photomultiplier tube.

Principle

A scanning electron microscope is a kind of specific microscope (**figure 19.3**). In this microscope, the image is produced by using scanned electron beams of the sample. The image is produced on both top-down sides. There are different ways by which the primary electron beam interacts with the sample.The sample of primary electrons generates low energy secondary electrons which fall onto the sample and produce an image. The primary electrons can also produce images. The X-rays are produced by relaxed ionized atoms which also help in producing the image of the sample.

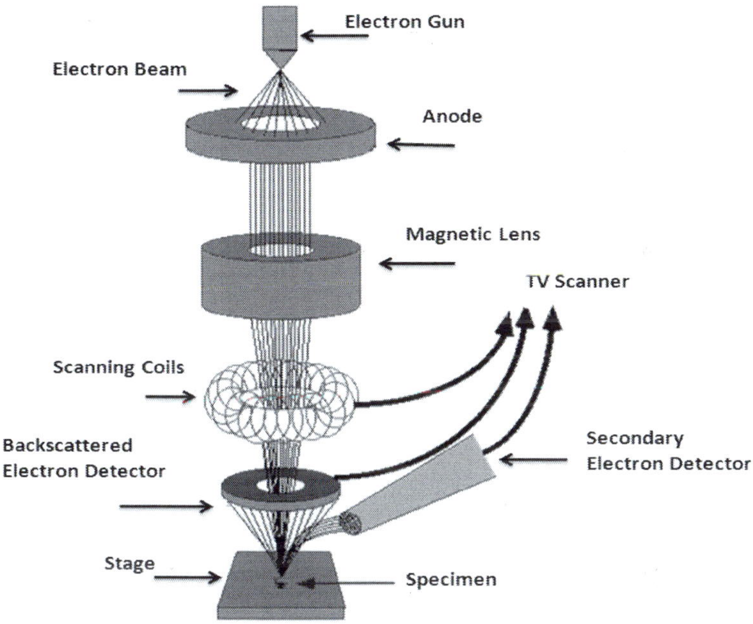

Figure 19.3. A complete schematic diagram of the working principle of SEM

How to work

Sample preparation

- The sample needs to be electrically connected to the sample holder to prevent the electron beam from "charging" the sample and distorting the image is called sample grounding.
- The samples for SEM were sputtered with gold and platinum coating is called sample coating.
- After sample preparation, the SEM software is opened.
- The sample is placed on an SEM sample mount.

- When the sample is properly placed on the SEM stage, close the door of the chamber gently.
- Now click OK in the box on the screen asking "Press OK to pump."
- After press OK to pump click on the "Vacuum" tab to view the vacuum level in the chamber and the gun.
- The voltage and resolution aresaved according to the nature of the sample.
- Activate the SCM (sample current monitor) window to check whether it is on or off.
- Remove the sample and collect your results from the system.

Usage

There are multiple usages of SEM, some are mentioned below:

Life Sciences

Biomedical engineering: For designing a variety of technologies such as support matrices for cell growth, artificial tissue, and implantable biomedical device. Cell and tissue morphology: For investigating cellular and tissue structure with high resolution (**Figure 19.4**). Typical applications involve observing shape changes of grooves, pores, blebs or microvilli on the cellular in response to the changes in the extracellular environment. Microbiology: High-resolution imaging of the microbial surface using SEM helps to have a better understanding of the morphology of microbial populations, bacteria communication, and biofilm formation. It helps researchers visualize microbial populations with great focal depth and high resolution. Pharmaceuticals: SEM offers very high-resolution imaging and superb analytical performance, which make them perfect and reliable tools for quality control of new preparations.

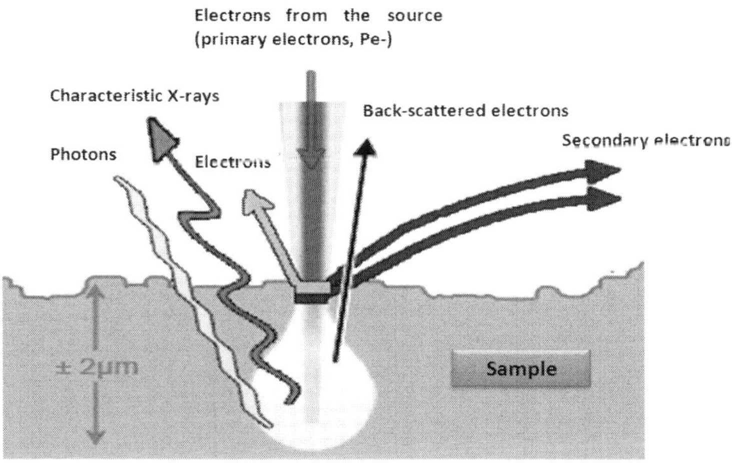

Figure 19.4. An image of electrons scatter during sample inhaled.

Earth Sciences

Magnetic and metamorphic petrology and mineralogy: Petrology and mineralogy benefit from the electron microscopy technique because of several reasons. The high magnification factor, as well as the phase-contrast provided by the back, scattered electron imaging play a crucial role in these disciplines. The analytical capabilities of the technique make the SEM an ultimate solution in this field. Ore Processing: Scanning electron microscopy helps to characterize the mineral of interest as well as the surrounding gangue. Such a technique is capable of both phase identification and evaluation of textural relationships of individual minerals. Oil and Gas: Scanning electron microscopy is an integral part of reservoir rock characterization workflow. The elementary imaging function of SEM is used when searching for basic rock properties such as porosity or grain shape. Paleontology: Scanning electron microscopy is a key tool in microfossil characterization helping to distinguish features that are otherwise irresolvable by means of optical microscopy. Sedimentary Petrology: Electron microscopy is helpful in tackling a wide range of petrological problems. Rock characterization mostly relies on optical microscopy, however, there is growing demand to characterize the rocks with greater detail nowadays. Recycling: Automated mineralogy approach is currently finding its way to other areas than just ore processing. This technique can be applied e.g., in the field of electronic scrap recycling where it is used to characterize the composition and structure of comminuted e-scrap.

20. EVOLUTION OF CELL THEORY, CELL STRUCTURE, ORGANELLES AND THEIR FUNCTION, TYPES OF CELLS, EUKARYOTES/PROKARYOTES CELLS AND ITS ORGANIZATION PLANT AND ANIMAL CELLS.

Introduction

The cell theory states that all organisms are composed of similar units of an organization called as cells.

There are **3** major parts of the cell theory:

1. All organisms are made of cells.
2. All existing cells are produced by other living cells.
3. The cell is the most basic unit of life.

The cell theory states that all organisms are composed of similar units of organization called cells.There are **3** major parts of the cell theory: All organisms are made of cells. All existing cells are produced by other living cells. The cell is the most basic unit of life.

There are **5** contributors to the cell theory

1. Robert Hooke
2. Anton van Leeuwenhoek
3. Matthias Schleiden
4. Theodor Schwann
5. Rudolf Virchow

Each of the above contributors produced evidence through investigations and experiments that led to today's cell theory. Before the cell theory people believed in spontaneous generation which means living things come from non-living thing. Two scientists disproved spontaneous generation. Francisco Redi proved that microorganisms do not spontaneously generate. People thought that maggots came from meat. Redi disproved this in a controlled experiment.

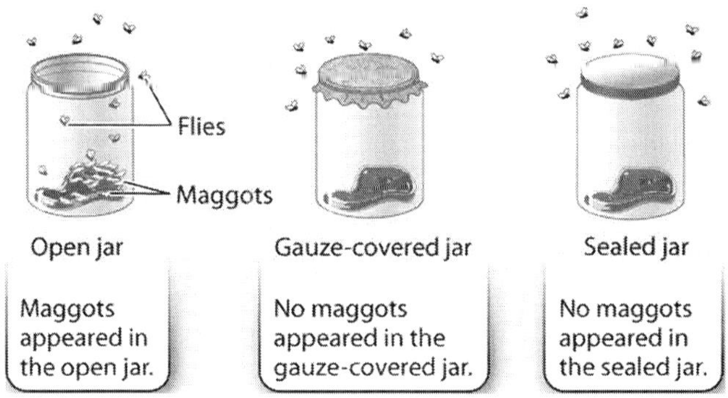

Pasture proved that microorganisms do not spontaneously generate.

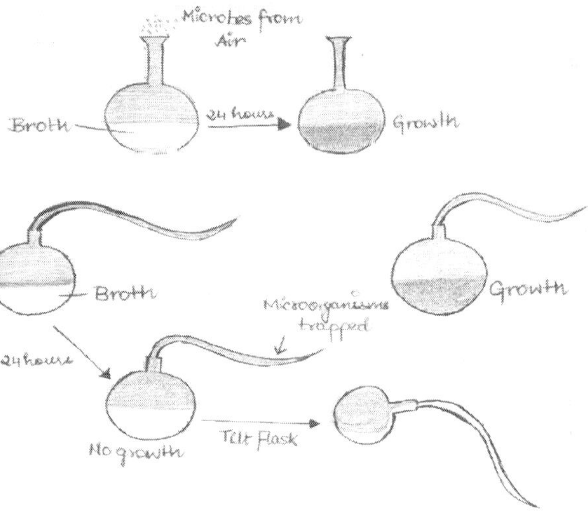

Theodor Schwann and Matthias Schleiden are credited with coming up with the cell theory.

- Schwann was discovered animal cells.
- Schleiden was discovered plant cells.
- Janssen first compound microscope was invented.

Robert Hooke

In 1663 an English scientist, Robert Hook improved Janssen's invention. Used the compound microscope to observe cork. Hooke observed that cork is composed of small, hollow compartments. The parts prompted Hooke to think of small rooms (cells) in a monastery, so he gave them the same name. **Cells** was investigated cork through experimenting with the compound microscope and came up with the name cells. **Key Note:** Hooke discovered the cell in 1665, which started formulating the cell theory.

Anton van Leeuwenhoek

Dutch Tradesman

The study new approaches for creating lenses to observe cloth. Leeuwenhoek's microscope was more powerful than Hooke's compound microscope. Leeuwenhoek became one of the first scientists to refer to living cells when he observed an abundant number of single-celled organisms, which he called animalcules (plant & animal), swimming in a drop of pond water.

Matthias Schleiden (German Scientist)

It is involved with plant cells, Schleiden used the compound microscope and studied plant cells. From investigating and experimenting with plants, projected plant parts were made of cells. He was discussed what he observed with his dear friend, German scientist Theodor Schwann.

Theodor Schwann

Studied plant & animal cells, and was intrigued by the similarities between the two. Schwann was able to determine that all animals are made of cells. Schwann published the 1st statement of the cell theory: all living things are made of cells and cell products.

Rudolf Virchow

Based on his investigations and experiments, he stated that all cells come from preexisting cells, which is the 2nd part of the cell theory: all existing cells are produced by other living cells. **Key Note:** After Virchow's completed his work, the cell theory was finally formulated in 1838.

Types of cells

Prokaryotic cells are mainly single celled organisms.

Eukaryotic cells are mainly multicellular organisms, although some are single celled (amoeba, algae, yeast).

Prokaryotic cell

Prokaryotes have no membrane bound organelles. **Example** – Bacteria cell

Eukaryotic cell

Eukaryotes are contain membrane bound organelles. **Example** – Plant and animal cells

Prokaryotic cells are single-celled entities that are primitive in structure and function as they lack a membrane-bound nucleus and other organelles. The term "prokaryote" is derived from two Greek words, 'pro' meaning 'before' and 'karyon' meaning 'nucleus'. Prokaryotes are considered to be the first living organisms of the earth as they are the simplest form of life.

- Single-celled.
- Reproduce by binary fission (other copy by dividing).
- No cell nucleus or any other membrane-bound organelles. DNA travels openly around the cell.
- All bacteria are prokaryotes.
- Eu = "true"; karyon = "nucleus"

Most organisms that we can see, such as trees, grass, worms, flies, mice, humans, mushrooms and yeast are eukaryotes. It can either be single-celled or multi-celled. It can reproduce in one of several ways (Ex. meiosis and mitosis). It has a cell nucleus within containing its DNA. The nucleus is the most evident distinction between these cell types.

A plasma membrane: an outer covering that separates the cell's interior from its surroundingenvironment. **Cytoplasm:** a jelly-like cytosol within the cell in which other cellular components are found. **DNA:** the genetic material of the cell. **Ribosomes:** where protein synthesis occurs.

However, "prokaryotes differ from eukaryotic cells in several ways. A prokaryote is a simple, single-celled (unicellular) organism that lacks an organized nucleus or any other membrane-bound organelle. We will shortly come to see that this is significantly different in eukaryotes. Prokaryotic DNA is found in a central part of the cell: the nucleoid. Most prokaryotes have a peptidoglycan cell wall and many have a polysaccharide capsule. The cell wall acts as an extra layer of protection, helps the cell maintain its shape, and prevents dehydration. The capsule enables the cell to attach to surfaces in its environment. Some prokaryotes have flagella, pili, or fimbriae.

Flagella are used for locomotion. Pili are used to exchange genetic material during a type of reproduction called conjugation". Fimbriae are used by bacteria to attach to a host cell.

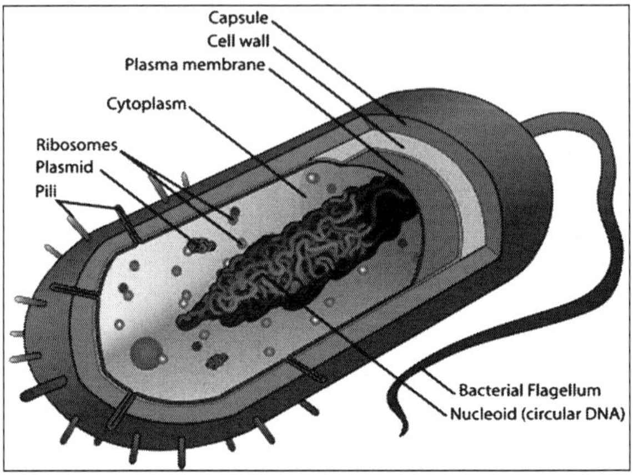

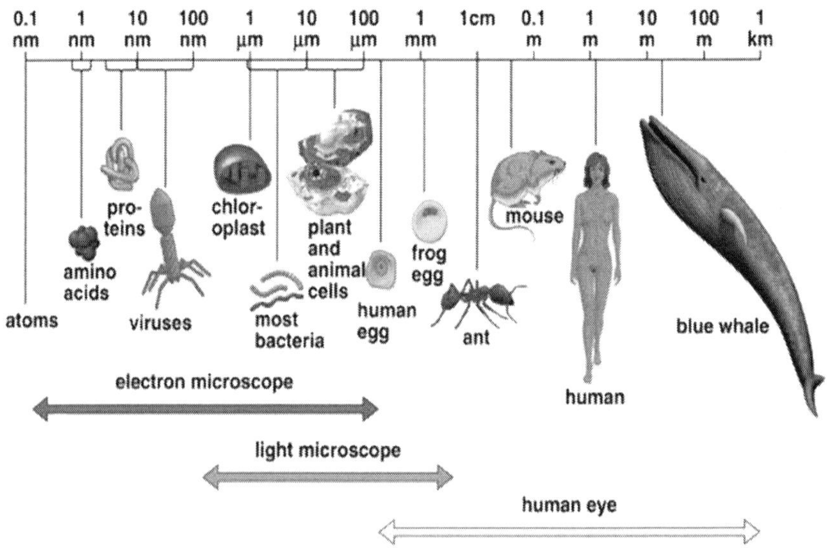

Prokaryote Genetics

Nucleoid

The region of cytoplasm where prokaryote's genome (DNA) is located and it usually a singular and circular chromosome.

Plasmid

The small extra piece of chromosome or genetic material is called plasmid. It has 5 - 100 genes. It can provide genetic information to promote.

- Antibiotic resistance
- Virulence factors

The molecules are produced by pathogens specifically influence the host's function to allow the pathogen to thrive. It promote conjugation (transfer of genetic material between bacteria through cell-to-cell contact).

Cytoplasm

It is also known as protoplasm. Gel-like matrix of water, enzymes, nutrients, wastes, and gases and contains cell structures. The site of growth, metabolism, and replication.

Granules

Bacteria's is the way of storing nutrients and staining of some granules aids in identification.

Cytoskeleton

The cellular "scaffolding" or "skeleton" is present within the cytoplasm. The major advance in prokaryotic cell biology in the last decade has been the discovery of the prokaryotic cytoskeleton.

Ribosomes

It found within the cytoplasm or attached to the plasma membrane. It is made of protein & rRNA. It composed of two subunits.The cell may contain thousands of ribosomes.

Plasma membrane

It separates the cell from its environment. Phospholipid molecules oriented so that hydrophilic water-loving heads directed outward and hydrophobic water-hating tails directed inward Proteins embedded in two layers of lipids (lipid bilayer). Membrane is semi-permeable.

Plasma membrane as a barrier

Osmosis

The diffusion of water across a semi-permeable membrane is called osmosis. The environment surrounding cells may contain amounts of dissolved substances (solutes).

Tonicity and Osmosis

- *Isotonic*: equal concentration of a solute inside and outside of cell.
- *Hypertonic*: a higher concentration of solute.
- *Hypotonic*: a lower concentration of solute.
- Water will always move towards a hypertonic environment.

Active Transport

The molecules move across the plasma membrane. It is pump moving water uphill. Types of active transport are classified by type of energy used to drive molecules across membranes. ATP driven active transport. Energy from adenosine triphosphate (ATP) drives substances across the plasma membrane with the aid of carrier molecules.

Prokaryotes: Cell wall

All bacteria are very similar inside the peptidoglycan wall, the bacterial world is divided into two major classes gram positive and gram negative.

Gram-positive Gram-negative

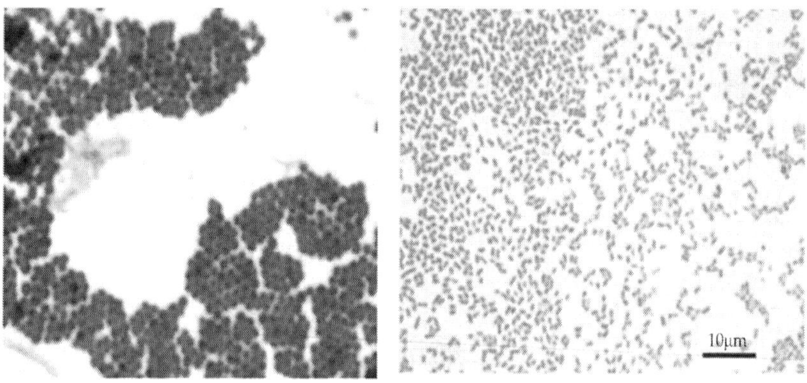

Bacterial cell wall

Peptidoglycan is a huge polymer of interlocking chains of alternating monomers. It provides a rigid support while freely permeable to solutes. The peptidoglycan molecule is composed of two amino sugar derivatives of glucose. The "glycan" part of peptidoglycan.

- N-acetylglucosamine (NAG)
- N-acetlymuramic acid (NAM)

NAG / NAM strands are connected by interlocking peptide bridges. The "peptid" part of peptidoglycan is shown in **figure 20.1**.

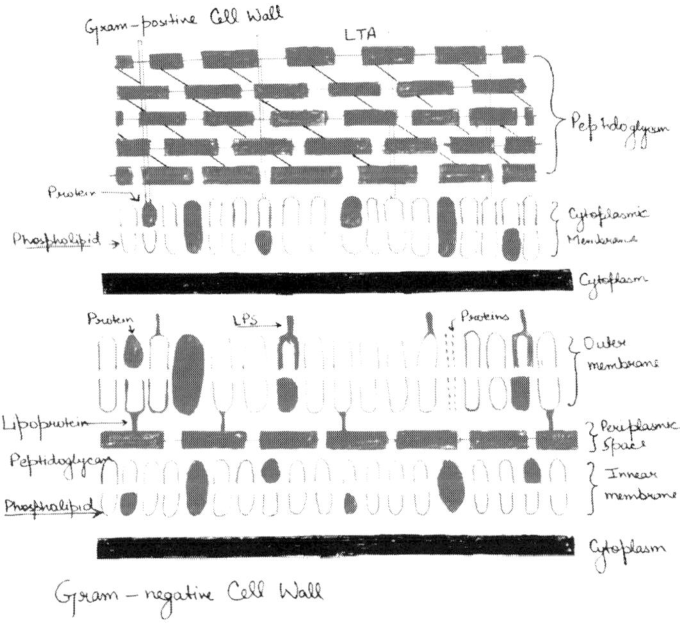

Figure 20.1. Prokaryotes Cell Wall

Prokaryotes – Glycocalyx

Some bacteria have an additional layer outside of the cell wall called the glycocalyx. This additional layer can come in one of two forms:

1. Slime Layer

Glycoproteins are loosely associated with the cell wall. Slime layer causes bacteria to adhere to solid surfaces and helps prevent the cell from drying out. The slime layer of Gram+ *Streptococcus mutans* allows it to accumulate on tooth enamel (yuck mouth and one of the causes of cavities). Other bacteria in the mouth become trapped in the slime and form a biofilm & eventually a buildup of plaque.

2. Capsule

Polysaccharides are firmly attached to the cell wall. The capsules adhere to solid surfaces and to nutrients in the environment. Adhesive power of capsules is a major factor in the initiation of some bacterial diseases. The capsule also protects bacteria from being phagocytized by cells of the host's immune system.

Endospores

It is dormant, tough, non-reproductive structure produced by small number of bacteria. It is resistant to radiation, desiccation, lysozyme, temperature, starvation, and chemical disinfectants. Endospores are commonly found in soil and water, where they may survive for very long periods of time.

Surface appendages

Some prokaryotes have distinct appendages that allow them to move about or adhere to solid surfaces. They consist of delicate stands of proteins. Flagella: It is long, thin extensions that allow some bacteria to move about freely in aqueous environments. Axial filament (endoflagella): Wind around bacteria, causing movement in waves. Fimbriae: Most Gram-negative bacteria have these short, fine appendages surrounding the cell. Gram positive bacteria don't have. It has no role in motility. It helps bacteria adhere to solid surfaces. They have Major factor in virulence. Pili: Tubes that is longer than fimbriae, usually shorter than flagella. Use for movement, like grappling hooks, and also use conjugation pili to transfer plasmids.

Prokaryotes – Cell Shapes

Most bacteria are classifies according to shape:

- Bacillus *(bacilli)* = rod-shaped
- Coccus *(cocci ... sounds like cox-eye)* = spherical
- Spiral shaped

  Spirillum *(spirilla)* – spiral with rigid cell wall, flagella

- Spirochete *(spirochetes)* = spiral with flexible cell wall, axial filament.

There are many more shapes beyond these basic ones. A few examples

- Coccobacilli = elongated coccal form
- Filamentous = bacilli that occur in long threads
- Vibrios = short, slightly curved rods
- Fusiform = bacilli with tapered ends

Bacteria has sometimes occur in groups, rather than singly. Bacilli divide along a single axis, seen in pairs or chains. Cocci divide on one or more planes and producing cells in:

- Pairs (diplococci)
- Chains (streptococci)
- Packets (sarcinae)
- Clusters (staphylococci).

Size, shape and arrangement of cells often are the first signs in the identification of a bacterium. Many "look-alikes", so shape and arrangement are not enough for identifying the genus and species. Like prokaryotes, and most living things, the genome is made of DNA. It may include several to many linear chromosomes within a membrane-bound nucleus.

Cytoplasm

The cytoplasm is present inside the plasma membrane; the nucleus is surrounded by cytoplasm. A water-like substance is that fills inside the cells. Cytosol is made up of water, salts, organic molecules and many enzymes that catalyze reactions.

Membrane-bound organelles

Eukaryotic cells have many organelles. Prokaryotes only have ribosomes, which are not bound by a membrane. Membrane-bound eukaryotic organelles organize functions within the cell. The system is present of internal membranes within eukaryotic cells that divide the cell into compartments, or organelles. Transport system, for moving molecules through interior of cell, as well as interactive surfaces for lipid and protein synthesis. The membranes of the endomembrane system are made of a lipid bilayer, with proteins.

Endomembrane System Organelles: Nucleus

It is a control center and separates the genetic material (DNA) from the rest of the cell. DNA, the genetic material, is a blueprint, or code for making proteins. The plasma membrane/cell membrane

- The flexible boundary of a cell
- separates a cell from its surroundings

It allows nutrients to enter the cell and waste to be removed

- This is referred to as selective permeability (Selective=Chooses, Permeability=filter through).
- Keeping a healthy balance of nutrients and water within the cell is called homeostasis

Overview of Organelles

Nucleus

Largest organelle in the cell and it is the most inner compartment of the cell; contains chromatin (DNA); genetic information on strands called chromosomes "control centre" for cell metabolism and reproduction.

Chromatin

Chromatin is help to make proteins.

Nucleolus

It is found inside nucleus and ribosomes are made here.

Ribosomes

It make proteins (made up of RNA and protein) and also called as protein "factories"

Cytoplasm

It is clear gel like fluid inside the cell, which suspends all organelles.

Endoplasmic reticulum

It is two types

- Rough ER: with ribosomes.
- Smooth ER: with no visible ribosomes.
- Golgi apparatus: Sorts proteins made by the ribosomes and sends them to needed places in the cell.
- Lysosomes- Organelles that are filled with digestive enzymes to remove waste and invading bacteria.

Mitochondria: It referred to as the "powerhouse" of the cell release energy for the cell. It converts the energy stored in glucose into ATP for the cell.

Vacuoles are fluid which filled organelles enclosed by a membrane. It is store materials such as food, sugar, water, and waste products.

Eukaryotic plant cell

Plant cells are also eukaryotic cells, but plant cells contain some organelles that are not found in animal cells.

Plant cell organelles

Cell wall

It is a rigid wall outside the plasma membrane. It provides the cell with extra support.

Chloroplasts

It captures light and energy; and converts it into chemical energy.

Chlorophyll

It is the green pigment found inside the chloroplast.

Plastids

The organelles that store things such as food in the plant cell.

21 CELL ORGANELLE PLASMA MEMBRANE, ENDOPLASMIC RETICULUM, MITOCHONDRIA, CYTOSKELETON, NUCLEUS AND GOLGI APPARATUS.

Introduction

The basic structural and functional unit of life. The smallest units that display the characteristics of life, i.e. reproduction, metabolism, response to stimuli. Plasma membrane is selectively permeable boundary between the cell and the environment. Nucleus is regulatory center of the cell. Cytoplasm is everything between the plasma membrane and the nucleus (fluid + organelles).

Plasma Membrane

Structure is phospholipd bilayer with proteins embedded in, and attached to, the inner (intracellular) and outer (extracellular) surfaces.

Function

- Selectively permeable barrier: controls what enters and leaves the cell.
- Phospholipids are liquid at body temperature, so proteins float around in the membrane

Functions as a Fluid Mosaic

- Main responsibility: ensure the composition of extracellular fluid is not the same as the composition of the intracellular fluid.
- Water-soluble substances (salts, nutrients) cross membrane with aid of protein channels, which are selective about what can pass through.
- Lipids can pass directly through bilayer by diffusion (the random walk of molecules)
- Attachment site of cytoskeleton, the internal support of the cell.

Glycocalyx is protein and carbohydrate coat covering the extracellular surface of the plasma membrane. Allows attachment to other cells. Allows the cell to interact with the environment. Gives each person's cell a distinctive surface. Allows your body to recognize foreign tissues. As different from your own (i.e., blood type, transplant rejection). Plasma Membrane =

selectively permeable boundary between the cell and the environment. Nucleus = regulatory center of the cell. Cytoplasm = everything between the plasma membrane and the nuclear compartment.

Nucleus

Nuclear envelope (membrane)

Phospholipid bilayer with nuclear pores. Controls what enters/leaves the nucleus. Things only go in or out by passing through. Protein channels, which are selective. Encloses are all the chromosomes chromatin = all the chromosomes, which are long strands of the molecule DNA. DNA regulates all cell activities, yet never leaves the nucleus.

It produces RNA, short messenger molecules that exit through nuclear pores. RNA carries instructions out into the cytoplasm. Nucleolus – site of ribosome synthesis. The nucleus is build all the proteins in the cell.

- Compartment in the nucleus where ribsomes are assembled.
- Ribosomes are then moved out into cytoplasm through nuclear pores.
- Ribosomes and RNA work together outside.

Ribosomes

Ribosomes is the site of protein synthesis which assembled in the nucleolus and exported into the cytoplasm.

- Free – It is unbound in the fluid cytoplasm, produce proteins for use in the cell.
- Bound – It is attached to the endoplasmic reticulum (ER) which produce proteins for export, or for the plasma membrane.

Cytoplasm + organelles

- Fluid portion is called as "cytoplasm"
  - Water with dissolved salts, nutrients and enzymes.
  - Site of many metabolic reactions
- Cytoskeleton is a network of protein fibers extending throughout the fluid cytoplasm.
  - Support and structure for the cell.

- Very dynamic and always remodeling itself.
- Critical for cells to divide and copy themselves.

Organelles

The separate compartments within the cytoplasm formed by membranes. Mitochondrion is just "thread granule", major source of cell's energy. Energy is taken from sugar, stored in molecule called ATP (adenosine triphosphate). It requires oxygen to make this exchange (aerobic metabolism). It is contained within double membrane. The Cytomembrane system = system of tubes and chambers formed by membranes. It is extensively distributed throughout the fluid cytoplasm. It is involved in the synthesis, modification, processing & packaging of cellular lipids and proteins.

Cytomembrane System

Endoplasmic reticulum (ER) is present "within the cytoplasm network", a system of tubes and sacs formed by membranes (an enclosed space).

a) Rough = with bound ribosomes (Modifies proteins produced by the ribosomes).

b) Smooth = without bound ribosomes (Doesn't modify proteins, Functions in lipid synthesis, drug, detoxification and carbohydrate metabolism).

Mitochondria

Mitochondria were first seen by kollicker in 1850 in muscles and called them 'sarcosomes. Flemming (1882) described these organelles as 'filia'. Altmann (1890) observed these structures and named them 'bioblasts'. Benda (1898) stained these organelles with crystal violet and renamed them 'mitochondria'. Michaelis (1900) used Janus green B as a vital stain to observe mitochondria in living cells. The shape of mitochondria is highly variable ranges from short rod shape to elongate filamentous form. The size of mitochondria is variable, they generally measures about 0.5 to 2um in diameter. Mitochondria have an average length of 3 to 4um. The number of mitochondria varies from one cell type to another. Mitochondria are not found in prokaryotes.

Ultra-structure

Mitochondria are enclosed by a double membrane envelope composed of lipid and protein. The two membranes are separated by a narrow fluid – filled space called the outer compartment. The outer membrane is smooth; it is more permeable to small molecules, contains some enzymes but is poorer in proteins. The inner membrane surrounds a central cavity or matrix (inner compartment) filled with a fluid. Folds of inner wall of mitochondria are called cristae.

Inner membrane

It is the site of the e^- transport chain, across which the proton pump occurs and contains ATP synthase. Inner membrane is highly folded called cristae increasing the surface area on which the above reactions can take place. The inner mitochondrial membrane is compartmentalized into numerous cristae, which expand the surface area of the inner mitochondrial membrane, enhancing its ability to generate ATP. In typical liver mitochondria, for example, the surface area, including cristae, is about five times that of the outer membrane. Mitochondria of cells which have greater demand for ATP, such as muscle cells, contain more cristae than typical liver mitochondria.

Outer mitochondrial membrane

The outer mitochondrial membrane, which encloses the entire organelle, has a protein-to-phospholipid ratio similar to the eukaryotic plasma membrane (about 1:1 by weight). It contains numerous integral proteins called porins, which contain a relatively large internal channel (about 2-3 nm) that is permeable to all molecules of 5000 daltons or less. Larger molecules, for example most proteins, can only traverse the outer membrane by active transport.

Chemical Composition

Mitochondria consists of protein-70 percent & lipids -25 -30 percent. Mitochondria contain 0.5 percent of RNA & traces of DNA. Mitochondrial DNA comprises about 1 percent of total cell DNA.Mitochondria contain enzymes for oxidation phosphorylation & electron transfer. It produces most of a cells ATP – acetyl groups in the Kreb's cycle producing CO_2 and NADH. NADH donates the e^- to the electron transport chain and becomes oxidized to NAD^+. e^- transfer promotes proton pump and ATP synthesis in process called oxidative phosphorylation. Cells that require large amounts of energy such as the heart have large numbers of mitochondria. It contain their own copies of DNA and RNA along with transcription and translation system (ribosomes). It is able to regenerate themselves without the whole cell undergoing division. The shape and size are dependent on the cell's function.

Golgi Apparatus

The series of flattened sacs is formed by membranes, functions in final protein processing prior to use by the cell. Proteins get shuttled from the ER to one end of the Golgi. In each sac, different modifications are made (proteins get individually tailored). Proteins get sorted and shipped off to their destination (like the post office of the cell). Cis face is nearer the center of cell. Cis face is often adjacent to ER. Trans face is nearer the cell membrane. Activity is directional. Starts at cis face and moves to trans. Synthesis in ER but modification in Golgi. Similar to an assembly line. Options packages are added in Golgi. Sequential passage through cisternae (cis to trans). Each cisternae contains different enzymes. Examples of golgi function. Replace sugars placed on glycoproteins. Modification of phospholipid acylgroups and head groups. Molecule is targeting and assembly. The production of vesicles to delivery membrane is associated molecules and excreted molecules to cell surface.

Cell renewal

Cells and their molecules age is become less effective. An important cellular function is renewal.

- Old molecules and organelles are recycled.
- New phospholipids are produced for the membrane.
- Toxic molecules are need to be collected and detoxified.

Sometimes is a cell needs to commit suicide. It is called autolysis or apoptosis.

Vesicles is small membrane-bound structures that transport proteins and lipids around the cell. Little transporters that shuttle their contents from one organelle to another. When they contact the appropriate organelle, they fuse with its outer membrane and dump Their contents inside Same for plasma membrane, allowing the export of materials from the cell (exocytosis).

- It is perform many of the renewal functions
- Lysosome
- Peroxisome
- Glyoxisome

Lysosome is membrane-bound vesicle that contains digestive enzymes or toxic chemicals merges with vesicles containing food particles, invading bacteria. Harsh chemicals and enzymes are degrade the food or bacteria, without harming rest of cell. It contains digestive enzymes used to degrade macromolecules or organelles. It originates in ER but enzymes are activated in Golgi. Important in macrophages for degradation of particles acquired via phagocytosis. It used to degrade organelles-autophagy.

Cell death

Sometimes used in development of organism like hand development. The certain cells are programmed to die at appropriate times. The defense against viral or intracellular bacterial infections. The self-destruct. program is started by T cells. Apoptosis is often accomplished by leaky or popping lysosomes.

Peroxisomes

It contains enzymes to collect free hydrogen and oxygen and combine into H_2O_2 (peroxide). Peroxide is also toxic so they contain enzymes to degrade into water and oxygen molecules. Catalase experiment in lab. Peroxisomes are formed by aggregation of lipids and proteins (not formed in ER). Many are found in liver cells. Degrade alcohols into peroxide then into water and oxygen. Also responsible for degradation of fatty acids in cytosol. The modify fatty acids and phospholipids and incorporate into peroxisome membrane. It is divided by binary fission.

Glyoxysomes

It is specialized peroxisomes which found in plant seeds. It is recognize appropriate signals and begin to degrade stored fat in seed. Fat is converted to sugars which allows seedling to sprout. Once photosynthesis starts and glyoxysomes degrade.

22. CELL DIVISION, DIFFERENTIATION –MITOSIS MEIOSIS INTER AND INTRACELLULAR COMMUNICATION.

Introduction

Cells are specialize to achieve their required functions are called cell differentiation. The cell division involves splitting of a cell and it will express specific genes.

Mitosis

The division of somatic cells (non-reproductive cells) in eukaryotic organisms. A single cell divides into two identical daughter cells. The daughter cells have the same number of chromosomes as parent cells.

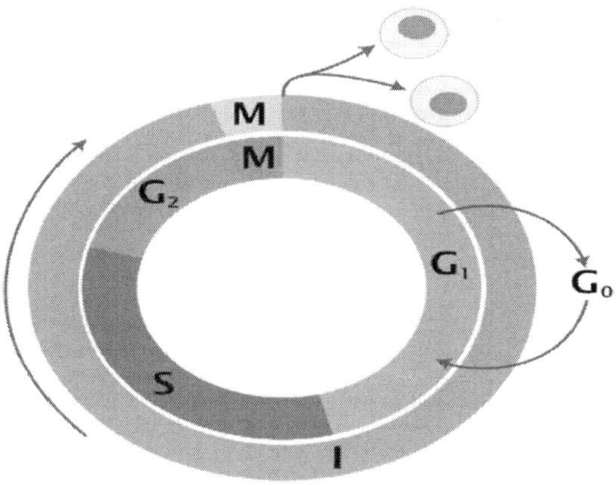

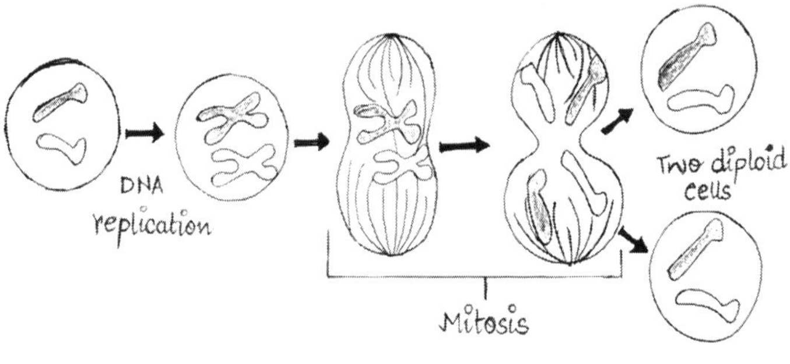

Cell is not dividing

DNA molecules in extended. Uncondensed form is called chromatin. A cell can only replicate and transcribe DNA when in an extended state.

Cell is preparing for division

DNA molecules condense to form chromosomes prior to division. Each chromosome is a single molecule of DNA. It is easier to sort and organize the replicated DNA into daughter cells.

Phases

- 1st – Prophase
- 2nd – Metaphase
- 3rd – Anaphase
- 4th – Telophase
- Cytokinesis

Prophase

It contains 3 major events Chromosomes condense. Spindle fibers form (Spindle fibres are specialized microtubules radiating out from centrioles). Chromosomes are captured by the spindle.

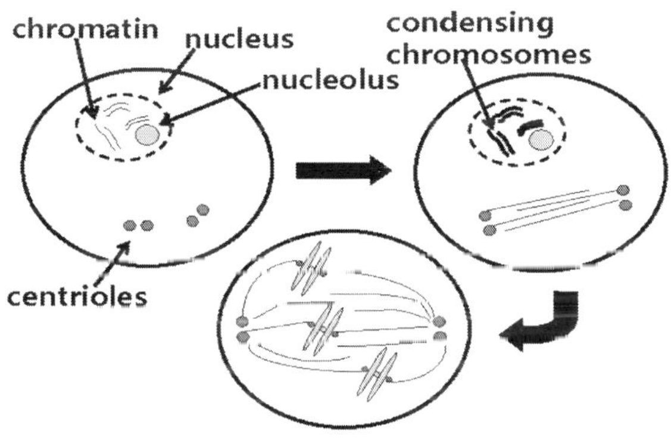

Prophase

Metaphase

Chromosomes align alongthe equator of the cell, with one kinetochore facing each pole.

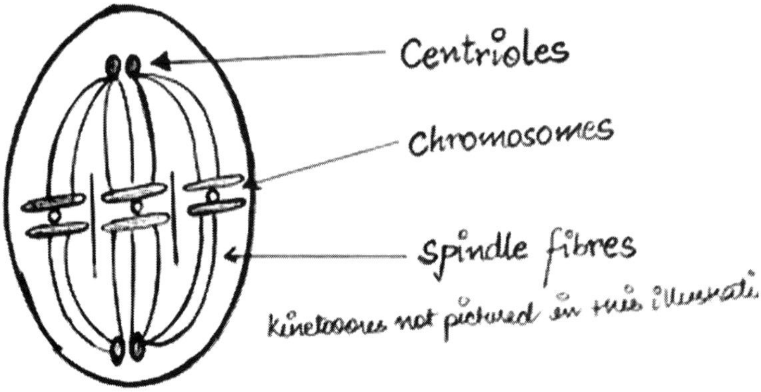

Metaphase

Anaphase

It has sister chromatids separate. Spindle fibers attached to kinetochores shorten and pull chromatids towards the poles. It is free spindle fibers lengthen and push poles of a cell apart.

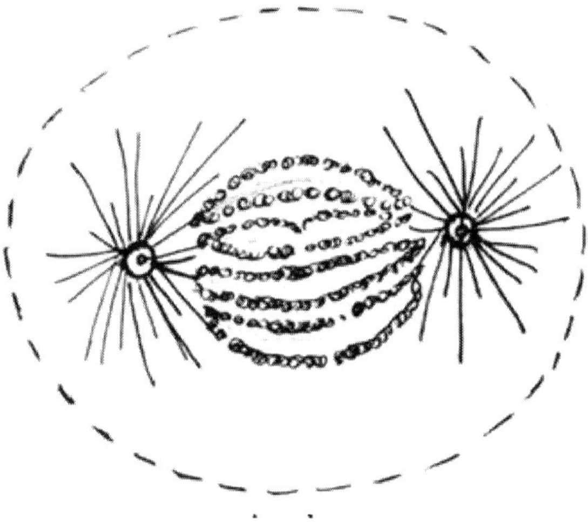

Anaphase

Telophase

The spindle fibres are disintegrate. Nuclear envelopes form around both groups of chromosomes. Chromosomes revert to their extended state. Cytokinesis occurs, enclosing each daughter nucleus into a separate cell.

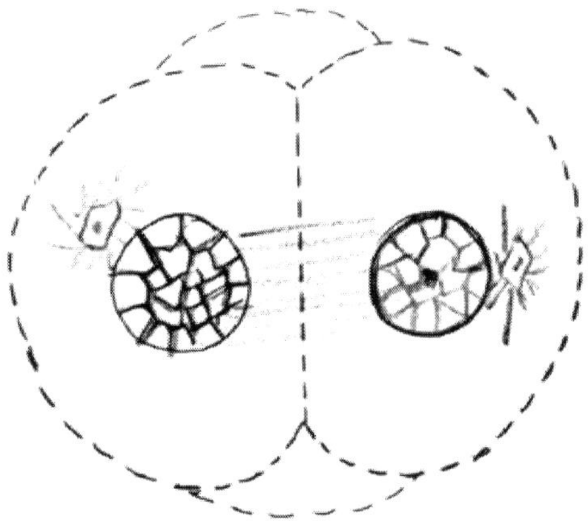

Telophase

Cytokinesis

- *Plant cells:* It undergoes cytokinesis by forming a cell plate between the two daughter nuclei.
- *Animal cells:* It undergoes cytokinesis through the formation of a cleavage furrow. A ring of microtubules contract, pinching the cell in half.

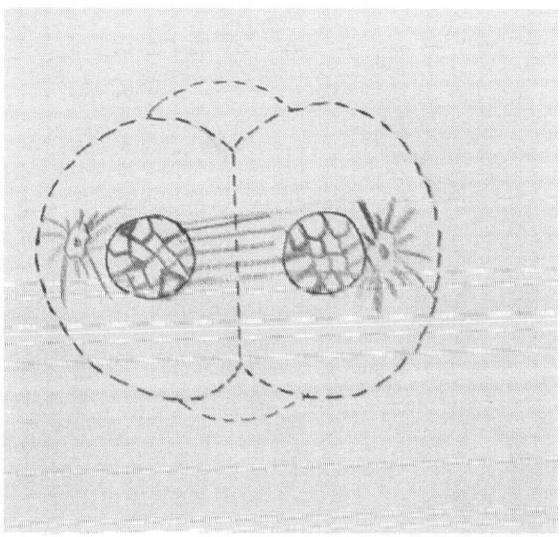

Cytokines

Stages of Mitosis

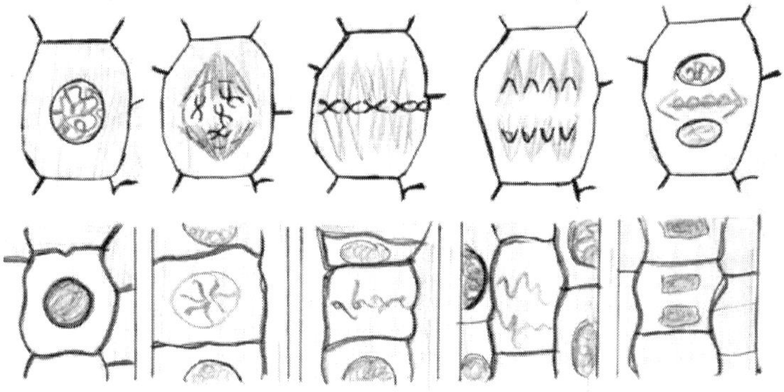

Stage of Mitosis

The cell division of gametes called Meiosis

A single germ cell divides into four unique daughter cells. Daughter cells have half of the chromosomes as the parent cell, so they are considered haploid.

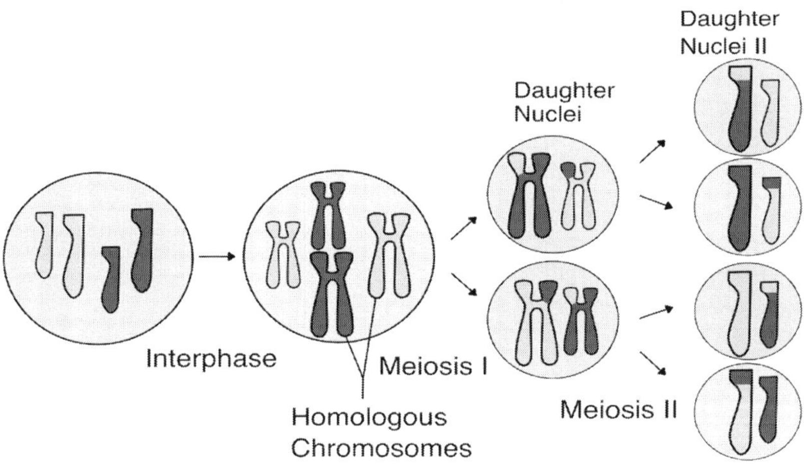

Stage of Meiosis

Haploid

One copy of each chromosome. Designated as "n", the number of chromosomes in one "set" gametes.

Diploid

Two sets of chromosomes. Two of each chromosome. Designated as "2n"

Somatic cells

Diploid organisms receive one of each type of chromosome from female parent (maternal chromosomes) and one of each type of chromosome from male parent (paternal chromosomes)

Homologous

Chromosomes exist in homologous pairs in diploid (2n) cells.

Exception: Sex chromosomes (X, Y).

Other chromosomes, known as autosomes, they are homologus.

Asexual reproduction

Many single-celled organisms reproduce by splitting, budding. Some multicellular organisms can reproduce asexually; produce **clones** (offspring genetically identical to parent).

Sexual reproduction

A fusion of two gametes to produce a single zygote. Introduces greater genetic variation, allows genetic recombination. With exception of self-fertilizing organisms, the zygote has gametes from two different parents.

Reproduction in Humans

It is donated by each parent *(*total = 46 or 23 pairs*)*. Gametes (sperm/ova):Contain 22 autosomes and 1 sex chromosome. Are haploid (haploid number "n" = 23 in humans).

Fertilization results in the diploid zygote. Diploid cell; 2n = 46. (n = 23 in humans)

Meiosis

Sex cell (gamete) formation. In meiosis, there are 2 divisions of the nucleus: meiosis I and meiosis II. Meiosis &sexual reproduction life cycle.

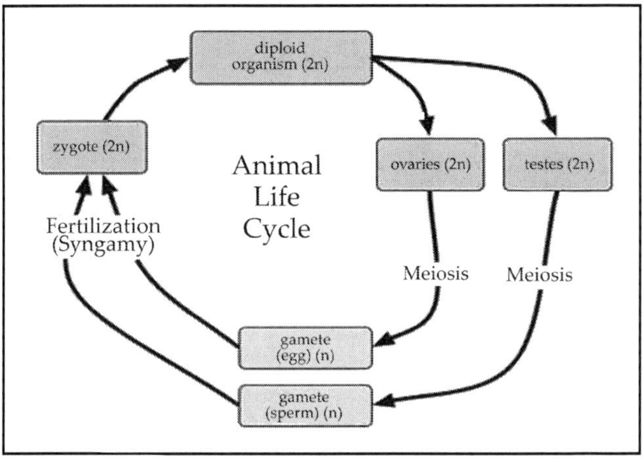

Genetic Variation in Diploid Organisms

The fusion of sperm and egg results in unique offspring. But not only because the young are a product of two individuals with different genetic makeup. Meiosis "shuffles" the genes so that the individual's gametes are genetically different from one another.

Genetic shuffling of Meiosis I

In addition to a new combination of chromosomes resulting from fertilization, there are also events in Meiosis I that shuffle the genes.

1. Crossing over in Prophase I.
2. Independent assortment in Metaphase I.

Crossing over

Homologus break at identical locations and then rejoin opposite partners. This creates new combinations of the alleles on each chromosome. It occurs randomly several times on every chromosome. Results in mixing of the genes you inherited from your parents.

Spermatogenesis

Males produce sperm throughout life, after the onset of puberty, about 1,500 sperms per second.

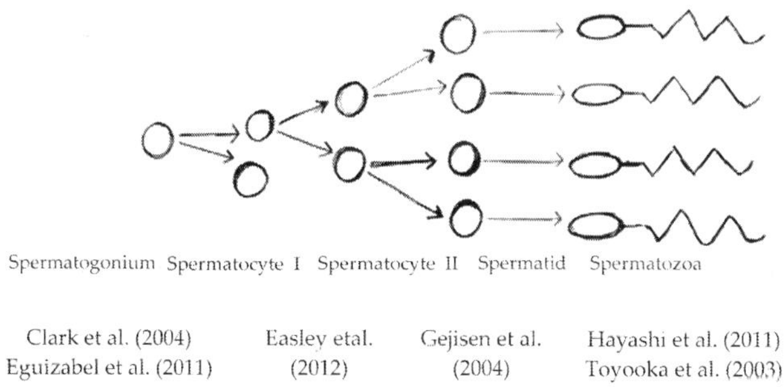

Structure of spermatogenesis

Oogenesis

Oogenesis in females is probably complete either before or shortly after birth. During oogenesis, three polar bodies develop as the mature ovum is generated. Polar bodies contain little cytoplasm and eventually degenerate.

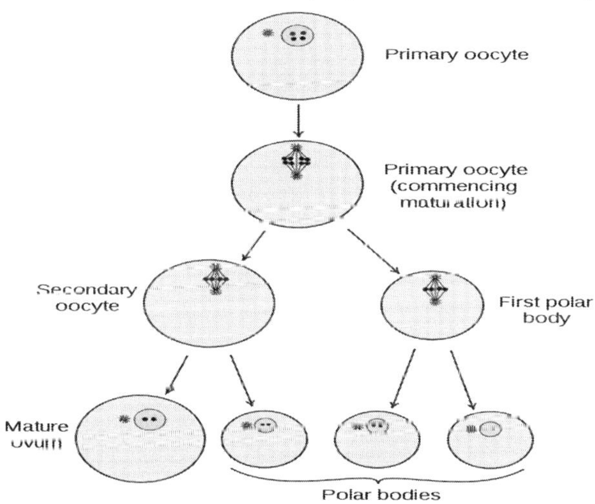

Oogenesis

Cell signaling

- General principles of cell signaling (signaling cell => signaling molecule binds to a receptor molecule on target cell)
- Endocrine
- Paracrine
- Neuronal
- Contact dependent

Receptor

Each target cell responds to a limited set of signal (origin of complexity). Must have receptor molecule for the signaling molecule (thousands of receptors). Different receptors for the same signal on different cells.

Signalling Pathways

Same receptor molecule can interact with many intracellular relay systems so the same signal and same receptor of different effects in different cells. Same relay system acts on many different intracellular targets. Receptors sense diverse stimuli but initiate a limited repertoire of cellular signals. Receptors contain a ligand-binding domain. An effector domain. Receptor modularity allows a wide variety of signals to use a limited number of regulatory mechanisms. Cells may express different receptors for the same ligand. The same ligand may have different effects on the cell depending on the effector domain of its receptor.

Receptors are catalysts and amplifiers: Receptors act by increasing the rates of key regulatory reactions. Receptors act as molecular amplifiers. Ligand binding changes the receptor conformation. Receptors can exist in active or inactive conformations. Ligand binding drives the receptor toward the active conformation. Signalling pathways usually have multiple steps and can diverge and/or converge. Divergence allows multiple responses to a single signal. Convergence allows signal integration and coordination. Signalling networks are composed of groups of biochemical reactions. The reactions function as mathematical logic functions to integrate information. Combinations of such logic functions combine as signalling networks to process information at more complex levels.

- Dephosphorylation
- Phosphorylation

Diverse receptors recruit protein tyrosine kinases to the plasma membrane. Receptors that bind protein tyrosine kinases use combinations of effectors similar to those used by receptor tyrosine kinases.These receptors often bind directly to transcription factors.

23. CELL GROWTH AND PROLIFERATION, MOLECULAR CONSTITUENTS OF CELLS.

Introduction

Cells in a multicellular organism communicate through chemical signals are called cell growth. Hormones act over a long-range. Local mediators are secreted into the local environment. Some cells communicate through direct cell-cell contact. Cells are stimulated when extra-cellular signalling molecules bind to a receptor. Each receptor recognizes a specific protein (ligand). Receptors act as transducers that convert the signal from one physical form to another.

Signalling Molecules

Most signalling molecules cannot pass through the cell membrane. Their receptors are in the cell membrane. Small hydrophobic signal molecules can diffuse directly into the cell cytoplasm. Their receptors are cytoplasmic or nuclear. Hormones are insulin and cortisol. It also contain local mediators such as Epidermal Growth Factor (EGF), Platelet Derived Growth Factor (PDGF), Fibroblast Growth Factor (FGF), TGFβ and Cytokines, e.g. Interferons, Tumour necrosis factor (TNF).

Receptors

There are three main classes of receptors. Ion-channel-linked receptors. G-protein-linked receptors. Enzyme-linked receptors. Ion channel-linked receptors are important in neural signalling. G-protein and enzyme linked receptors respond by activating cascades of intracellular signals. These signals alter the behavior of the cell.

G-protein-linked receptors

G-protein-linked receptors activate a class of GTP-binding proteins (G-proteins). G proteins are molecular switches. They are turned on for brief periods while bound to GTP. They switch themselves off by hydrolyzing GTP to GDP.

G proteins

Some G proteins directly regulate ion channels. Others activate adenylate cyclase, thus increasing intracellular cyclic AMP. Some activate the enzyme Phospholipase C, thus increasing intracellular inositol triphosphate (IP_3) and Diacylglycerol (DAG).

Enzyme-linked receptors

Many receptors have intracellular domains with enzyme function. Most are receptor tyrosine-kinases. They phosphorylate tyrosine residues in selected intracellular proteins. These receptors are activated by growth factors, thus being important in cell proliferation.

Receptor Tyrosine Kinases

Receptor tyrosine kinase activation results in the assembly of an intracellular signaling complex. This complex activates a small GTP-binding protein and Ras. Ras activates a cascade of protein kinases that relay the signal to the nucleus. Mutations that make Ras hyperactive are a common way of inducing increased proliferation in cancer.

Signaling: cytoplasm to nucleus

Many signaling cascades culminate in the activation of nuclear transcription factors. Transcription factors alter gene expression. C-jun and C-fos (that form an AP1 complex) and C-myc are three important transcription factors.

Signaling pathway interactions

There are many signaling molecules and receptors. A given cell expresses only a subset of receptors. Different intracellular signaling pathways interact. This enables cells to respond appropriately to complex combinations of signals.

Cell signaling and proliferation

Animal cells proliferate when stimulated by growth factors. These bind mainly to receptor tyrosine kinases.These signaling pathways override the normal brakes on proliferation. These brakes are part of the cell cycle control system. This ensures that cells divide only under appropriate circumstances.

The Cell Cycle

The eukaryotic cell cycle consists of distinct phases. The most dramatic events are nuclear division (mitosis) and cytoplasmic division (cytokinesis). This is the M phase. The rest of the cell cycle is called interphase which is deceptively and uneventful. During interphase, the cell replicates its DNA, transcribes genes, synthesizes proteins and grows in mass.

Phases of the cell cycle

- S phase – DNA replicates in this phase
- M phase – Nucleus divides (mitosis) and thecytoplasm divides (cytokinesis)
- G1 phase – Gap between M and S phase
- G2 phase – Between S and M phase

Cell cycle control

Cell cycle machinery is subordinate to a cell cycle control system. The control system consists mainly of protein complexes. These complexes consist of a cyclin subunit and a Cdk subunit. The cyclin has a regulatory function, the Cdk catalytic function. Cdk expression is constant, but cyclin concentrations rise and fall at specific times in the cell cycle.The Cdks are cyclically activated by cyclin binding and by phosphorylation status. Once activated, Cdks phosphorylate key proteins in the cell. Different cyclin-Cdk complexes trigger different cell cycle steps. Some drive the cell into the M phase, others into the S phase. The cell cycle control system has in-built molecular breaks (checkpoints). The checkpoints ensure that the next step does not begin until the previous one is complete.

The G1 checkpoint has been widely studied. The retinoblastoma (Rb) protein plays a key role at this checkpoint. The Rb protein function is determined by its phosphorylation status. The S phase cyclin-Cdk complexes phosphorylate Rb. This checkpoint is influenced by the action of cyclin-dependant kinase inhibitors (CKIs, e.g. p21, p16). e.g. p53 senses DNA damage and induces p21 expression. CKIs inactivate cyclin-Cdk complexes. Cells must respond to a variety of stimuli that may be hormonal, paracrine or through direct cell contact. These stimuli may arise under physiological or pathological conditions. The way that cells adapt in terms of growth and differentiation depends in part on their ability to divide. Tissues can be classified according to the ability of their cells to divide. Some tissues contain a pool of cells that move rapidly from one cell cycle to the next. These are labile cells. Some cells dismantle their cell cycle control machinery and exit the cell cycle. These cells are said to be in $G_{0.}$ Some of these cell scan re-enter the cell cycle when stimulated, e.g. by growth factors. These are stable cells. Others are unable to re-enter the cell cycle. These are permanent cells.

CHAPTER 4

INTRODUCTION TO BIOTECHNOLOGY

24. INTRODUCTION DEFINITION AND CLASSICAL VS MODERN BIOTECHNOLOGY.

Introduction

The biotechnology term was coined in 1917 by a Hungarian Engineer, Karl Ereky, to describe a process for the large scale production of pigs. According to him all types of work are biotechnology by which products are produced from raw materials using living organisms. During the end of the 20th century biotechnology emerged as a new discipline of biology integrating with technology, but the route of biotechnology from biology. There was no sudden sprout of this discipline, but some of the methods for the production of products were developed centuries back. Therefore, biotechnology is concerned with the exploitation of biological components for the production of useful products. Biotechnology is defined by different organizations in different ways. It has been broadly defined as "the development and utilization of biological processes, forms and systems for obtaining maximum benefits to man and other forms of life". Biotechnology is "the science of applied biological process" (Biotechnology A Dutch Perspective, 1981). Biotechnology is the "controlled use of biological agents such as microorganisms or cellular components for beneficial use" (U.S. National

Science Federation). The definition was given by OECD "scientific and engineering principles" refer to microbiology, biochemistry and genetics ,etc. Biological agents: we can call biological agents such as microorganisms, enzymes, plant and animal cells. The meaning of these three definitions and others given by many organizations are more or less similar. The definition of genetic engineering has been given by Smith (1996) as "the formation of new combinations of heritable material by the insertion of nucleic acid molecules produced by whatever means outside the cell, into any virus or bacterial plasmid or other vector systems so as to allow their incorporation into host organisms in which they do not naturally occur but in which they are capable of continued propagation".

Importance

Biotechnology is widely used in pharmacy to create more efficient and less expensive drugs. Recombinant DNA technology is used for the production of specific enzymes, which enhance the rate of production of a particular range of antibodies in the organism. The hormones such as somatostatin, insulin and the human growth hormone can be synthesized easily and cheaply. The first human hormone to be synthesized by genetic engineering was somatostatin. Somatostatin is the brain hormone originating from the hypothalamus. It acts to inhibit the release of human growth hormone and insulin is related to the treatment of diabetes, pancreatitis and few other conditions. Genetech, a California based company, has produced human growth hormone (hGH) from genetically engineered bacteria. Human insulin is the first genetically engineered pharmaceutical product, developed by Eli Lilly and company in 1982. Bovine Somatotropin (BST) is produced for a large quantity of milk production in cows.

Antibiotics are chemical substances produced by several microorganisms. Recombinant DNA technology has helped in increased production of antibiotics for example; the rate of penicillin produced at present is about 150,000 unit/ml against about 10 unit/ml in the 1950s. Antibiotics produced using such technology has very specific effects and causes fewer side effects. Currently, scientists are working on vaccines for fatal illnesses such as AIDS, hepatitis, malaria, flu, and even some forms of cancer. Interferon, an anti-viral protein, is prepared from mammalian cells by recombinant DNA technology. By cloning cDNA to genes for human interferon, it has been found that there isa large number of interferon differing in amino acid sequences and properties. A large number of interferon is prepared in yeast cells by the fermentation process. Expects that in the near future vaccines will come in more convenient ways "some will come in the form of mouthwash; others will be swallowed in time-release capsules, avoiding the need for boosters".

Biotechnology Century

Biotechnology is the technologies applied to biology, molecular biology, genetics, and many other subfields of biology. Biotechnology utilizes cellular and biomolecular processes to create technologies and products that help improve our lives and nature. By making useful food, such as bread and cheese, and preserving dairy products, we have done these for many years by now. Recent biotechnology develops breakthrough products and technologies to fight diseases, reduce our environmental harm, feed the hungry, and useless and cleaner energy, and have safer, cleaner and more efficient industrial manufacturing processes. More than 250 biotechnology health care products and vaccines have been made available to patients, many for previously untreatable diseases. More than 13.3 million farmers around the world use agricultural biotechnology to increase yields prevent damage from insects and pests and reduce the damage done to the environment due to farming and more than 50 biorefineries are being built across North America to test and refine technologies to produce biofuels and chemicals from renewable biomass, which can help reduce greenhouse gas emissions. Biotechnology is the third wave in biological science and represents such an interface of basic and applied sciences, where the gradual and subtle transformation of science into technology can be witnessed. Biotechnology is defined as the application of scientific and engineering principles to the processing of the material by biological agents to provide goods and services. Biotechnology comprises a number of technologies based upon an increasing understanding of biology at the cellular and molecular level. The Bible already provides numerous examples of biotechnology. Namely, it deals with the conversion of grapes to wine, of dough to bread and of milk to cheese.

The oldest biotechnological processes are found in microbial fermentation, as born out by the Babylonian tablet dated circa 6000 B.C explaining the preparation of beer. The Sumerians were able to brew as many as twenty types of beer in the third millennium B.C. In about 4000 B.C. leavened bread was produced with the aid of yeast. During the Vedic period (5000-7000 B.C). Aryans had been performing daily *Agnihotra* or *Yajna*. In *Ayurved*, production of 'Asava' and 'Arista' using different substrates and flowers of mahua (*Madhuca indica*) or dhataki (*Wodfordiafructicosa*) has been well characterized till today since the Vedic period. One of the materials used in *Yajna* **is animal fat** (i.e. ghee) that is the fermented product of milk. The term 'biotechnology' was described in a Bulletin of the Bureau of Biotechnology published in July 1920 from the office of the same name in Leeds in Yorkshire. The articles in this bulletin described the varied roles of microbes in the leather industry to pest control.

Sub-fields of Biotechnology

Red biotechnology is biotechnology applied to medical processes. Some examples are the designing of organisms to produce antibiotics, and the engineering of genetic cures to cure diseases through genomic manipulation.

White biotechnology, also known as grey biotechnology, is biotechnology applied to industrial processes. An example is the designing of an organism to produce a useful chemical. White biotechnology tends to consume less resources than traditional processes when used to produce industrial goods.

Green biotechnology is biotechnology applied to agricultural processes. An example is the designing of an organism to grow under specific environmental conditions or in the presence (or absence) of certain agricultural chemicals. One hope is that green biotechnology might produce more environmentally friendly solutions than traditional industrial agriculture. An example of this is the engineering of a plant to express a pesticide, thereby eliminating the need for external application of pesticides. Whether or not green biotechnology products such as this are ultimately more environmentally friendly is a topic of considerable debate.

The term blue biotechnology has also been used to describe the marine and aquatic applications of biotechnology, but its use is relatively rare.

Broadly Biotechnology can be Divided Into Two Major Branches

Non-gene biotechnology deals with whole-cell, tissues or even individual organisms. Gene biotechnology involves gene manipulation, cloning, etc. Non-gene biotechnology is a more popular practice, and plant tissue culture, hybrid seed production, microbial fermentation, production of hybridoma antibodies or immunochemical are wide-spread biotechnology practices. For centuries humans have used microorganisms to produce foods and drinks without understanding the microbial processes underlying their production.

In recent years the understanding of the biosynthetic pathways and regulatory control mechanisms used by microorganisms for the production of several metabolites has been increased by developing the knowledge of the biochemistry of industrially important organisms. Notable biotechnologies for food processing include fermentation technology, enzyme technology and monoclonal antibody technology. Beneficial microbes participate in fermentation processes, producing many useful metabolites such as enzymes, organic acids, solvents, vitamins, amino acids, antibiotics, growth regulators, flavors and nutritious foods. Some leading food bioprocessing technologies are dairy processing, alcohol and beverage processing. Production of alcoholic beverages include wine, beer, whiskey, rum, shake, etc. utilizing microorganisms like *Clostridium acetobutylicum*, *Lecuonostoc mesenteroides*, *Aspergillus oryzae*, *Saccharomyces cerevisiae*, *Rizopus sp.* and *Mucor sp.* Biotechnologically produced organic acids like citric acid, acetic acid, gluconic acid, D-Lactic acid, fumaric acid also havea very high market value.The application of biotechnology can result in (a) new ways of producing existing products with the use of new inputs, and (b) new ways of producing new products. Examples of the former include the production of gasoline from ethanol which in turn is produced from sugar; the production of insulin using recombinant DNA technology; the production of hepatitis B vaccine using recombinant DNA technology and the extraction of copper using mineral leaching bacteria.

The alternative inputs are oil for gasoline, porcine pancreases for insulin, human blood for hepatitis vaccine, and the conventional mining techniques for copper. Examples of the latter include possible medicinal substances thatare produced in minute quantity in the human body and which cannot be synthesized such as insulin, interleukin or Tissue Plasminogen Activator (TPA).

A wide variety of microorganisms are now being employed as tools in biotechnology to produce useful products or services. Raw materials can be converted to useful finished products both by ordinary chemical processes and by biological means. Generally, the costs of chemical conversion are quite high as the reactions require high temperature or pressure. In contrast, biological alternatives, using microbes or cultured animal or plant cells, operate at physiologically normal conditions of temperature, pressure, pH, etc. During the next few decades, biotechnology would have overtaken chemical technology and many such chemicals thatare today produced chemically would be made through biotechnology.

Enzyme technology is an area of considerable current interest and development. Enzymes are biological catalysts and have been used for many years as isolated agents particularly in food e.g. rennin, papain and invertase. These enzymes have increasingly replaced plants and animal enzymes thus amylase from *Bacillus* and *Aspergillus* has substituted those of malted wheat and barley in brewing, baking and biscuit-making and also in the textile industry, etc. Today, enzyme technologies have four distinct areas of application in cosmetics, therapy, the food and feed industry, and for diagnostic purposes. One very important recent application is the production of foodstuffs from non-traditional raw materials ,for instance, the development of the sweetener, high fructose corn syrup (HFCS), also called isoglucose. Another recent application is the use of phytase in animal feed. Nowadays, interest in the traditional fermentation technology for food processing has greatly increased because of emphasis placed upon plant materials as human foods.

Single-cell protein (SCP) is the term generally accepted to mean the microbial cells (algae, bacteria, actinomycetes and fungi) grown and harvested for animal or human food. During World War II, when there were shortages in proteins and vitamins in the diet, the Germans produced yeasts and a mold (*Geotrichum candidum*) in some quantity for food. Research on SCP has been stimulated by a concern over the eventual food crisis or food shortages that will occur if the world's population is not controlled. Many scientists believe that the use of microbial fermentation and the development of an industry to produce and supply SCP are possible solutions to meet a shortage of protein if and when the amount of protein produced or obtained by agriculture and fishing becomes insufficient. The roots of molecular biology were established only after the British biophysicist Francis Crick and the American Biochemist James Watson, in 1953, proposed the structure of DNA (deoxyribonucleic acid) molecule which is well known as the chemical bearer of genetic information of most of the organisms. We really began understanding and utilizing molecular biotechnology (or gene biotechnology) only after recombinant DNA technology was developed in the 1970's.

Daniel Nathans (in 1971) of John Hopkins University utilized the restriction enzyme to split DNA of monkey tumor virus, Simian Virus (SV_{40}). Recombinant DNA technology, often referred to as genetic engineering or gene manipulation, involves the extraction of a particular gene of interest from one organism and then the insertion of the gene into other organisms. Genetic manipulation may be defined as the extracellular (i.e. *in vitro*) creation of new forms of arrangements of DNA in such a way as to allow the incorporation or continued propagation of altered genetic conditions in nature. Among the first scientist to attempt genetic manipulation was Paul Berg of Stanford University who is1971 along with his co-workers opened the DNA molecule of SV_{40} and spliced it into a bacterial chromosome and constructed the first recombinant DNA molecule.

Genetic engineering techniques are useful tools for genetic research. They can help to gain in the structure, function and regulation of genes. They also help to prepare the physical maps of the viral genome. Maps of several viruses have been made available like SV_{40}, Polyoma virus and adenovirus. Another goal in genetic engineering is to design super bugthatcan degrade most of the major hydrocarbon components of petroleum. The different strains of *Pseudomonas putida* contain a plasmid that has genes coding for enzymes that digest a single family of hydrocarbons. By crossing the various strains of this bacterium, a superbug has been created. The multiplasmid bacterium is able to grow on a diet of crude oil. The superbug has the potential for clearing up oil spills. One of the best-known applications of genetic engineering is that of the creation of genetically modified organisms (GMOs).

There are potentially momentous biotechnological applications of GM, for example, oral vaccines produced naturally in fruit, at very low cost. This represents, however, a spread of genetic modification to medical purposes and opens an ethical door to other uses of the technology to directly modify human genomes. A genetically modified food is a food product derived in whole or part from a genetically modified organism (GMO) such as a crop plant, animal or microbe such as yeast. Genetically modified foods have been available since the 1990s. The principal ingredients of GM foods currently available are derived from genetically modified soybean, maize and canola. Between 1996 and 2001, the total surface area of land cultivated with GMOs had increased by a factor of 30, from 17,000 km^2 (4.2 million acres) to 520,000 km^2 (128 million acres). The value for 2002 was 145 million acres (587,000 km^2) and for 2003 was 167 million acres (676,000 km^2).

Future applications of GMOs include bananas that produce human vaccines against infectious diseases such as Hepatitis B, fish that mature more quickly, fruit and nut trees that yield years earlier, and plants that produce new plastics with unique properties. Now scientists have transformed Tobacco Mosaic Virus (TMV) to infect host plants and produce immunizing proteins rather than debilitating leaf shrivel, turning greenhouse tobacco into a biofactory for plague vaccine.

Genetic diseases could be treated through the use of genetic engineering. Defective genes in an organism cause genetic disorders. If a defective gene could be identified and located in a particular group of cells – it could be replaced with a functional one. The transgenic cells are then planted into the organism, resulting in a cure of the disorder. Cloning is a relatively new sector of biotechnology, but it promises answers to very important problems related to surgery. Tissues and organs could be cloned for surgical purposes. If scientists could isolate stem cells and then direct their development, they would be able to create any kind of tissue, organ or even a whole part of a body. Another revolutionizing tool of biotechnology is DNA fingerprinting. DNA fingerprints are useful in several applications of human health care research, as well as in the justice system. DNA fingerprinting is used to diagnose inherited disorders in both prenatal and new born babies in hospitals around the world. These disorders may include cystic fibrosis, hemophilia, Huntington's disease, familial Alzheimer's, sickle cell anemia, thalassemia, and many others. Early detection of such disorders enables the medical staff to prepare themselves and the parents for proper treatment of the child. In some programs, genetic counselors use DNA fingerprint information to help prospective parents understand the risk of having an affected child. DNA fingerprint information can also help in developing cures for inherited disorders. DNA fingerprints help to link suspects to biological evidence blood or semen stains, hair, or items of clothing found at the scene of a crime and help in solving crime. Another important use of DNA fingerprints in the court system is to establish paternity in custody and child support litigation. The U.S. armed services have just begun a program to collect DNA fingerprints from all personnel for use later, in case they are needed to identify casualties or persons missing in action or for suspect verification. Due to the revolutionary development of biotechnology during the last couple of decades agriculture has drastically advanced. Sensational achievements were made in both plant cultivation and animal husbandry.

Plants have been Improved in four Different Ways

Enhanced potential for more vigorous growth and increasing yields. Increased resistance to natural predators and pests, including insects and disease-causing microorganisms. Production of hybrids exhibiting a combination of superior traits derived from two different strains or even different species. Selection of genetic variants with desirable qualities such as increased protein value increased content of limiting amino acids, which are essential in the human diet, or smaller plant size, reducing vulnerability to adverse weather conditions. Another important area of biotechnology is an improvement of livestock. Improvement in disease control, the efficiency of reproduction, yields of livestock products i.e. meat, milk, wool, eggs, the composition of livestock products i.e. leaner meat, feed value of low-quality feeds i.e. straw, are some of the applications of biotechnology. One of the major scientific revolutions of the twentieth century was the breaking of the genetic code and the development of tools that enable scientists to probe the molecules of life with incredible precision.

Now, in the twenty-first century, these developments in biology are being married with the use of ever-increasing computer power to help us face the challenges that the new century brings. Bioinformatics is the name given to the new discipline that has emerged at the interface of biology and computing. A huge amount of genetic data (DNA, RNA, amino acid and protein sequences) of various organisms, from bacteria to humans, being generated worldwide is stored in a computer database. Specialized software programs are used to find, visualize, and analyze the information, and most importantly, communicate it to other people. Various computer tools are used to predict protein structure which is valuable information for the development of vaccines, diagnostic tools as well as more effective drugs. Bioinformatics can help in easy and early detection of various diseases like cancer, diabetes and many more with the help of microarray chips (microarrays are miniature arrays of gene fragments attached to glass slides).

Bioinformatics also helps scientists to construct a phylogenetic tree based on molecular biology and ultimately contribute to the study of evolution. Computer simulations model such things as population dynamics or calculate the cumulative genetic health of a breeding pool (in agriculture) or endangered population. One very exciting potential of this field is that entire DNA sequences or genomes of endangered species can be preserved. Biotechnology has a promising future. In the future, biotechnology will be accredited for some revolutionary technology. Recent advances in bioenergy, bioremediation, synthetic biology, DNA computers, virtual cell, genomics, proteomics, bioinformatics and bio-nanotechnology have made biotechnology even more powerful. A recent discovery of the conduction of electricity by DNA and its behavior as a superconductor has opened a new realm in modern science. In the future, biotechnology will have a profound impact on the world economy.

Biotechnology is a golden tool to solve some of the key global problems like a global epidemic, fatal diseases, global warming, rising petroleum fuel crisis and above all poverty. For all the positive effects of biotechnology, there are some possible side effects. Nobody knows what ecological hazards could be caused by transgenic organisms. Some even speculate that some transgenic organisms could fall into wrong hands to develop bioweapons. The opposition of genetic engineering says that "the science is very young and needs a lot more research". The path from a test tube to the field is not a straight highway. Both intellectual and financial resources should be realized before new discoveries pave their way to industrial applications.

In conclusion, biotechnology has also proved to be extremely productive and innovative and the 21st century should be the century of biotechnology.

The branches of microbiology are categorized into pure and applied sciences such as:

- Bacteriology: It is the study of bacteria.
- Mycology: It is the study of fungi.
- Protozoology: It is the study of protozoans.
- Virology: It is the study of viruses.
- Algology or Phycology: It is the study of algae.
- Parasitology: It is the study of parasitism or parasites.
- Microbial Ecology: It is the study of the interrelationship between the environment and microbes.
  - Microbial Morphology: It is the study of the detailed structure of microbes.
  - Microbial Taxonomy: It is concerned with the classification, naming and identification of different microbes.
  - Microbial Physiology: It is the study of the metabolism of microbes at the cellular and molecular levels.
  - Microbial Genetics: It is the study of the genetic material of microbes.
  - Molecular Microbiology: It is the study of structure and function and the biochemical reactions of microbial cells involved in metabolism and growth.
  - Industrial Microbiology: It is concerned with the industrial uses of microbes in the production of vitamins, alcoholic

beverages, amino acids, enzymes, antibiotics and other drugs.

- Food Microbiology: It is the study of the interaction between food and microbes in relation to food bioprocessing, food spoilage and their prevention.
- Agricultural Microbiology: It is the study of the interrelationship of crops and microbes with an emphasis on control of plant diseases and improvement of yield.
- Dairy Microbiology: It deals with the production of and maintenance ofquality control of dairy products.
- Aquatic Microbiology: It is the study of microorganisms and their activity concerning human and animal health in fresh, estuarine and marine waters.
- Air Microbiology: It deals with the role of aerospora in contamination and spoilage of food and discrimination of plant and animal disease by air.
- Exomicrobiology: It is the study of exploration for microbial life in outer space.
- Medical Microbiology: It deals with the fundamental principles and techniques involved in the study of pathogenic organisms as well as their application in the diagnosis of infectious diseases.
- Immunology: It deals with the immune system that protects against infections and attempts to understand the many phenomena that are responsible for both acquired and innate immunity.
- Biotechnology: It is the scientific manipulation of living organisms especially at the molecular and genetic level to produce useful products.
- Public Health Microbiology: It is concerned with monitoring, control and spread of diseases in communities.

Fermentation

Lactic acid production got increased when he added chalk powder to the fermentation medium. Pasteur showed the presence of lactic acid in curd by using a polarimeter. In 1860, Pasteur provided a detailed report on the use of a synthetic medium for microbiological studies. He concluded that there are many steps for fermentation such as: Fermentation is carried out anaerobically by the living cells.During the alcoholic fermentation in a synthetic medium the yeast increased the weight with an increase in carbon and nitrogen contents of an overall batch. The increase in yeast protein in the synthetic medium was accompanied by a related decrease in the ammonium nitrogen in the medium. Fermentation sugar was required for the multiplication of yeast cells.A similar phenomenon occurred in the fermentation of lactic acid, tartaric acid and butyric acid ,etc.The pure culture is not using; some of the fermentation processes were stopped.Growth and physiology of yeast differ when they are grown under anaerobic growth conditions. A large amount of sugar was converted into alcohol under anaerobic growth conditions, while in the aerobic conditions a large amount of sugar was converted into yeast cell mass. Pasteur suggested that the high percentage of the microbial population is killed by heating the juice at 62.8° C which is now called as Pasteurization. Robert Koch in 1881 gave the method for establishing the relationship of a pathogen with a disease which is known as "**Koch's**" postulates or "pathogenicity test". He proved that the anthrax disease is caused by "***Bacillus anthracis***".

Role of Microorganisms In Fermentation

The causes of fermentation could be discovered after observing microoorganisms using a microscope Antony van Leeuwenhoek (1673 1723) at Delf (Holland). During the 18th century, a significant contribution was done by the chemists on the process and products of fermentation. In 1757, it was demonstrated that a milky precipitate could be obtained when the gas evolved from fermentation was passed through the lime water. It was called the lime water test. A similar gas comes out from the burning of charcoal. Henry Cavendish demonstrated that the gas evolved from brown sugar in water treated with yeast was absorbed by sodium hydroxide solution. Schele, J.Priestly (1772-1774) confirmed the identity of oxygen gas present in the air. Using an analytical technique for carbon estimation, Antoine Lavoisier gave the chemical basis of alcoholic fermentation.The study of microbiology was started withthe first report of Louis paster (1857) on lactic acid fermentation from sugar. He isolated the microorganisms (lactic yeast) that were associated with lactic acid and formed curd. The cells of lactic yeast were smaller than that of beer yeast Lactic acid production got increased when he added chalk powder to fermentation media. Pasteur showed the presence of lactic acid in curd by using a polarimeter.In 1860, Pasteur provided a detailed report on the use of the synthetic medium.

He concluded that fermentation is carried out anaerobically by the living cells. During alcoholic fementation in synthetic yeast increased the weight with an increase in C and N contents of the batch. The increase in yeast protein in the synthetic medium was accompanied by a related decrease in anmonium nitrogen in the medium. Fermentation of sugar was required for the multiplication of yeast cells. Similar phenomenon occurred in the fermentation of lactic acid, tartaric acid, butyric acid, etc. because of using pure culture, some of the fermentation processes were stopped growth and physiology of yeast differ when they are grown conditions was later on called as a large amount of sugar was converted into alcohol, while under aerobic conditions a large amount of sugar was converted into yeast cell mass. Pasteur suggested that a high percentage of microbial population is killed by heating the juice under aerobic and anaerobic (Pasteur Effect) under anaerobic growth conditions a 62.8° C (which is now called Pasteurization).

Modern Biotechnology

There are two major features thatdifferentiate the modern biotechnology from classical biotechnology such as Capability of science to change the genetic material for getting new products for specific requirement through recombinant DNA technology. The ownership of technology and its socio-political impact. Now the conventional technology industries, pharmaceutical industries, agro-industries ,etc are focusing their attention to produce biotechnology-based products.

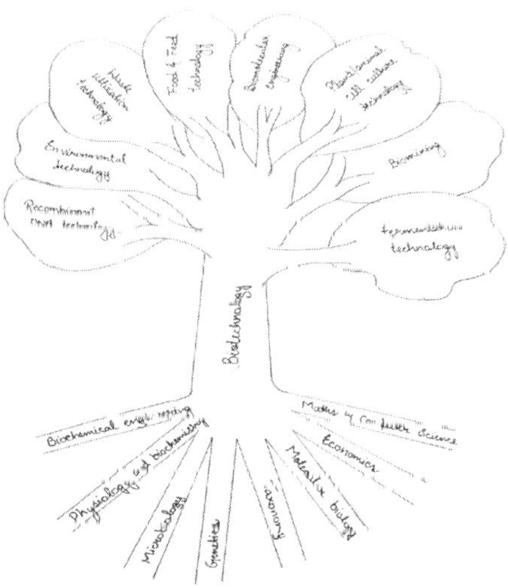

Figure 24.6.A tree of biotechnology

Many areas of biotechnology have arisen through the interaction between various parts of biology and engineering, biochemistry, biophysics, cell biology, colloid chemistry, embryology, ecology, genetics, immunology, molecular biology, medical chemistry, pharmacology, polymer chemistry, thermochemistry and virology. Modern biotechnology has developed several technologies extracting the basic knowledge from biology.

25. HISTORY OF BIOTECHNOLOGY AND DIFFERENT BRANCHES OF BIOTECHNOLOGY

Introduction History of Biotechnology

The origin of biotechnology, it is as old as human civilization. The development of biotechnology can be studied considering its growth that occurred in two phases

- Traditional Biotechnology (old)
- New Biotechnology (Modern)

The origin of biotechnology can be traced back to prehistoric times when microorganisms were already used for processes like fermentation. Although a molecular biologist may consider the cloning of DNA to be the most important event in the history of biotechnology, later has actually been rediscovered in the 1970's for the third time during the present century. In the 1920's *Clostridium acetobutylicum* was used by Chaim Weizmann for converting starch into butanol and acetone, the latter was an essential component of explosives during World War I. This raised hopes for commercial production of useful chemicals through biological processes and may be considered as the first rediscovery of biotechnology in the present century. Similarly, during World War II (in the 1940's), the production of penicillin (as an antibiotic discovered by Alexandar Fleming in 1929 on a large scale from cultures of *Pencillium notatum,* the second rediscovery of biotechnology. This is the beginning of an era of antibiotic research. The third rediscovery of biotechnology is its recent reincarnation in the form of recombinant-DNA technology, which leads to the development of a variety of gene technologies and is considered to be the greatest scientific revolution of the century. Thus, even though biotechnology had its origin in the past, it has assumed special significance only in the1970s and 19s. Biotechnology is the product of the interaction between the science of biology and technology.

This relationship between science and technology has been observed to be complex, so that not only science has influenced technology, but technology has also influenced science. Because of this complex relationship and its major impact on human welfare, it is believed that biotechnology in the future may become a major force for human existence. Already, the products of biotechnology (including, diagnosis, prevention and cure of diseases; new and cheaper chemical products e.g. pharmaceutical drugs and the new food sources; devices for environment protection and energy conservation ,etc are playing a very important role in employment, productivity, trade, economics and the quality of human life throughout the world. Biotechnology being such an important area of study; attempts have been made to define it.

It has been recognized that a complete definition of biotechnology is difficult due to such a wide range of its usage. Fermentation, by some microorganisms, the formation of yogurt (curd) and cheese from penicillin from certain fungi and the processes of baking and brewing are often included in describing what is called old biotechnology. The other example of biological processes involving the techniques of recombinant DNA and polymerase chain reaction (PCR)and bioprocessing which become possible only through the researches in molecular biology have been described as new biotechnology. However, in recent literature, no effort as made to distinguish between old and new biotechnology. Following are some available definitions of biotechnology, and one of them may be really adequate.

Traditional Biotechnology

It is kitchen technology developed by our ancestors using fermenting bacteria. It is as old as human civilization. The Aryans are performing daily **Agnihotra or Yojna**during the *Vedic* period (5000-7000 BC). The material used in Yajna is animal fat (ghee) which is fermented product of milk.Similarly, the divine "soma" had been offered to God (a fermented microbial product used as a beverage). Summarians and Babylonians (6000 BC) are drinking the beer. Egyptians were baking leavened bread by 4000 BC. The preparation of curd, ghee, wine, beer and vinegar ,etc. were kitchen technology. In spite of all development, preparation of curd, ghee, vinegar and alcoholic beverages, idli, jalebi, dosa (Table 25.1) have become an art of the kitchen of all Indians. Traditional biotechnology refers to conventional biotechnology thathas been used for many centuries. Beer, cheese, wine and other many foods have been produced using traditional biotechnology. Thus the traditional biotechnology includes the processes that are based using traditional biotechnology. It includes the process that is based on the natural capabilities of microorganisms. It has established a huge and expanding world market. In monetary terms, it represents a major part of all biotechnology financial profits.

Table 25.1. Fermented foods prepared in different parts of India

Types of foods	Substrate	Regions	Quality and Uses
Wari	Black gram	North	Spongy cake used as a snack
Vadai (wada)	Black Gram	South	Deep fried cake used as snack
Tharra	Mahua	North	Sweet alcoholic beverage obtained through distillation
Tari	Date Palm	East	Sweet alcoholic beverage
Sinki	Radish tap root	Himalaya	Sun-dried sour soup pickles
Srikhand	Milk	West	Concentrated sweetened preparation
Rabdi	Mixture of butter-milk/ wheat/ barley/ peralmillet	West	Cooked paste used as Staple food
Paneer	Milk	North	Soft cheese used as a fried curry
Papad	Black gram flour (besan)	North	Circular wafers used as a snack
Nan	Wheat Flour (maida)	North	Leavened flat baked bread used as staple food
Mishti dahi	Milk	East	Thick sweet gel
Mesu	Bamboo shoot	Himalaya	Sour pickles
Khalpi	Cucumber	Himalaya	Sour pickles
Khaman	Bengal gram	West	Spongy cake, used in breakfast
Jalebi	Wheat Flour (maida)	North	Crispy, deep-fried used as sweet confectionery
Idli	Rice and Black gram	South	Steeamed Spongy cake used in Breakfast
Gundrum	Leafy Vegetables	Himalaya	Sun-dried, sour taste,used as-soup or pickles
Dhokla	Bengal-gram	West	Spongy cake used as a snack
Dahi	Milk	North	Sour, thick gel with Whey
Dosa	Rice and black gram	South	Spongy, shallow fried used as staple food
Chhurpi	Milk	Himalaya	Cheese like mild sour, soft mass used as curry

Ambali	Millet and rice	South	Steamed cake used as asnack
Bhatura	Wheat flour (maida)	North	Flat fried, leavened bread used with chhola

The Egyptians during the 2000 BC period used to prepare vinegar from crushed dates by keeping for longer time. But the crushed dates produce intoxicants at first. In Egypt, Mesopotamia and Palestine during 1500 BC the art of production of wine from crushed grapes, and beer from the germinated cereals (malt) using a bread leaven (a mass of yeast) was established. In Indian Ayurved, production of "Asava" using different substrates and flowers of mahua (*Madhuca indica*) or dhataki (*Wodfordia fructicosa*) has been well characterized till today since the Vedic period. In these methods, various substrates are transformed into a number of products.

Hence the odor, color and taste of final products are changed. Moreover the use of salts for preservations of different foods has been earlier developed even in Europe. The preservation methodology is still of the mummies of Egypt are noteworthy. The possible mummification have been done through dehydration of dead body followed by the use of a mixture of salts largely sodium carbonate. Traditional technology is an art rather than a science.The cause of fermentation could be discovered after observing microorganisms using a microscope by Antony van Leeuwenhoek (1673-1723) at Delt (Holland). Asignificant contribution was done by chemists on the process and the products of fermentation during the 18th century. The milky precipitate was obtained when the gas evolved from fermentation was passed through the lime water in 1757. It was called lime water test.

A similar gas was comingout from the burning of charcoal. Henry Cavendish demonstrated that the gas evolved from brown sugar in water treated with yeast was absorbed by sodium hydroxide solution. Schele and J. Priestley in 1772-1774 confirmed the identity of oxygen gas present in the air. Using an analytical technique for carbon estimation, Antoine Lavoisier gave the chemical basis of alcoholic fermentation. A French Man, Nichola Appert in 1810 was described the method of food preservation. In the same year, Peter Durand also gave the use of a tin container for food preservation. It was done by putting an air-tight vessel containing food material in the boiling water. It increased the importance of the canning industry. Lack of oxygen in such a closed and heated vessel was reported by Gay Lussac. He concluded the oxygen required for initiation of alcoholic fermentation, but not for further progress of fermentation.Charles B Astier gave the concept that air is the carrier of all kinds of germs after 1830. Theodore Schwann after a series of experimentsdemonstrated that the development of the fungus (sugar fungus) on fruit juice causes fermentation. He was the first to observed and describe the yeast in the growing process. Charles Cagiard-Latour (1838) observed yeast budding using a microscope allowing 300-400 power magnification. Justus von Liebig in 1839 a well-known chemist proclaimed that all the activity of yeast cells was the result of chemical and physical reactions going on in the medium.

Microbes also called microorganisms are invisible creatures too small to be seen with the unaided eye and a microscope is needed to study them. They are diverse, present everywhere and exist in a range of environments from mountains and volcanoes to deep sea vents and hot spring. *Thermus aquaticus* is an example of an unusually heat-stable bacterium isolated from hot springs from which heat resistant enzyme *Taq polymerase* was isolated.

Biotechnology Impact on Health Care

The maximum benefits of biotechnology has been utilised by health care. Biotechnology derived proteins and polypeptides form the new class of potential drugs. For example, insulin was primarily extracted from slaughter animals. Since 1982, human insulin (Humulin) has been produced by microorganisms in a fermenter. Similarly, hepatitis B vacines. Recombivax HB" (from Merk), Guni (from Shantha Biotechnics Ld, Hyderabad), Shanvac (Biological E. Laboratory), etc, are the genetically engineered vaccines produced biotechnologically. Since 1987, the number of biotechnology-derived new protein drugs has surpassed the new chemical drugs (Padh, 1996).

Some of the Important Products Produced Through Genetically Modified Organisms

With the advancement of gene manipulation in organisms the science has led to a new revolution in biology called gene revolution. Obviously, it is a third revolution in science after the industrial revolution and computer revolution. Thus, the roots of today's biotechnology lie in chemistry, physics and biology. Currently there are about 35 biotechnology-derived therapeutics and vaccines approved by the USFDA alone for medical use, and more than 500 drugs and vaccines reach in market. Similarly, about 600 biotechnologies diagnostics practices. About 130 gene therapy protocols have been approved by the US authorities imports of many immunodiagnostic are worldwide available in clinical. The major landmark in human history is the human genome sequence. The HGP is an international research program. Almost the whole human genome has been sequenced and chromosome map has been developed in various laboratories world wide through co-ordinated efforts. Human chromosome mapping was completed by March 2003. There are about 3,000 functional genes in human. More than 97% genes are non-functional. They do not encode for any polypeptide chain. Objectives of human genome project are to: construct the detailed genetic and physical map of human genome, (i) determine the complete nucleotide sequence of human DNA, (ii) tore information in database, (iii) locate the estimated 50,000-10,000 genes within the human genome. (iv) Address the ethical legal and social issues LELSI) that may arise from the project, and (v) perform similar analysis on the genomes of several other organisms.

Biotechnology Impact on Environment

The natural biodegradability of pollutants present in the environment has increased with the use of biotechnology. Bioremediation technologies have been found successful to combat the pollution problems. Bioremediation is the use of microorganisms to detoxify pollutants, present in the environment usually as soil or water sediments. The pollutants cause several health problems. Microorganisms that show potential to degrade of oil, pesticides and fertilizers belong to the geniera of bacteria Pseudomonas, Micrococcus, Bacillus, and fungi Candida, Cladosporium, torulopsis, Trichoderma, etc. Computer-based study and designing of genome is called genomics. Genomics deals with the sequencing of the complete genome of a particular organism. Similarly, study of proteins present on genome using computer is called proteomics. The proteomics can be defined as the study of all the proteins present in the genome of an organism. With the help of proteomics and genomics , new molecules that can interact with the other partners could be identified. This gives us deep insights into biological pathways. Now the whole genome is available to biologists for scrutiny. There may be new small molecules as potential drug candidates. Therefore, interactions of some molecules can be studied in greater detail.The 33,000 genes of human beings are on a microchip. It has helped to design specific drugs for genetic diseases, for which there is no cure so far For example, a specific gene expresses in breast cancer patients. A designed drug (Herceptia) is good for the treatment of breast cancer. Thus, the field of genomics has helped the growth of phamacologic, toxicologic and protein studies on animals, therefore, and the new areas are called phamacogenomics, toxicogenomics and proteogenomics, respectively.

Bioinformatics

It is a new field of biotechnology linked with information technology. Bioinformatics may be defined as the application of information sciences mathematics, statistics, and computer sciences) to increase the understanding of biology, biochemistry and biological data. The most remarkable success of bioinformatics to date has been is use in the 'shotgun sequencing' (breaking of a large piece of DNA into smaller fragments) of the human genome.

Biotechnology impact on agriculture

Biotechnology is making new ground in the food/agriculture area. Current public debate about BSTC bovine somatotropin (a hormone administered to cows to increase milk production) example of biotechnology product testing public acceptance. Similarly the FlavrSavr tomato (produced by transgenic plants engineered antisense technology to preserve flavour. texture and quality) is a new breed of value added foods. Food biotechnology offers valuable and alternative to food problems and a solution to nutritionally influenced diseases such as diabetes, hypertension, cancer, heart diseases, arthritis etc. A transgenic golden rice has been produced by introducing three genes for the pduction of **vitamin A in Taipei' rice**. Several insect resistant transgenic Bt plants have been produced insecticidal toxin producing Bt gene of ***Bacillus thuringiensis*** into the desired plants. A transgenic cotton named Ingaurd' was released in Australia which contained Bt gene and provided resistance against insects. We can recall the Bt cotton prepared for Andhra Pradesh but sown inGujarat in 2001 which raised debate throughout the country. Biopesticides are coming to thel market and their sales are increasing. Molecular Pharming is a new concept where therapeutic drugs are produced in farm animals, for example, therapeutic proteins secreted in goat milk. There are about a dozen companies that produce lactoferin. tPA, haemoglobin, melanin and interleukins in cows, goat and pigs (Padh, 1996). However it is not surprising those vegetables producing vaccines, insulin. Interferon and growth hormones would be available inl market in 2Ist century, beside, human clones and several other miracles.

26. TOOLS OF GENETIC ENGINEERING: CLONING VEHICLES, RESTRICTION ENZYMES, MODIFYING ENZYMES, DNA LIGASE, POLYMERASE ETC. CLONING VECTORS: PLASMIDS, LAMBDA PHAGE, PHAGEMIDS, COSMIDS, ARTIFICIAL CHROMOSOMES (BACS, YACS), SHUTTLE VECTORS, VIRUS BASED VECTORS, METHODS OF GENE TRANSFER: TRANSFORMATION, TRANSDUCTION, PARTICLE GUN, ELECTROPORATION, LIPOSOME MEDIATED, MICROINJECTION, AGRO-BACTERIUM MEDIATED GENE TRANSFER, PREPARATION AND APPLICATION OF MOLECULAR PROBES: DNA PROBES, RNA PROBES, RADIOACTIVE LABELING, NON-RADIOACTIVE LABELING, USE OF MOLECULAR PROBES, DNA FINGERPRINTING, ANALYSIS AND EXPRESSION OF CLONED GENE IN HOST CELLS: EXPRESSION VECTORS, RESTRICTION ENZYME ANALYSIS, SOUTHERN BLOTTING, NORTHERN BLOTTING, WESTERN BLOTTING, IN-SITU HYBRIDIZATION. COLONY AND PLAQUE HYBRIDIZATION, FACTORS AFFECTING EXPRESSION OF CLONED GENES, REPORTER GENES, FUSION PROTEINS, GENE LIBRARIES - CDNA SYNTHESIS, GENOMIC DNA LIBRARIES, AMPLIFICATION OF GENE LIBRARIES, IDENTIFYING THE PRODUCTS OF CDNA CLONES.

Introduction

Recombinant DNA technology is the set of techniques that enable the DNA from different sources to be identified, isolated and recombines so that new characteristics can be introduced into an organism. Paul Berg developed the first recombinant DNA molecules that combined DNA from the SV40 virus and lambda phage. Later in 1973, Herbert Boyer and Stanley Cohen developed recombinant DNA technology showing that genetically engineered DNA molecules may be cloned in foreign cells. The enzyme used in the recombinant DNA technology fall into broad categories.

Cloning vehicles

Cloning vehicles are used to clone foreign DNA whereas expression vectors are engineered so that any foreign DNA can be transcribed into RNA and translated into protein.

Exonuclease

These enzymes act upon the genome and digest the base pairs on 5' or 3' ends of a single-stranded DNA or at single-strand nicks or gaps in double-stranded DNA: **figure 26.1**.

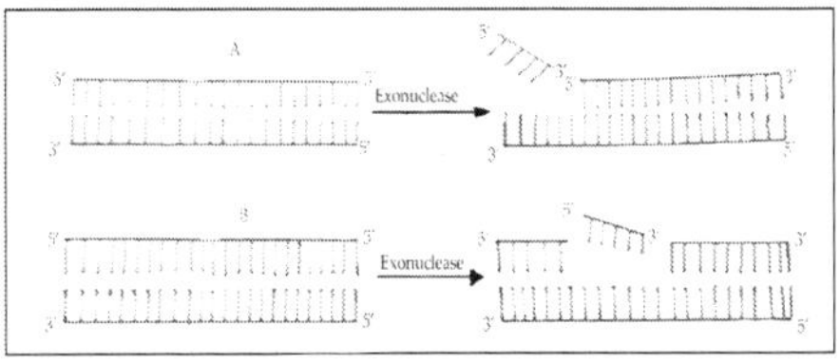

Figure 26.1. Action of exonuclease (a) endonuclease (b).

Endonuclease

They act upon genetic material and cleave the double-stranded DNA at any point except the ends, but their action involves only one strand of the duplex figure 26.2.

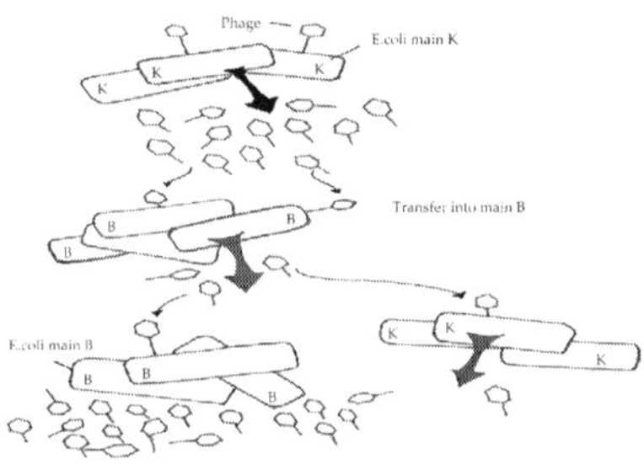

Figure 26.2 Host controlled restriction and modification of phage I in *E.coli* strain K and strain B.

Restriction Endonucleases

The restriction enzymes are called as 'molecular scissors'. These act as foundations of recombinant DNA technology. These enzymes are present in bacteria and provide a type of defense mechanism called the 'restriction-modification system. Molecular basis of these systems was elucidated first by Werner Arber in 1965. However, there was no understanding about cutting and joining of DNA molecules using DNA enzymes, before 1970. During the early 1950s, it was discovered when phage A infects *E. coli*, phage DNA is delivered inside the cell. Host-control restriction modification system provides a defense to *E.coli*. Bacterium protects itself through (i) restriction mechanism (i.e. identifies the introduced foreign DNA and cuts into pieces called restriction endonucleases, and (ii) modification system (i.e. methylation of certain bases of its own DNA by methylase so that phage A-expressed restriction endonuclease could not cleave its DNA).

Nomenclature

The restriction enzymes are named based on some of the following principles "Name of the organism is identified by the first letter of the genus name and the first two letters of the species name to form a three-letter abbreviation in the italic, for example, *E. coli* = *Eco* and *H. influenzae* = Hin, etc. A strain or type identified is written as subscript e.g. *EcoK* for *E. coli* strain K. Hind for *H. influenzae* strain Rd. In such cases where the restriction and modification systems are genetically specified by the plasmid, the extra chromosomal element is identified by a subscript e.g. *Ece RI, EcoPI*, etc. When a strain has several restriction and modification systems, these are identified by Roman numerals, for example, *HindI, HindII, HindIII for H. influenzae* strain Rd. ,etc These Roman numerals should not be confused with those in the classification of restriction enzymes into Type I.

Type III Restriction Enzymes: The Type III enzyme is made up of two sub-units, one specifies site recognition and modification and the other for cleavage. In a reaction, it moves along the DNA and requires ATP as the source of energy and Mg as a co-factor. ATPase activity is lacking in these enzymes. Some examples of Type III enzymes are *Hpal, MbolI, Fokl*, etc. They have symmetrical recognition sites and cleave DNA at double-stranded DNA two sites in opposite orientation must be present. One strand of double-stranded DNA is cleaved about 25-27 bp away from the recognition site which is located in its immediate specific non-palindromic sequences. Since the cleavage products of these enzymes are a homogeneous population of DNA fragments, they cannot be used for genetic engineering experiments.

There are some other restriction enzymes that do not fall under Type I-III. Hence there is a need for reclassifying restriction enzymes under the other categories from Type I to Type IV. Examples of the other restriction enzymes are *Eco 571*, MCR (modified cytosine reaction), etc. *Eco571* is made up of a single polypeptide. It shows both nuclease modification and nuclease activity. S-adenosyl methionine stimulates the nuclease activity of this enzyme. It bears some other properties that resemble Type II enzymes. Therefore, *Eco 571* should be put under Type II to Type IV.

S1 Nuclease

It degrades the single-stranded DNA or single strand protrusion of double-stranded DNA cohesive ends. As a result of the action of nuclease cohesive ends are converted into blunt ends. Therefore, if nuclease is used to remove the incompatible ends so that overlapping ends may be developed for annealing of two fragments of DNA molecules, mode of action of S1 nuclease.

DNA ligases

Mertz Davis (1972) for the first time demonstrated that DNA cleaved DNA molecules could be covalently sealed with E. coli DNA ligase and were able to produce recombinant DNA molecules that have 5'-3 '-OH (hydroxyl) termini. There are two enzymes extensively used for covalently joining restriction fragments the ligase from that encoded by T4 phage. The main one of DNA phage, hence, the enzyme is known as T4 DNA ligase.

Alkaline phosphatase

When a plasmid vector, for joining a foreign DNA fragment, is treated with a restriction enzyme, the greatest difficulty arises at the same time. Because the cohesive ends of broken plasmids, instead of joining with foreign DNA join the cohesive end of the same DNA molecules and get recircularized. To overcome this problem, the restricted plasmid (i.e. plasmid treated with restriction enzymes) is treated with an enzyme, alkaline phosphatase that digests the terminal 5 'phosphoryl group. The restriction fragments of the foreign DNA are not treated with alkaline phosphatase. Therefore, the 5 'end of the foreign DNA fragment of the plasmid. The hybrid ligase will only join 3 'and 5' ends of recombinant DNA together if the 5 'end is phosphorylated. Thus, alkaline phosphatase and ligase prevent recircularization of the vector and increase the frequency of production of recombinant DNA molecules. The nicks between two 3 'ends of DNA fragment and vector DNA are repaired inside the bacterial host cells during the transformation can covalently join to 3 'end or recombinant DNA obtained has a nick with 3' and 5' hydroxyl ends: **figure 26.3**.

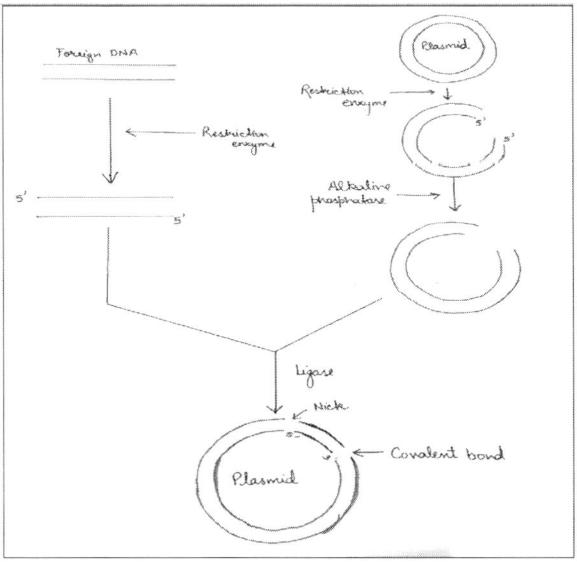

Figure 26.3. Alkaline phosphates

Reverse Transcriptase

In addition to these enzymes, reverse transcriptase is used to synthesize the copy DNA complementary DNA (cDNA) by using an mRNA template. Reverse transcriptasc is vcry useful in the synthesis of cDNA and construction of or as a cDNA clone bank. Until recently, it was known that the genetic information of DNA pass to protein through mRNA. During the 1960s, Temin and co-workers postulated that in certain cancer-causing-animal viruses that contain RNA as genetic material, the transcription of cancerous genes (on RNA into DNA) takes place probably by DNA polymerase directed by virus RNA. Then DNA is a template for the synthesis of many copies of viral RNA in a cell. Baltimore discovered that information can also pass back from RNA to DNA. They found that retroviruses (possessing

RNA) contain an RNA-dependent-DNA polymerase, which is also called as reverse transcriptase. This produces single-stranded DNA, that in turn functions as a template for a complementary long chain of DNA.

DNA Polymerase

This enzyme polymerizes the DNA synthesis on a DNA template (or cDNA template) and also catalyzes a 5'- 3' and 3'-5' exonucleolytic degradation of DNA (Kornberg, 1974).

T4 polynucleotide kinase

T4 polynucleotide kinase catalyzes the transfer of y PO, of ATP to 5' terminus of a depho phorylated DNA (i.e. DNA lacking -PO, residue) or RNA as below: 5'GGOH 3 3'-CCTTAAp 5'-GGOH Polynucleotide kinase -3' 3'-CCTTAA5' .S- 7P-ATP. This enzyme shows both phosphorylations as well as phosphatase activities; therefore, two types of reactions can be used accordingly. In phosphorylation , the y-PO is transferred to the 5'end of dephosphorylated DNA. When ADP is in excess, it causes the T4 polynucleotide kinase to transfer the terminal phosphate (attached to a nucleotide) from phosphorylated DNA to ADP. By using radio- labeled gamma-phosphate (PO) from P-ATP. This shows that the enzyme plays a dual role of phosphorylation activity and phosphatase activity. It was found that DNA is rephosphorylated by the transfer of the phosphate group. Due to the presence of both properties, this enzyme is used in radiolabeling of the 5' end of duplex during DNA sequencing (see DNA sequencing method), constructing the terminally labeled DNA, phosphorylating the synthetic linkers/oligonucleotide linkers/other DNA fragments lacking terminal 5'phosphates, etc.

Terminal Transferase

This enzyme is from calf thymus. It has a property to add an 'oligodeoxynucleotide tail' to the 3'-OH end of double-stranded DNA fragments. Thus it extends homopolymer tails and this phenomenon is called homopolymer tailing deoxynucleotidyl transferase. It has been isolated and purified.

Use of linkers and adaptors

The DNA with Type II restriction enzymes results in the formation of fragments having sticky ends or blunt ends. Such DNA fragments cannot be ligated with cloning vectors. Therefore, artificially cleavage sites at the blunt end can be added as linker molecules. Chemically synthesized double-stranded oligonucleotides are called linkers. They are added double-stranded DNA molecules that act as adaptor molecules are added recognition sites for restriction enzymes. For example, the following oligonucleotide possess a recognition sequence for *EcoRI* (asterisk nucleotides). It can be ligated to the blunt end of any double-stranded DNA and cut by *EcoRl* to generate sticky ends: **figure 26.4**.

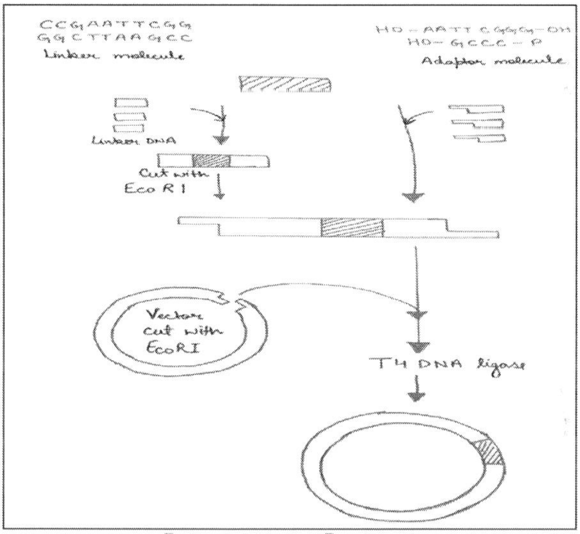

Figure26.4 Linker and adaptor molecules.

Tools of Genetic Engineering (Cloning Vectors)

Cloning and expression vectors

Techniques for manipulating prokaryotic, as well as eukaryotic DNA, have witnessed a remarkable development. This has allowed the breakage of a DNA molecule at two desired places to isolate a specific DNA segment and then insert it in another DNA molecule at a desired position. The product thus obtained is called recombinant DNA and the techniques are often called genetic engineering. Using this technique we can isolate and clone a single copy of a gene or a DNA segment into an indefinite number of copies, all identical. This became possible because vectors like plasmids and phages reproduce in a host in their usual manner ever after insertion of foreign DNA so that the insertion DNA will also replicate faithfully with the parent DNA. A variety of vectors have been developed which not only allow multiplication but may also be manipulated in such a way that the inserted gene may express in the host.

Cloning vector for recombinant DNA

One of the most important uses of recombinant DNA technology in the cloning of (i) random DNA or cDNA segments, often use as a probe or (ii) specific genes, which may be either isolated from the genome or synthesized organochemically or in the form cDNA from mRNA. This other DNA molecule is often used in the form of a vector, which could be a plasmid, a bacteriophage, a derived cosmid or phagemid, a transposon or even a virus. Techniques should also be available, which will allow the selection of chimeric genomes obtained after the insertion of foreign DNA from a mixture of chimeric and the original vector. Another critical desired feature of any cloning vector is that it should possess a site at which foreign DNA can be inserted without disrupting any essential function. Therefore, in each case, an enzyme will also have to be selected which will cause a single break. Sometimes vectors are modified by inserting a DNA segment to create a unique site for one or more enzymes to facilitate its use in gene cloning. This inserted DNA with restriction sites for several enzymes is sometimes called a **polylinker.**

Plasmids as vectors

Plasmids are defined as autonomous elements, whose genomes exist in the cell as extrachromosomal units. They are self-replicating circular duplex DNA molecules, which are maintained in a characteristic number of copies in a bacterial cell, yeast cell or even in organelles found in eukaryotic cells. These plasmids can be single-copy plasmids that are maintained as one plasmid DNA per cell or multicopy plasmids, which are maintained as

10-20 genomes per cell. These are also plasmids, which are under relaxed replication control, thus permitting their accumulation in very large numbers. These are the plasmids that are used as cloning vectors, due to their increased yield potential. Circular plasmid DNA used as a vector can be cleaved at one site with the help of a restriction enzyme to give a linear DNA molecule. A foreign DNA segment can now be inserted, by joining the ends of broken circular DNA to the ends of foreign DNA, thus regenerating a bigger circular DNA molecule that can now be separated by electrophoresis on the basis of its size. Selection of chimeric DNA is also facilitated by the resistance genes, which the plasmid may carry against one or more antibiotics. If a plasmid has two such genes conferring resistance against two antibiotics and if the foreign DNA insertion site lies within one of these two genes, then the chimeric vector loses resistance against one antibiotic, the gene for which has foreign DNA inserted within its structure. In such a situation, the parent vector in bacterial cells can be selected by resistance against two antibiotics and the chimeric DNA can be selected by retention of resistance against only one of the two antibiotics.

pBR322 and pBR327 vectors

The naturally occurring plasmids may not possess all the above and other essential properties of a suitable cloning vector. Therefore, one may have to restructure them by inserting genes of relaxed replication and genes for antibiotic resistance. This has been done and suitable plasmid vectors have been obtained. One is pBR322, which is 4,362 bp DNA and was derived by several alterations in earlier cloning vectors. It has genes for resistance against two antibiotics an origin of replication and a variety of restriction sites for cloning of restriction fragments obtained through cleavage with a specific restriction enzyme. Another vector pBR327 was derived from pBR322, by deletion of nucleotides between 1427 to 2516. These nucleotides are deleted to reduce the size of the vector and to eliminate sequences that were known to interfere with the expression of the cloned DNA in eukaryotic cells. pBR327 still contains genes for resistance against two antibiotics. Both pBR322 and pBR327 are very common plasmid vectors; **figure 26.5.**

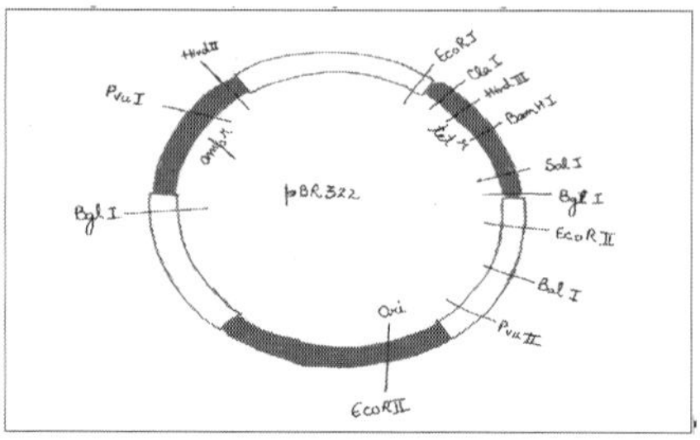

Figure 26.5. pBR322 vector

pUC vectors

Another series of plasmids that are used as cloning vectors belong to the pUC series. These plasmids are 2700 bp long and possess (i) ampicillin resistance gene, (ii) the origin of replication derived from pBR322 and (iii) the lac Z gene derived from *E.coli*. Within the lac region is also found a polylinker sequence having unique restriction sites. When DNA fragments are cloned in the region of pUC, the lac gene is inactivated. These plasmids when transformed into an appropriate *E.coli* strain, which is lac, and grown in the presence of IPTG and X-gal , will give rise to white or clear colonies. On the other hand, pUC having no inserts and transformed into bacterial will have an active lac Z gene and therefore will produce blue colonies, thus permitting identification of colonies having pUC vector with cloned DNA segments. The cloning vectors belonging to the pUC vector with cloned DNA segments. The cloning vectors belonging to the pUC family are available in pairs with reversed orders of restriction sites relative to lac Z promoter. pUC 8 and pUC 9 make one such pair. Other similar pairs include pUc12 and pUC 13 or pUC 18 and pUC 19: **figure 26.6.**

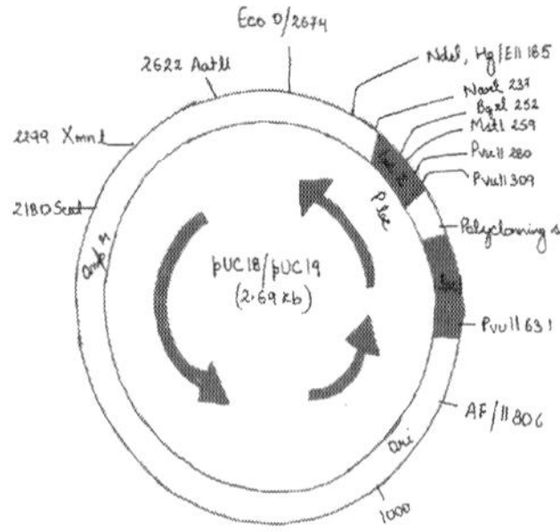

Figure 26.6. pUC 19 cloning vector

As discussed above, in pBR322 and pBR327, the DNA is inserted at a site located in one of the two genes for resistance against antibiotics, so that it will inactivate one of the two resistance genes. The insert bearing plasmid can be selected by their ability to grow in a medium containing only one of the two antibiotics and by their failure to grow in a medium containing both the antibiotics. The plasmids carrying no insert, on the other hand, will be able to grow in media containing one or both the antibiotics. In this manner, the presence of lac Z gene in pUC and resistance genes against ampicillin and tetracycline in pBR322 and pBR327 allows the selection of *E.coli* colonies transformed with plasmids carrying the desired foreign cloned DNA segment.

Yeast Plasmid Vectors

Special vectors are also known for introducing DNA segments in yeast cells, a eukaryotic system that has been used extensively for developing genetically engineered yeast cells. Although *E.coli* plasmids or phages can be are used for the transfer of genes to yeast cells, the frequency of transformation is rather low. For instance, yeast cells that are LEU could be transformed using chimeric plasmid carrying LEU + gene yeast. This involves integrative transformation having low frequency. To overcome this difficulty of low-frequency transformation, an effort was made to prepare a variety of vectors for yeast. The possible examples of yeast plasmid vector are (i) Yip or yeast integrative plasmids, which allow transformation by crossing over and have no replication origin (ii) Yep or yeast episomal plasmids, which carry 2 micron DNA sequences including the origin of replication and rep genes (iii) YRp or yeast replication plasmids, which carry autonomously replicating sequences; this sequence is very common with many yeast genes so that stable transformation can be achieved by crossing over (iv) YCp or yeast centromere (CEN) containing plasmids, which function as true chromosomes and segregate during mitosis and meiosis (v) pYAC or yeast artificial yeast chromosome vector carries both centromere and telomere sequences and, therefore, can be used to obtain artificial chromosomes.

The above vectors have been extensively utilized for a study of yeast genome and are also used as shuttle vector, which allow gene sequences to be routinely transferred back and forth between yeast and *E.coli* cells, provided they contain origins of replication active in both yeast and *E.coli*. These vectors also have markers enabling the selection of *E. coli* cells or colonies transformed with these vectors. The vectors can therefore be amplified in *E.coli* and then used to transform yeast cells. Retriever vectors are other class vectors, which are used to retrieve specific genes from the normal chromosome of an organism like yeast through recombination. They will multiply both in *E.coli* and yeast. In such a vector the gene in question is removed by enzymes to produce a gapped vector having flanking sequences. Through recombination, they acquire lost sequences from the host. Retriever vectors are very useful in the isolation of genes from yeast for further molecular studies including sequencing.

Ti and Ri plasmids as a vector for higher plants

Among higher plants, Ti plasmid of *Agrobacterium tumefaciens* or Ri plasmid of *A.rhizogenes* is the best-known vector. **T-DNA,** from Ti or Ri plasmid of *Agrobacterium*, is considered to be a very potential vector for cloning experiments with higher plants. This will involve the following steps: Foreign DNA has to be first cloned into T-DNA of Ti or Ri plasmids. Modified hybrid T-DNA can be transferred to the genome of plant cells by *Agrobacterium* infection. Recombinant Ti plasmid can also be obtained by cloning of a large fragment of Ti plasmid in pBR322. These recombinant Ti plasmids can then be used for the transformation of higher plants. This system can be widely used, since *Agrobacterium* infects nearly all dicotyledonous plants. Such cloning in plants has proved to be of immense use to modify agricultural plants to increase their productivity. Monocotyledonous plants to which our cereal crops belong are however, difficult materials to be used for improvement through such techniques; some success, however, has been achieved in cereals also in recent years: **figure 26.7.**

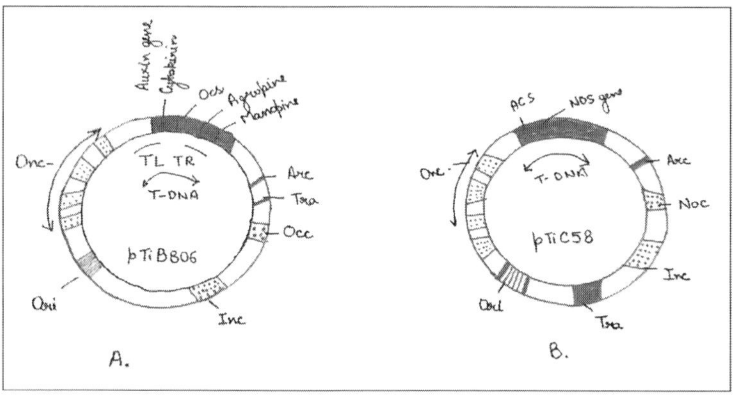

Figure 26.7. Octopine plasmid (a) Nopaline plasmid (b).

Bacteriophages as Vectors

Bacteriophages provide another source of cloning vectors. Since usually, a phage has a linear DNA molecule, a single break will generate two fragments, which are later joined together with generating two fragments, which are later joined together with foreign DNA to generate a chimeric phage particle. The chimeric phage can be isolated after allowing it to infect bacteria and collecting progeny particles after a lytic cycle. The use of phage particles as vectors imposes a limitation on the size of foreign DNA is too long, the size of phage DNA may not be accommodated in the phage head. In order to overcome this problem, those segments of DNA, which do not contain essential genes, may be removed. Such a technique has been followed in phage **lambda** to create a smaller vector genome having a single restriction site for the enzyme *Eco RI*. Since the reduced size also fails to be adequately packed in phage head, this automatically provides a selection method, in which only the chimeric particles will be obtained in the phage progeny and the vector particles lacking cloned segment will be eliminated due to its reduced size.

Lambda Phage Vectors

Plasmid vectors described in the previous section are often used for cloning DNA segments of small size. However, while preparing a genomic library in a eukaryote, the cloned fragments should be large enough to contain a whole gene. Therefore, it can be screened without serious difficulty. The above properties and other requirements of cloning the whole genome in eukaryotes are fulfilled by the phage lambda and cosmid vectors, the former permitting cloning of segments up to 20-25kb long and the latter accommodating segments up to 45kb long. Phage lambda is easier and more efficient for making genomic and cDNA libraries.

(a) lambda gt 10 and lambda 11

Lambda 10 and lambda gt 11 are modified lambda phages designed to clone cDNA fragments. The major difference between those two vectors is that gt11 is an expression vector, where inserted DNA is expressed as a B- galactosidase fusion protein. Lambda gt 10 is a 43 kb double-stranded DNA for cloning fragments that are only 7 kb in length. The insertion of DNA Inactivation cl+ gene generating a cl- bacteriophage. Non-recombinant lambda gt 10 forms cloudy plaques an appropriate *E.coli* host, while recombinant plaques. Further, in an *E.coli* strain carrying hf La150 mutation only cl- phage will form plaques, become cl+ will forms lysogens and will not undergo lysis to form any plaques. Recombinant Lamba gt 10 plaques can thus be easily selected. Lambda gt 11 is a 43.7kb double-stranded lambda phage for cloning DNA segments, which are less than 6 kb in length. Foreign DNA can be expressed as beta-galactosidase fusion proteins. Recombinant lambda gt 11can be screened using either nucleic acid or antibody probes. The recombinant lambda gt 11becomes gal-, while non –recombinant lambda gt 11 remains gal+ so that an appropriate *E.coli* host, and with recombinant phage will form white or clear colonies and that with recombinant phage (gal+) will form blue colonies permitting screening in the presence of IPTG (inducer) and X gal (substrate).

(b) EMBL3 AND EMBL4 (replacement vectors)

EMBL3 and EMBL4 are two vectors that are designed so that a central non-essential part of the 44 long phage can be replaced by a foreign DNA. Cleavage of the phage with an appropriate enzyme generates three fragments. The central fragment representing 40% of the phage genome is nonessential for the propagation of the phage and can be replaced by foreign DNA that may be as long as 20-30kb. Non-recombinant DNA resulting from the fusion of the left arm and right arm will give a DNA that is too small for packaging so that there is automatic selection against non- recombinant phages. EMBL3 and EMBL4 are two such replacement vectors used for preparing genomic libraries in eukaryotes with cloned fragments, 15-25 kb in size. The two vectors have polylinkers with reverse orders of restriction sites with respect to each other.

(c) Charon 34 and Charon 35

Charon 34 and Charon 35 differ from each other only in their central fragments and will accept fragments 9-20 kb long. They have a more extensive range of restriction targets within their polylinkers than do EMBL3 and EMBL4. In these vectors, physical separation of lambda arms from the central fragments is required before ligation to donor fragments. In EMBL3 and EMBL4, removal of central fragments is not required, because there is a mechanism that ensures that 98% of packageable DNA are recombinants.

M13 phage as a cloning vector for DNA sequencing

M13 is a filamontous bacteriophage of *E.coli* and contains a 7.2 kb long single-stranded circular series of cloning vectors which can be used for cloning of a wide variety of DNA. M13 phage has been variously modified to give rise to an M13 mp series of cloning vectors that can be used for cloning of a wide variety of DNA fragments particularly for the purpose of sequencing through Sanger's method of dideoxy chain termination. Cloning vectors of M13 mp series have a lac Z gene that complements gal host giving blue colonies. But if foreign DNA segments are inserted at one of the polylinker sites associated with the lac Z gene, it inactivates the lac Z gene and no complementation is possible. Thus on transformation, only white or clear plaques are obtained permitting the selection of recombinant M13mp plaques. Once the foreign DNA is cloned in the M13 vector, commercially available oligonucleotide primers are used for copying the insert in presence of dideoxynucleotides, so that fragments of different sizes with known termini are produced, permitting the determination of the sequence. Reversed order of the restriction sites in polylinker is present in a pair of vectors like M13mp8 and M13mp9 permitting sequencing from both the ends of a double-stranded DNA molecule. The extension of the commercial primer or cloned fragments in the M13 vector in the presence of radioactively labeled nucleotides will also allow the generation of radioactivity single-stranded probes.

Cosmids as vectors

Cosmids are plasmids particles, into which certain specific DNA sequences, namely those for cos sites are inserted. Since these cos sites enable the DNA to gel packed in lambda particle, cosmids allow the packaging of DNA in phage particle *in vitro*, thus permitting their purification. Like plasmids, these cosmids perpetuate in bacteria and do not carry the genes for lytic development. The advantage of the use of cosmids for cloning is that its efficiency is high enough to produce a complete genomic library of 10/6 -10/7 clones from a mere 1 micro g of insert DNA. The disadvantage, however, is its inability to accept more than 40-50 kbp of DNA. Bacteriophage P1 system and F- factor based vectors can allow cloning of DNA segments, as large as 100 to 1000 kbp length.

Phagemids as Vectors

Phagemids are prepared artificially by combining features of phages with plasmids, as the name suggests. One such phagemid, which is commonly used in molecular biology laboratories, is pBluescript 11 KS, which is derived from pUC19 and 2961 bp long. The KS designation indicates the orientation of polylinker, such that the transcription of lac Z gene proceeds from the restriction site for KpnI to that for SacI. The detailed structure of pBluescript II KS: It may be noted that it has the following features: (i) a multiple cloning site (MCS) flanked by T3 and T7 promoters to be read in opposite directions on the two strands, (ii) an inducible lac promoter (Lac I), upstream of lac Z region, which complements with *E.coli* and provides the facility for selection of chimeric vector DNA using the criterion of white colonies (iii) fl (+) and fl(-) origins of replication derived from a filamentous phage for recovery of sense (+) and antisense (-) strands of lac z gene, when the host is coinfected with a helper phage; (iv) an origin of replication derived from the plasmid, and used in the absence of helper phage; (iv) a gene for ampicillin resistance for antibiotic selection of chimeric phagemid vector.

P1 cloning vectors for cloning large DNA segments

The bacteriophage P1 cloning vectors can allow cloning of 100 kbp long DNA segments with an efficiency of 10^5 clones per micro gm of insert DNA. Therefore, in their capability, they fall between YACs and cosmids. The vector with insert DNA is amplified in *E.coli* and several microgarams of cloned DNA can be recovered from 5-10 ml of exponential phase of *E.coli* cells.

F- factor based vector (which behave like a phage)

F-factor based vectors have recently been developed for cloning large DNA segments in *E.coli*. The cloning of large DNA segments is achieved by a method called **chromosomal building,** in which through repeated recombination the size of cloned segments can be increased. These bacterial vectors will complement the YAC vectors for cloning segments larger than 100 kbp in length and offer some advantages over the YAC system. These vectors have already been used for cloning large DNA segments from the **bithorax** gene of Drosophila.

Plant and Animal Viruses as Vectors

A number of plants and animal viruses have also been used as vectors both for introducing foreign genes into cells and for gene amplification and expression in host cells. We may be interested only in getting the increased quantity of the gene product and may not be interested in the integration of the foreign gene in the host genome. Some of the viruses that are commonly used as vectors will be described below.

Plant viruses (Cauliflower mosaic virus or CaMV and germiniviruses)

Cauliflower mosaic viruses, tobacco mosaic virus (TMV) and geminiviruses are three groups of viruses that have been used as vectors for cloning of DNA segments. CaMV infects particularly the members of Cruciferac and has a double-stranded DNA molecule, 8kbp in size, whose sequence is now known. Following infection, the virus spreads systematically throughout the plant in a very high copy number reaching up to 10^5 virus particles per cell. These features make CaMV a suitable vector for the transformation of higher plants although there are instance of use of other viruses also as vectors leading to the production of transgenic plants. Geminiviruses comprise a group of single-stranded DNA plant viruses causing important diseases in cassava, maize and other cereals. They replicate via double-stranded DNA forms and have been subdivided into two groups, one infecting monocots and transmitted by leafhoppers and the other infecting dicots and transmitted by the whitefly. The dicot geminiviruses have two DNA segments in their genome, DNA A and DNA B, both are essential for infection by mechanical inoculation. DNA-A carries all genes responsible for replication. It has been shown that deletions in the coat protein gene do not destroy the infectivity of the virus, which means that long deletions in this region can be used to insert foreign DNA to obtain gene constructs that will be extremely useful as amplification and expression vectors in plants. Further, since the coat proteins are the most abundant protein synthesized by the virus, promoters of coat protein genes can be used for high-level expression of cloned genes.

Transposons of higher plants (Ds, Ac or Mu 1 of maize).

Ac and **Ds** are popular transposons in corn, and were earlier known to represent a transposon with short terminal repeats enclosing a long DNA segment, which measures more than 4500bp in **Ac** and about 400bp in Ds. Each possesses genes including the gene for transposase enzyme responsible for transposition. Part of this region can be deleted and the transposon can be used for cloning of foreign DNA segment in much the same way as in other cases.

Transposons of Drosophila

P elements of *Drosophila* consist mainly of 31bp terminal inverted repeats enclosing a 3 kilobase proteins coding region, coding for transposase and transposase derived from other complete P element help their mobilization. In these deleted P elements, the gene for transfer can be inserted to yield recombinant P element, which can be microinjected into the fertilized egg, along with normal P element. The inserted gene is transposed onto an embryonic chromosome.

Artificial chromosome (YAC and MAC) vectors for cloning large DNA segments

We know that in plasmids, sequences up to 10-15 kbp; in lambda phage sequences up to 22kbp and in cosmids sequences up to 40kbp can be cloned. Yeast artificial chromosome vectors mentioned earlier allow cloning of sequences that are several hundred kilobase pairs long. These long molecules of DNA which may represent whole chromosomes can be cloned in yeast by ligating them to vector sequences that allow their propagation as linear artificial chromosomes. These long molecules of DNA which may represent whole chromosomes can be cloned in yeast by ligating them to vector sequences that allow their propagation as linear artificial chromosomes. These long DNA molecules can be generated and allow the construction of comprehensive libraries in microbial hosts. With the isolation of mammalian telomere and centromere, mammalian artificial chromosomes (MACs) will also be produced further. YACs have two disadvantages. (i) Cloning efficiency is low (1000 clones/ micro g DNA against 10^6-10^7 clones/micro g DNA for cosmids), thus making it impractical to generate complete genomic libraries through the use of YACs (ii) It is not possible to recover a large amount of pure insert DNA from the individual clones; selective amplification of YACs DNA has recently allowed this problem to be overcome. Both these problems have been overcome by the bacteriophage P1 system,

Promoters

Promoters from non-plant sources express poorly in plants and promoters from non-animal sources express poorly in animals. Therefore, to obtain correct expression, genes transferred to plants should be linked to plant-specific promoters and those transferred to the plant should be linked to the plant-specific promoters and those transferred to animals should be linked to animal-specific promoters. These promoters can be interchanged so as to confer the transferred gene, a specific pattern of expression.

Nopaline synthase (nos) promoter from T-DNA

They are 200 bp long and contain several DNA sequence motifs that direct the expression of linked genes. The upstream element from -97 to –130 when duplicated increases the expression three-fold, suggesting that this may be an enhancer. The nos promoter is the most active in basal regions of plants, its activity slowing down in vegetative parts at the onset of flowering, even though in the flower its activity is increased.

Dual promoter of mannopine synthase (mas) genes 1 and 2

These promoters are present in T-DNA fragments 467 bp long. The expression seems to be developmentally regulated and the linked genes seem to be developmentally regulated and the linked genes to be most active in the basal region of the plant. The expression is also induced on wounding. Since they are closely linked, if a marker gene is linked to one of these two promoters and a foreign gene is linked to the other promoter, selection for high expression of marker gene will lead to a high level of expression of the foreign gene.

35S RNA promoter of CaMV

This is the most extensively used promoter in a wide variety of plants. The promoter is 343 bp long and contains a strong transcriptional enhancer, which on duplication results in 10-fold increase in expression, even at a distance of 2 kbp. High expression is mainly observed in leaf tissue.

Polyhedrin promoter from baculovirus

Baculovirus can synthesize polyhedron protein in infected cells, so that polyhedron makes up to 50% of the total protein in the cell, giving up to 1g of protein per liter of insect cell culture. In view of this, the promoter has been used to construct an expression vector to allow a high level of expression of any gene under the influence of this promoter. This is being used for developing biopesticides or even for the production of specific chemicals in industry.

Expression cassettes

The literal meaning of cassette is 'a device containing film such as magnetic tape for insertion into equipment like a tape recorder, VCR, etc.' Therefore, by expression cassette, we mean gene constructs, which allow the insertion of foreign genes, either as transcriptional or translational fusions, behind specific promoters. pRT plasmids are typical in this connection and have been derived from Puc18/19. They consist of a series of vectors, which differ in polylinker sequences; each flanked by the 35S promoter of CaMV on one end and a sequence for Poly-A addition on the other end. A variety of marker genes have been inserted into these cassettes and their expression studied, both in protoplasts and in stable transgenic tissues.

Baculovirus, an expression vector system for insect cells

Development of the baculovirus expression vector system is based on its life cycle which involves the following steps: Early phase virus particles are bubbled from infected cells and spread the virus. Late phase (2-5 days), occluded virus units accumulate in the host nucleus and get embedded in the protein polyhedral.The insect is decomposed releasing polyhedrals. Polyhedrals are ingested by another insect, where polyhedrals are dissolved releasing viruses for multiplication. The use of baculovirus as a cloning vector involves the following steps: Desired gene is cloned in a transfer vector under the control of polyhedron promoter; Chimeric transfer vector is co-transfected with wild type virus into the host cell; Polyhedron gene of the wild type is replaced by the cloned gene through recombination; Recombinant virus is plaque purified, since they will have polyhedral occlusion bodies; The selected recombinant virus is multiplied and used for the expression of heterologous proteins.

Virus expression vector for mammalian cells

A number of viruses have also been used as transfer or expression vectors for mammalian cells. These include the following: (i) The first viruses used for mammalian cells included DNA viruses like *papoviruses* and *adenoviruses,* but since their capacity for foreign genes is small, they can not be used for the expression of many genes which are fairly large in size; (ii) Retroviruses are a very useful vector, since they can infect a broad variety of cell types and can be used for cloning large genes; however, being single-stranded RNA, they can not replicate without being integrated so that we will need replicating cells for expression or alternatively high-titer for fully differentiated cells (iii) *Adeno-associated virus* (AAV) overcomes the limitations associated with retroviruses even though, like retroviruses, it is an integrating virus, with 5kb single-stranded DNA; it is non-pathogenic, and can be concentrated to high titers; (iv) *Herpes virus* , is another important non- integrating virus of large size (150 kb), which can be used for expression of large intact genes. Several of the above viruses have already been using and showed expression of cloned genes, but they need to further characterize their long term usage.

Binary and Shuttle Vectors

For the purpose of gene transfer in higher plants, a number of binary vectors were developed. These were based on the pPCV series of plasmids. These vectors contain a conditional mini-RK2 replicon, which is maintained and mobilized by transacting functions derived from the plasmid RK2 replicon. The plasmid RK2 was introduced into both *E.coli* as well as *Agrobacterium* to facilitate replication of binary vectors in both these hosts so that the vector can be maintained and shuttled between both hence the name shuttle vector also. Similar binary or shuttle vectors for maintenance and transfer between *E.coli* and yeast cells have also been designed: figure 26.8.

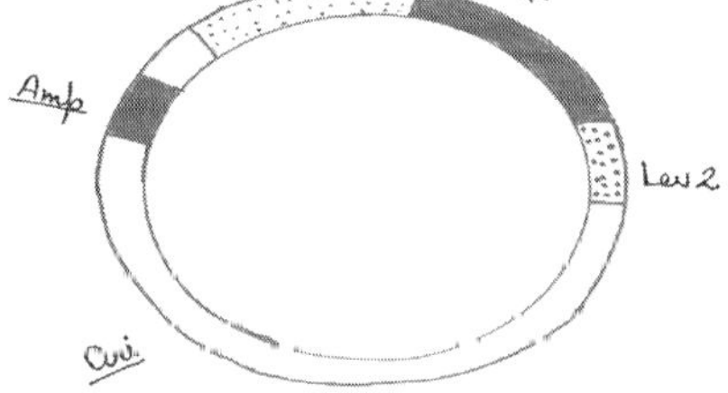

Figure 26.8. Shuttle Vector

Ac-Ds elements

The activator (Ac) and dissociation (Ds) elements are simple transposons of maize. It has 4.565 kb long autonomous sequence which transcribes 3.5 kb long GC-rich untranslated leader sequence. Four introns are spliced out from the primary transcript and a 3.312 nucleotide long mature mRNA is obtained. Consequently a 807 amino acid- long protein is translated. Within the Ac element, the terminal inverted repeats (TIR) of 11 bp are located at both the ends which play a key role in transposition. The protein synthesized by Ac is called transposase that causes transcription of Ac elements. The Ds (dissociation) is the defective element of Ac family. Due to deletion, non-autosomal elements are obtained from the autosomal elements. It activates the trans-acting transposase. The 11 bp long TIR is common to all Ds elements. The enzyme transposase cannot be expressed by the Ds element. But the transposase that is produced by the Ac element is diffused and trans-acting. It causes transposition of Ds elements. By detecting part of the region transposons can be used as a vector for cloning a foreign DNA.

P element

The transposons of Drosophila are known as P elements. The size of different P elements varies and the largest clement is 1907 bp long containing 31 bp long terminal inverted repeats at both ends. The complete elements move autonomously due to the expression of the transposase enzyme. The other P elements lack internal genes that produce transposase. The terminal and sub-terminal sequences cause transposition. A foreign DNA can be inserted into the transposase element to develop recombinant P elements which can be microinjected into the fertilized egg along with normal P elements. The normal P supplies transposase to the recombinant P element. Here the non-autonomous element acts as a transformation vector consequently the foreign gene inserted into an embryonic cell is transposed onto an embryonic chromosome of Drosophila. Gierl et al. (1989) have reviewed the transposable elements of maize. A recombinant P element was prepared by inserting a DNA fragment of the rosy gene (and a marker gene for identification of a transgene). It was inserted into a plasmid. Similarly, a P element, lacking the ability for integration was inserted into a plasmid to encode transposase. Both the plasmids were co-injected into the embryo of *Drosophila melanogaster* mutant in the rosy gene. It was found that 50% of progeny of *D. melanogaster* were rosy. This shows the successful transfer of a foreign gene into a cell of P element (Gierl et al., 1989).

Expression Vector

The main utility has been to clone a gene or a DNA sequence. In other words, they are used for obtaining millions of copies of the cloned DNA segments for further use in genetic engineering experiments or for further basic studies. With this objective in mind, the cloned genes in this vector need not express themselves either at the transcription level or at the translation level. But, when the cloned gene is used for transformation to generate transgenic plants or animals or the production of microbes to be used in industry, then the cloned gene must be expressed. Sometimes, a high level of expression of the gene is desirable, if the product of a cloned gene is to be recovered as a commercial product. The objective can be achieved through the use of promoters and expression cassettes.

Gene cloning

It is an essential process in the success of genetic engineering where insertion of a specific fragment of foreign DNA into a cell, through a suitable vector, so that it can replicate independently and be transferred to the progeny during cell division: **figure 26.9.**

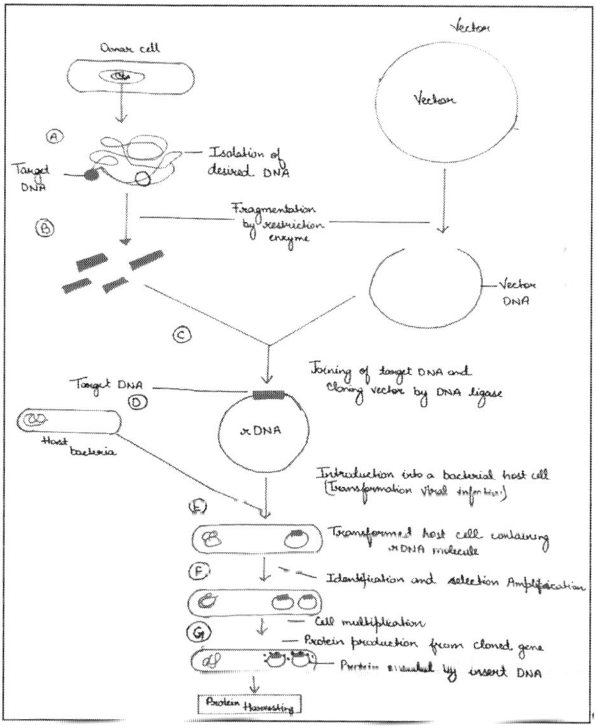

Figure 26.9 Gene Cloning

Cloning in Prokaryotes

The DNA fragment to be introduced into the host cell is called insert DNA, desired DNA, target DNA or foreign DNA. Such gene/DNA fragment is taken out from an organism or gene library i.e. a random collection of cloned DNA fragments in vectors that consist of all genetic information of that organism concerned. However, getting the desired DNA is not so easy because the cellular DNA is very large and the position of the desired DNA is not accurately known. Libraries of such cloned fragments prepared.

Strategies of recombinant DNA technology

There is no single method of recombinant DNA technology, but it involves several steps. Maniatis et al. (1976) have described the basic techniques of gene cloning. The principal steps of gene cloning :

- Isolation of DNA of known function from an organism (A)
- Enzymatic cleavage (B) and joining (C) of insert DNA to another DNA molecule (cloning vector) to form a recombinant DNA (i.e. vector + insert DNA) molecule (D).
- Transformation of a host cell i.e. transfer and maintenance of this rDNA molecule into a host cell (E).
- Identification of transformed cells (i.e. cells carrying rDNA) and their selection from non-transformants.
- Amplification of RDNA (F) to get its multiple copies in a cell.
- Cell multiplication (G) to get a clone i.e. a population of genetically identical cells. This facilitates each of the clones to possess multiple copies of foreign DNA.

Gene library

A gene library is the collection of different DNA sequences from a sequence that has been cloned into a vector for ease of purification, storage and analysis. There are two types of libraries on the basis of source of DNA used: (i) genomic library (where genomic DNA is used), and (ii) cDNA library (where cDNA or mRNA). A gene library should contain a certain number of recombinants with a high probability of containing any particular sequence.

Genomic library

It is a collection of clones that represent the complete genome of an organism. All fragments of DNA inserted into vectors for further propagation into suitable hosts represent the entire genome of an organism. For the construction of a genomic library, the entire genomic DNA is isolated from host cells/tissues, purified and broken randomly into fragments of the correct size for cloning into a suitable vector. There are two basic ways of fragmenting genomic DNA randomly: physical shearing (e.g. pipetting, mixing or sonication) and partial restriction enzyme digestion, (by using a limiting amount of restriction enzyme the DNA is not using these methods genomic DNA is broken randomly into smaller fragments. A total of 6 random DNA fragments obtained after physical shearing. The vector isolated from the bacterium is also digested with the restriction enzyme which digests genomic DNA. The fragments of genomic DNA are inserted into the vector. Each vector consists of different fragments of DNA. The recombinant DNA molecules are transferred into bacterial cells or bacteriophage particles are assembled.

The genomic library for organisms with smaller genome size (e.g. *E. coli*) can be constructed in a plasmid vector. Only 5,000 clones (of the average size of 5,000 bp) results in a 99% chance of cloning the entire genome of 4.6 x 10 bp. In addition, libraries from organisms with larger genomes are constructed using phage, Cosmid, BAC or YAC vectors.

cDNA library

The mRNAs are highly processed representatives of genes that express under specific conditions. The mRNAs cannot be directly cloned because they are unstable hence mRNAs are converted into cDNAs. The made from library complementary or copy DNA (cDNA) is called cDNA library. The cDNA library represents the DNA eukaryotic the of only organisms, prokaryotic ones. Because genomic DNA of eukaryotes contains introns, regulatory not regions, repetitive sequences. Therefore, the establishment of library genomic eukaryotes is not meaningful. The cDNA library can be constructed by using mRNA because mRNAs are the highly processed, intron-free representatives of DNA only coding having sequence.

Isolation of mRNA

It is much difficult to isolate a specific mRNA if it is in low amounts in the cell. The majority of mRNA sequences in eukaryotic cells contain a long polyadenylated tract (i.e. about 100A residues) at their 3' termini (Kates 1970). By this virtue, they can be separated from other types of RNA_S. Therefore, an mRNA binds with an oligo-dT cellulose affinity column or poly-U sepharose from which it can be eluted. Its integrity can be studied using gel electrophoresis that allows mRNAs to be size-fractionated by recovering chosen regions of the gel lane. Commercial preparations of both these substrates are now available. The mRNA preparations can be enriched for the desired molecules by fractionation according to their size by using sucrose density gradient centrifugation (Boffey, 1987).

Preparation of cDNA.

Reverse transcription

Reverse transcriptase is required for the synthesis of a DNA copy of an mRNA molecule. The mRNAs are reverse transcriptase enzymes and four deoxynucleotide triphosphates (dNTPS) i.e. dATP, dGTP, dCTP and dTTP. The oligo-dT primer binds to polyadenylated tail that contains –AAAA(n) residues and provides a free -OH site for reverse transcription. Reverse by (b) treated with oligo-dT primer, transcriptase adds complementary dNTPs to free 3' -OH site of primer, forms a single strand of DNA resulting in RNA-DNA hybrid. Cellular DNA and total RNA inhibit reverse transcriptase activity. Therefore, it is necessary that mRNA must be in the pure form before cloning. The oligo-dT segments contain between 10- 20 nucleotides in length that hybridize to the poly-A tract on mRNA. The oligo-dT primer ensures the initiation of cDNA synthesis at 3' terminus of RNA-DNA hybrid. Cellular DNA and total RNA inhibit reverse transcriptase activity.

Oligo-dG primer

Oligo-dG primer is added to the reaction mixture and (e) temperature is maintained at about 55°C. This facilitates the binding of dG to oligo-dC tails formed on cDNA. Template for the synthesis of double-stranded cDNA in the presence of now cDNA acts as a DNA polymerase I or AMV (Avian myeloblastosis virus) reverse transcriptase.

Synthesis of the second strand of cDNA

Klenow DNA polymerase I and dNTP precursors are added to the reaction mixture. This enzyme adds complementary nucleotides to the cDNA one by one to the 3' -OH site of the primer. Consequently, from one cDNA, a double-stranded DNA clone is synthesized the length of which is similar to that of the mRNA molecule initially used. This is used for cDNA cloning and construction of cDNA libraries and in gene cloning experiments. The double-stranded cDNA is identical to the gene which codes for Mrna (only in prokaryotic genes). However, the majority of eukaryotic genes contain introns (the non-interrupt their coding sequences. Methods of removing cDNA in eukaryotes have been described earlier. cDNA can be radioactively labeled by nick translation and used for the identification of DNA fragments containing the gene. By using these procedures, a cDNA clone bank is built up.

Insertion of DNA into Vector

The DNA thus isolated as above or procured from DNA or gene bank is fragmented by using the specific restriction enzyme to vector (e.g. PBR322) is also treated with the same restriction enzymes so that the cohesive ends generated may have complementary residues similar to foreign DNA. The plasmid contains amp and tet marker genes to confer resistance against antibiotics, ampicillin and tetracycline respectively.

Sometime after cleavage by restriction enzymes and ligation of foreign DNA fragment with the plasmid a few important genes are destroyed in develop specific cohesive ends Example, BamHI destroys tet genes. The sticky ends can form a phosphodiester linkage between free 5' phosphoryl and 3' OH group) is added along with ATP at 4-100C for a long incubation time. A number of products: highly heterogeneous mixture of recombinant DNA molecules together with parental plasmids are formed from this reaction that are highly heterogeneous mixtures of recombinant DNA molecules together with parental plasmids.

For insertion of double-stranded cDNA into a cloning vector it is necessary to add to both termini single-stranded DNA sequences which should be complementary linear vectors.

In order to get an efficient formation of recombinant DNA molecules, the addition of sticky to a tract of DNA at the termini of ends on both termini is necessary. There are two methods for the generation of cohesive ends on the double-stranded cDNA: use of linkers and homopolymer tails.

Use of Restriction Enzyme Linkers

Linkers are the chemically synthesized double-stranded DNA oligonucleotides containing on them one or more restriction sites for cleavage by restriction enzymes, e.g. Eco RI, Hind II, Bam HI, etc. Linkers are ligated to blunt end DNA by using T4 DNA ligase. Both the vector and DNA are treated with restriction enzymes to develop sticky ends. The staggered cuts i.e. sticky ends are then ligated with T4 DNA ligase and very high-efficiency molecules are produced.

Use of homopolymer tails

For the first time, Jackson et al. (1972) applied this method to construct recombinant DNA by inserting phage A DNA into SV40 DNA by using dA-dT homopolymers. Using terminal transferase, the synthesis of homopolymer tails of the defined at both 3' termini of double-stranded DNA and vector can be done. In the presence of length precursor DATP, terminal transferase helps to add poly-dA at 3' termini of vectors. Likewise, the same enzyme adds poly-T at 3' termini of DNA molecule, when precursor TTP is present. The linearized vector having tails is incapable of recircularization, unless ligated to a double- stranded DNA fragment. The vector and DNA tails are annealed.

Transfer of recombinant DNA into the bacterial cell

Several methods have been developed for the introduction of recombinant DNA molecules into host cells. According to types of vectors and host cells, the methods are gene transfer into host cells are briefly discussed.

Transformation

Once a mixture of recombinant DNA is obtained, it is allowed to be taken up by suitable bacterial cells.

Originally the transformation procedure was developed by Mandell and Higa (1979). The strains of *E.coli* usually do not have restriction systems, hence it degrades foreign DNA. To escape from degradation exponentially growing cells are pretreated with CaCl2, at low temperature; thereafter DNA is mixed up. The event of entering the plasmid containing foreign DNA fragment into a bacterial cell is known as transformation.

Transfection

Transfection is the transfer of foreign DNA into cultured host cells mediated through chemicals the charged chemical substances such as cationic liposomes, calcium phosphate is taken and mixed with DNA molecules.

The recipient host cells are overlayed by this mixture . Consequently, the foreign DNA is taken up by the host cells. For detail see transfer in the preceding section. DEAE dextran liposome-mediated gene For the first time a phage was used to transfer the foreign DNA into *E. coli* cell; therefore, it is often termed as transfection (a hybrid of transformation and infection). The efficiency of transformation is not high as it is influenced by bacterial strain and the size of foreign DNA. It has been found that the efficiency of this process could generate of a cloned circular plasmid (e.g. pBR322). It has not been possible to achieve efficiencies of over 10 transformants per g of plasmid. If found that the efficiency of this process could generate about 10 transformants per microgram (ug) of cloned circular plasmid (e.g. pBR 322). It has not been possible to achieve efficiencies of over 10^8 transformants per g plasmid. If linear DNA is transformed it is almost completely insufficient in transformation.

Selection (Screening) Of Recombinants

After the introduction of recombinant DNA into suitable host cells, it is essential to those cells which have received the recombinant DNA molecules. This process is called screening or selection. Selection of recombinant DNA cells is based on the expression or non-expression of certain characters or traits. The vector or foreign DNA present in recombinant cells expresses the characters, while the non-recombinants do not express the traits. Following which is used mainly for selection of recombinants in *E.coli* are some of the methods.

Direct selection of recombinants

The above features ease the direct selection of recombinant cells. If the cloned DNA itself codes for resistance to the antibiotic ampicillin (amp) the recombinants can be allowed to grow on a minimal medium containing ampicillin. Only such recombinants will grow and form a colony in the medium that contains an amp gene on its plasmid vector. Following this procedure, you cannot say whether the recombinants growing on such medium contain religated plasmid vector or contain recombinant plasmid plus foreign DNA fragment. Because the amp gene is present in both the recombinants on pBR322.

Insertional selection inactivation method

This is a more efficient method than direct selection, in this approach one of the genetic traits is disrupted by inserting foreign DNA Antibiotic resistance genes acts as a good insertion inactivation system. As discussed in the earlier section that plasmid PBR322 contains two antibiotic resistance genes, one for ampicillin (amp gene) and the other for tetra-cycline (ter" gene). If the target DNA is inserted into the tet" gene using BamHI, the property of resistance to tetracycline will be lost. Such recombinants would be test-sensitive. When such recombinants (containing target DNA in the tet gene) are grown on a medium containing tetracycline, they will not grow because their tet^R gene has been inactivated. But they are resistant to ampicillin because the amp gene is functional. On the other hand, the self-ligated recombinants will show resistance to tetracycline and ampicillin. Therefore, they will grow on medium with both antibiotics.To ascertain the presence or absence of the gene in the inserted DNA fragment in plasmid replica plating is done from the master plate. Bacterial colonies of the master plate are gently pressed with sterile velvet so that a few cells of each colony may adhere to it which is then pressed on another plate containing the nutrient medium amended with tetracycline. Plates are incubated for the growth of bacterial colonies. The appearance of colonies is compared with the master plate and those colonies that fail to grow on a plasmid that had the insert DNA in the tet gene of plasmid and had destroyed.

Blue-white selection method

It is also a powerful method used for screening recombinant plasmid. A reporter gene lacZ is inserted in the vector. The lacZ encodes the enzyme B-galactosidae and contains several recognition sites for restriction enzymes. Beta-galactosidase breaks a synthetic substrate, X-gal (5-bromo-4-chloro-indolyl-B-D-galacto-pyranoside) into an insoluble blue colored product. If a foreign gene is inserted into lacZ, this gene will be inactivated. Therefore, no blue color will develop, because B-galactosidase is not synthesized due to the inactivation of lacZ. The host cells containing rDNA will form white colored colonies on the medium containing X-gal, whereas e other cells containing non-recombinant DNA will develop the blue colored colonies. On the basis of colony color the recombinants can be selected.

Colony Hybridization (Nucleic Acid Hybridization)

After the growth of cells on the nutrient medium it is likely that a few cells may have specific DNA as desired among several thousand cells. The portion of host DNA possible to pick up those cells that have the desired DNA. To make it Separation of strand easy, the "colony hybridization technique" has been developed by Grustein and Hogness (1975). It is suitable for use with plasmids. An analogous hybridization technique has been given for use with phages (Benton and Davis, 1977). Single-stranded DNA Hybridisation plaque method, network bases 32 p-DNA Probe hybridizes the of (a) DNA Probes: The colony hybridization technique is based on Hybridisation of desired DNA with a DNA probe the availability of a radiolabeled DNA probe. A DNA probe is the radiolabeled (2PO,) fragment of DNA molecule (20-40 bp) which is complementary at least to one part of desired DNA Generally probe DNA is labeled with 3PO while I emit y-rays. In addition to using radioactive isotopic elements, isotopic sulfur or fluorescent molecules may also be used. Therefore, a DNA probe recognizes A, T, G and C nucleotides and combines them with the complementary sequences of the target DNA. The A and T of the probe combine with T and A of target DNA and vice versa, Similarly, G and C of probe combine with C and G of target DNA and vice versa. The use of the DNA probe in searching the functional target DNA is given in nucleic acid hybridization technique small or sometimes with 121. The 3P liberates B-particles the small-sized DNA probe easily catches the target DNA for hybridization than the long sized probe. Besides, a ^{32}P-DNA than the target DNA molecules. Moreover, the DNA probe is very specific to its target DNA; this specificity is called stringency.

In vitro translation

In vitro translation is now used as a method to confirm the identification of recombinant clones. Nagata et al. (1980) have used primary translation screened by isolating the interferon gene from total leukocyte poly A+RNA. Translation of poly A+RNA is done when it is microinjected into oocytes from the toad, Xenopus laevis with the result of secretion of interferon into the culture medium. Interferon is detected by its antiviral activity. Thereafter, DNA is immobilized on nitrocellulose or diazobenzyloxymethyl cellulose paper to isolate complementary mRNA by nucleic acid hybridization as described earlier. The RNA can be eluted from the cDNA-RNA hybrid and microinjected into oocytes and the translation products are detected by immunoprecipitation techniques.

Immunological tests

The immunological techniques are the final test analogous to the colony hybridization technique as described earlier. It is an alternative screening procedure that relies on expression, and generally applicable approach to potentially a very powerful method since the only absolute requirement is that the required mRNA encodes a protein for which a suitable antibody is available (Williams, 1981). The in-situ techniques given by Broome and Gilbert (1978), and Erlich et al. (1978) are in current use. In the immunological test, instead of radio-labelling of DNA molecules, antibodies (immunoglobulins) are used to identify the colonies or plaques developed on master plates that synthesize antigens encoded by the foreign DNA present in plasmids of the bacterial clones. For this purpose, a special vector, known ds expression vector, is designed where the foreign DNA is transcribed and translated within the bacterial cell (Glover, 1984). The growth medium containing specific anti-serum may help the detection of viable immunoprecipitate (precipitin) around the colonies or plaques.

Blotting techniques

DNA, RNA and proteins are separated by blotting techniques which described below:

Southern blotting techniques

The genes in a DNA restriction fragment is called as Southern blotting technique. Southern blots can easily provide a physical map of restriction sites within a gene located normally and reveal the number of copies of the gene in the genome, and the degree of similarity of the gene when compared with the other complementary genes. A molecular biologist **E.M. Southern (1975)** for analyzing the related on a chromosome.The procedure is starts with digestion of DNA population by one or many restriction enzymes Consequently, DNA fragments of unequal length are produced. This preparation is passed through agarose gel electrophoresis that results in the separation of DNA molecules based on their size. DNA restriction fragments present in the gel are denatured by alkali treatment. The gel is then put on top of the buffer saturated filter paper. The upper surface of the gel is covered with nitrocellulose filter and overlaid with dry filter paper. The dry filter paper draws the buffer through the gel. A buffer contains single-stranded DNA. Nitrocellulose filter binds DNA fragments strongly after baking at 80° C, DNA fragments are permanent and fixed to the when coming in contact with nitrocellulose filter. Then the filter is placed in a solution containing radio-labeled RNA or denatures DNA probe of known sequences. These are complementary in sequence to the blot-transferred DNA. The radiolabeled nucleic acid probe hybridizes the DNA on the nitrocellulose filter. The filter is thoroughly washed to remove the probe. The hybridized regions are autoradiographically detected by placing the nitrocellulose filter in contact with a photographic film. The images are show the hybridized DNA molecules.

Northern Blotting Technique

Southern blotting technique could not be directly applied to the blot transfer of mRNA separated by gel electrophoresis, because RNA was found not to bind with nitrocellulose filter Alwines et al (1979) devised a technique in which RNA bands are blot transferred from the gel to chemically reactive paper. An aminobenzyloxymethyl cellulose paper prepared from Whatman filter paper No. 540 after a series of uncomplicated reactions, is diazotized and rendered into the reactive paper and, therefore, becomes available hybridization with radiolabeled DNA probes. The hybridized bands are found by autoradiography. These blot transfers are reusable because of the firm covalent bonding of RNA to the reactive paper. Effective in binding the DNA as well small fragments of DNA can most effectively be transferred to the diazotized paper derivative to nitrocellulose. These techniques were more advanced and demonstrated that mRNA bands can also be blotted directly on nitrocellulose paper and appropriate conditions.

Western blotting

Towbin et al. (1979) developed the western blotting technique to find out the newly encoded protein by a transformed cell. Its working principle lies on antigen-antibody reaction: hence is an immunodetection technique. In this method, radiolabeled nucleic acid probes are not used. This technique follows the following steps: Extraction of protein from transformed cells. (I) Separation of protein by using SDS-PAGE (sodium dodecyl sulfate polyacrylamide electrophoresis) where SDS acts as a solvent for electrophoresis. Transfer of electrophoresed gel in a buffer at low temperature (40° C) for half an hour blotting of proteins on nitrocellulose filter paper of nitrocellulose filter, Whatman filter and coarse filter in transfer buffer. Soaking of nitrocellulose filter, Whatman filter and coarse filter in transfer buffer. Placing of the Whatman filter paper on a cathode plate followed by a stack of coarse filter Whatman filter, electrophoresed gel, nitrocellulose filter, Whatman filter paper, coarse filter stack, Whatman filter and anode plate transferred. Putting the complete set up in a transfer tank containing sufficient transfer buffer. Application of an electric field (30 V overnight for 5 hours) to cause the migration of proteins from the gel to nitrocellulose filter and binding on its surface. Hybridization of proteins by using radiolabeled antibodies (antibodies) of known structure, isolated from the rabbit. Washing of nitrocellulose filter with a wash solution (Tris-buffered saline + tween 20) to facilitate the removal of unhybridized antibodies. Detection of hybridized protein image by autoradiography.

Shine-Dalgarno sequence

Prokaryotic mRNA, have a sequence upstream from the initiation codon that plays a role in attachment of the 30S ribosomal subunit to the mRNA. This sequence (about 3-9 bases long) is about 3-12 bases upstream from the initiation codon which is also complementary near the 3' terminus of 16S tRNA (Shine and Dalgarno, 1975). The name of this sequence was coined as the Shine-Dalgarno sequence after the name of the discoverer. For the expression of eukaryotic genes within *E. coli*, the presence of a Shine-Dalgarno sequence in bacteria is to a sequence.

Detection of nucleic acids

There are two approaches for labeling the nucleic acids to be used as a DNA probe: radioactive labeling and non-radioactive labeling.

Radioactive labeling

Methods for *in vitro* introduction of radioactivity in nucleic acids were developed during the 1970s through metabolic labeling. In this method, a large amount of radioactivity is required. However, this method was time-consuming and laborious. The methods developed later on utilized phage T4 polynucleotide kinase (which transfers y-PO of ATP to 5'-OH terminus in DNA or RNA) or *E. coli* DNA polymerase I (which replaces normal nucleotides). The nucleic acids are labeled *in vitro* by using the following methods: a normal nucleotide of double-stranded DNA with α - ^{32}P labeled nucleotide.

Nick translation

This method is used to label the double-stranded DNA probes. If there is no nick, a nick is made on double-stranded DNA by using DNase (in a buffered solution containing dNTPs) which exposes the 3'-OH group on one strand of DNA. All the four nucleotides are the alpha position of the phosphate group. (c-32P-dNTPs) at radiolabel generally, only one nucleotide i.e. a- P-dCTP is radiolabeled, whereas the other nucleotides are unlabelled. *E. coli* DNA polymerase I is used in this method. This enzyme shows 5' 3' hydrolyses the nucleotides from the 5' side of the nick. Exocatalytic activity and either remove or Nucleotides are eliminated from 5'end and simultaneously added at 3' side resulting in movement of the nick along with the DNA. Therefore, this method is called the nick translation. The reaction mixture includes probe DNA to be labeled. 2-P-dTP, of the other three dNTPs, DNase 1 and *E. coli* DNA polymerase I. It is incubated at 14°C for 2-4 hours. The hybridized DNA is separated from the rest of dNTPs and used to hybridize the DNA.

Random primed radiolabeling of probes

This technique was discovered by Vogelstein in (1983, 1984). The restriction fragments purified by gel electrophoresis from denatured linear or circular DNA are used to prepare probes. In this method, primers are used for DNA synthesis but DNA nicking is not done. The purified DNA to be labeled is mixed in a buffer and heated at 100°C so that it may be denatured. Then it is put on ice to lower down the temperature. The reaction mixture (containing the Klenow fragment of *E.coli*, DNA Polymerase I, one radiolabeled nucleotide e.g. O- P-dCTP and three unlabelled nucleotide triphosphates) is incubated at 37°C for 30 minutes to 16 hours. Since, Klenow DNA Polymerase I lacks 5'3' exonuclease activity, the radiolabeled probe is synthesized by primer extension but not nick translation as described earlier. The hybridized DNA is separated from unincorporated nucleotides and used as a probe. Probes Developed by PCR: This method is advantageous because very small amount of template DNA can be used. The method of operation is the same as described. In this method, the reaction mixture includes linear template DNA (to be labeled), amplification buffer **Taq** polymerase, primers, one radiolabeled and three unlabeled dNTPs. The hybridized DNA is isolated and used as a probe.

Non-radioactive labeling

Mostly, non-radiolabeled DNA probes are also used to detect the desired DNA. There are three systems of non-radioactive routes such as using horseradish peroxidase, biotin probe and streptavidin probes as described below

Horseradish peroxidase (HRP) method

The HRP is a plant-derived enzyme covalently linked to DNA. Chromogenic is substrates (such as chloronaphthol) or chemiluminescent substrates (probe enzyme e.g. luminol) are used for detection of HRP-linked DNA. Chloronaphthol yields an insoluble purple product in the presence of peroxide and HRP. Consequently, the presence of DNA is detected. Similarly, HRP oxidizes luminol and emits luminescence by using X-ray film; hence luminescence can be detected.

Digoxigenin (DIG) Labelling System

Digoxigenia is a cardenolide steroid derived from a plant digitaria as a hapten. The presence of digoxigenin can be detected by an antibody associated with an enzyme (anti digoxigenin-alkaline phosphatase conjugate). The probe is labeled with digoxigenin (11) dUTP which is a nucleotide tri- phosphate analog. This analog containing digoxigenin moiety is incorporated into DNA by nick translation or random prime labeling. The DNA acts as probe. Anti-DIG (antibody to digoxigenin) conjugated with alkaline phosphatase detects the DIG-labeled probe through enzyme-linked immunoassay. A chromogenic substrate [such as 5-bromo-4-chloro-3-indolyl phosphate (BCIP) and nitrobutane tetrazolium chloride (NBT)] is added which reacts with alkaline phosphatase and produce purple blue color. A chemiluminescent substrate such as dioxetane is also used with digoxigenin streptavidin systems. There are several derivatives of digoxigenin that emit light after enzymatic activation by alkaline phosphatase. The denatured DNA present on Southern blots is hybridized by labeled probe. The membrane is placed in a detection buffer containing anti-digoxigenin (20 pg/m) and bovine serum albumin (BSA) (5% w/v) and incubated at 37°C for 1 hour. After incubation membrane is washed thrice. Enzyme activity of alkaline phosphatase is detected by using BCIP (0.17 mg/ml) and nitrobutane tetrazolium chloride (NBT) (0.33 mg/ml) activation purple/blue color develops.

Biotin-Streptavidin Labeling System

Biotin is vitamin H and avidin is found in egg white. The principle of this system is based on the interaction between biotin and the glycoprotein avidin. A biotin-containing nucleotide analog (biotin-141ATP) or probe is prepared through nick translation or random priming methods. Streptavidin is conjugated to alkaline phosphatase. The biotinylated probes detected by specific binding of streptavidin-conjugated alkaline phosphatase. Thereafter, a chromogenic or chemiluminescent substrate is added that reacts with alkaline phosphatase and detects the hybrids. The intensity of color is proportional to the amount of biotin present in hybrid DNA. This method detects the hybrids much faster than the radioactive probes. But a very small number of probes (20 nucleotides) consists of only a small number of biotinylated sites. Therefore, the intensity of signals is limited or biotin as a dye.

Gene cloning in eukaryotes

Most studies on gene cloning have been carried out in bacteria. In addition, many difficulties are associated with them, for example

(i) Correction of introns of eukaryotic mRNAs,

(ii) Failure of transfer of an equal number of plasmids into daughter cells during cell division and yield of two types of cells, one with plasmid and the second without plasmid, post-translational modification (e.g. inhibition of proteolysis, the addition of oligosaccharides to specific sites on the polypeptide chain, the glycosylation),

(iii) Threat of hazardous effects.

(iv) Expression to ensure the presence of a stable plasmid in the bacterium for commercial applications.

With these prospects the attraction of the use of eukaryotic cells is obvious. Considerable efforts have been made for the improvement of crop plants through genetic engineering. Many works on various aspects of building up of vectors and expression of insert genes in eukaryotic cells are in progress.

Transformation in Filamentous Fungi

Transformation through plasmid of *E coli* was first described in protoplasts of *S. cerevisiae* (Himnen et al., 1978) and then in *N. crassa*. However, the situation in filamentous fungi is different from the others as there is no convincing evidence for autonomously replicating plasmids. But sequences having functions similar to ARS have been isolated. Method of transformation is similar to for plant cells, Following arc the examples where transformations in protoplasts have been carried out: *A. nidulans, A. niger, A. oryzae, Cephalosporium acremonium, Coprinus cinereus, Glomerella cingulata, Gaeumannomyces gruminis, Mucor sp. Penicillium chrysogenum, Septoriu nodorum*, etc.

Application of transformation of fungal protoplasts

Mainly the transformation approaches are used for the analysis of promoters and construction of expression vectors, and (ii) biotechnological application involves the construction of vector which has different pieces of promoters. These can be cut by restriction enzymes and joined with DNA ligase. By this method changes in the promoter region are also possible and this can be achieved through mutagenesis. Thus, a strain can be improved in the expression of foreign genes in fungi. Analysis of promoter overproduction of the products (Peberdy, 1989).

Gene transfer in dicots by using agrobacterium TI-DNA as Vector

It is not possible to transfer a Ti-plasmid into a plant cell because of its large size, most likely up to 235 Kb. Therefore, T-DNA is excised and used as foreign DNA for expression in plant cells. From a bacteriological point of view opine production and development of gall are not required for the successful integration and expression of DNA. Hence, the replacement of nos and ops genes with a foreign DNA fragment retaining the opine synthase promoter for the expression of DNA fragment has no harm. Schell and Van Montague (1983) undertook some experiments where genes for resistance to kanamycin and methotrexate were expressed in cultured callus cells. Generally speaking, promoters for the expression of foreign DNA. It has become possible that specific tissues of plants when they are with specifically regulated promoters.

Gene transfer in monocots

The limitation of Ti-plasmid is that it is specific only to the dicot plants. *A. tumifaciens* does not induce a tumor in monocot plants. However, most of the major crops including cereals are monocots. Hooykaas Van et al. (1984) discovered that *A. tumifaciens* could transfer T-DNA into certain monocots with the consequences of expression of opine gene within the plant cell without inducing the tumor. This discovery made it possible that T DNA can be expressed into cells of monocot plants.

Plant cell transformation

Plant cell suspension culture coupled with DNA transfer techniques has become a new field of gene manipulation in plants. Cell suspension cultures have been the traditional source of plant material for biochemical selection of plant cell mutants, such as (i) numerous cell lines resistant to various amino analogs (Maliga, 1980). Suspension of protoplasts constitutes an identical plant material from one laboratory to other. The approaches made for plant protoplast transformation are (ii) introduction of the DNA sequence via integration into T-DNA of Ti-plasmid of *A. tumifaciens*. (iii) Transfer of naked DNA into protoplast including bacterial spheroplasts and microinjection and (iv) Transformation with plant viruses carrying inserts e.g. *Gemini and caulimoviruses*.

Plant cell transformation by ultrasonication

Ultrasonication is done for various biological experiments. But in some plants, this technique has been done for successful gene delivery. The Biotechnology Research Centre, Beijing (China) has used this technique for gene transfer in plants like wheat, tobacco and sugarbeet. When the cultured explants were sonicated with plasmid DNA carrying market genes such as cat, nptll, and gus, and sonicated calli transferred of the selective medium, shoots were grown successfully. The calli in the control set of experiments that were not sonicated with plasmid did not grow on selective medium but they died.

Liposome-mediated gene transfer

Liposomes are microscopically small sized particles. They contain phospholipid bilayers that are concentric in nature. Liposomes enclose an aqueous chamber and can entrap water-soluble molecules. Therefore, they are called lipid bags. Many plasmids are enclosed in them. By using polyethylene glycol (PEG) they may be stimulated to fuse with protoplast. In several plants like carrot, tobacco, petunia ,etc. this technique has been used for a successful transfer of genes. Due to endocytosis of liposomes, DNA enters the protoplast. It gets adhered first to the protoplast surface and gets fused with the protoplast surface. Consequently, plasmids are released inside the cell. There are several advantages in using this technique such as low toxicity, (i) protection of DNA/RNA from nucleases that lyse them, (ii) long stable storage of DNA fragments in a liposome, (iii) applicability in various cell types, and (iv) high level of reproducibility.

Electroporation

Electroporation is now used to transfer the foreign DNA into the fragile cells. A brief pulse of high voltage electricity (about 350 V) is applied to a protoplast suspension containing naked or recombinant plasmids. The electric pulses induce the formation of large pores in the cell membrane. These pores give a passage through which the foreign DNA can enter into the protoplasts and thus, increase the transformation frequency. The transformed protoplasts are cultured for about one month. These develop microcalli which were plated on solid medium containing selective marker (e.g. kanamycin). After 37-45 days, the calli are analyzed for the presence of differences in transformed cells. By using electroporation, genes of choice have been successfully transferred in protoplasts of wheat, rice, petunia, sorghum, maize and tobacco. Moreover, transformation frequency can be improved by using linear DNA instead of a circular one, by giving heat sock to protoplast at 45°C for 5 minutes, by adding PEG and using 1.25 kV/em voltages.

Particle bombardment gun

It was developed by **Prof. Stanford** and coworkers of the Cornell University (USA) in 1987. As the term denotes, it shoots foreign DNA into plant cells or tissue at a very high speed. This technique is also known as particle bombardment, particle gun method, biolistic process, microprojectile bombardment or particle acceleration. This technique is most suitable for those plants which hardly regenerate and does not show sufficient response to gene transfer through *Agrobacterium*, for example, rice, wheal corn, sorghum, chickpea and pigeon-pea. The apparatus consists of a chamber connected to an outlet to create a vacuum. At the top, a cylinder is temporarily sealed off from the rest of the chamber with a plastic rupture disk. Helium gas flows into the cylinder. A plastic microcarrier is placed close to the rupture disk.

Pollen transformation through particle bombardment

Genetic improvement in *Alfalfa* has been made by plant tissue culture methods. This method is time-consuming, requires special techniques and the efficiency of getting stable regeneration is low. Ramaiah and Skinner (1997) produced transgenic alfalfa through direct delivery of DNA into pollen grains by particle bombardment method. Plasmid PBI121 bearing the GUS reporter gene was transfered into pollen grains. The microprojectile bombarded a liquid pollen suspension and tripped the recipient female flower. Pollinated flowers set seeds in about a month. Thirty percent of plants derived from fertile seeds showed the integration of GUS plasmid. It was confirmed by Southern analysis. After 10 vegetative generations, some were allowed to fertilize the tagged transgenic plants lost the integrated GUS plasmid, whereas, in few others, copies of the GUS gene decreased due to unknown reasons.

Microinjection

Microinjection is a technique of delivering foreign DNA into a living cell (a cell, egg, oocyte, embryos of animals) through a glass micropipette. One end of glass can be transformed by a plasmid carrying the early region of SV40 with cos cells express wild type large T antigen and can support the micropipette is heated until the glass becomes some forms to very fine tip at the heated end. The tip of the pipette attains to about 0.5 mm diameter which resembles an injection needle. The process of delivering foreign DNA is done under a powerful microscope. Cells are microinjected are placed in a container. A holding pipette is placed in the field of view of the microscope. The holding pipette holds a target cell at the tip when gently sucked. The tip of the micropipette is injected through the membrane of the cell.

Direct transformation

Like plant cells, mammalian cells can be transformed by foreign DNA fragments. It is necessary to precipitate the DNA with calcium phosphate and mix the cells to be transformed for transforming the mammalian cells. DNA molecule passes through the cell membrane and integrates randomly with mammalian chromosomes. The selective marker can be linked up with DNA fragments to be cloned and can be introduced into mammalian cells. The transformed cells are, thereafter, separated from the cell line after plating them on a selective medium. Many techniques transformed cell lines. **Perucho et al.** (1980) successfully isolated the chicken thyamidine gene (tk gene) by ligating with PBR322 and transforming the tk- (deficient for the thyamidine gene) mouse.

Gene DNA sequencing

The structure of DNA (e.g. gene insert, a recombinant plasmid or entire genome) can be analyzed by determining the nucleotide sequences. The information of nucleotide sequences is essential. In 1965, **Robert Holley** and his research group at the Cornell University completely sequenced nucleotides of tRNA (tRNA for yeast alanine). In 1977, the following two methods were DNA developed. **Allan Maxam and Walter Gilbert** developed the chemical method of sequencing. In this method, end- labeled DNA is subjected to base-specific cleavage reaction before gel separation. In routine sequencing of DNA, this method is not commonly followed. **Frederick Sanger** and co-workers developed in the same year (1977). It is also called the dideoxynucleotide chain termination method because dideoxynucleotides are used in an the enzymatic method of DNA sequencing.

Maxam and Gilbert's chemical degradation method

This method is not very popular because it is time-consuming and labor-intensive. In this method, the DNA molecule can be radiolabeled at either 5 'end by using polynucleotide kinase or 3' end by terminal transferase. One end of the radiolabeled double-stranded DNA is removed by using an endonuclease. A base is modified chemically followed by cleavage of sugar-phosphate backbone of DNA. No specific reaction for the four bases is carried out, except specific reaction to G only and purine specific reaction which removes A or G. A difference in these reactions indicates the presence of A.

Cleavage of purine

The mixture is separated in four sets, each treated with a different reagent that degrades only G or C or A and G or C and T. DNA is treated with acid followed by dimethyl sulfate in one set. This causes methylation of A, (at 3 'position and G (at 7' position). Subsequently, the addition of alkali (-OH) and pyrimidine results in DNA cleavage and removal of purines (A or G)

Cleavage of pyrimidine

Similar to the cleavage of purine, pyrimidine (C or T) is cleaved in the presence of 1-2 M NaCl solution.

It works only with C. Differences between these two indicate the presence of T in the DNA sequence. Cleavage of pyrimidines (C or T) through hydrazine hydrolysis. Partial chemical cleavage of DNA fragments as done above generates the populations of radiolabeled molecules extending from radiolabeled terminus to the site of chemical cleavage. These fragments of different sizes represent unique pairs of 5 and 3' cleavage products in the random collection. A complete set is formed by these products the length of each number is short by nucleotide. These can be separated by gel electrophoresis. The fragments containing labeled terminus can be observed by autoradiography of the gel.

Sanger method (Dideoxynucleotide chain termination method)

Twice Nobel Prize winner Frederick Sanger powerful method et al. (1977) developed DNA sequencing that utilizes single-stranded DNA as a template. This method is also called a dideoxynucleotide chain termination method. The requirements are: with free 3' -OH ends to primer start DNA synthesis, DNA polymerase and dNTPs. The presence of free 3'-OH group at 3' end in dATP and no 2',3' -OH in 2',3'ddATP. In 2',3', ddATPa hydrogen atom is attached at 2' and 3 carbons instead of -OH hydroxyl group. If any of four ddNTPs binds, the chain elongation is terminated. Because ddNTPs do not have free 3'-OH end which is required for chain elongation. Therefore, no phosphodiester bond will be formed.

Automatic DNA Sequencers

Automatic DNA sequencing machines were Sanger's method. In this new method, a different fluorescent dye is tagged to the ddNTPS. Using this technique, a DNA sequence containing thousands of nucleotides can be determined in a few hours. Each dideoxynucleotide is linked with a fluorescent dye that imparts different colors to all the fragments terminating in that nucleotide. All four labeled ddNTPs are added to a single capillary tube. It is a refinement of gel electrophoresis that separates fast. DNA fragments of different colors are separated by their respective size in a single electrophoretic gel. A current is applied to the gel. The negatively charged DNA strands migrate through the pores of gel towards the positive end. The small developed during the 1990s. It is an improvement of size.

DNA fragments migrate faster and vice versa. All fragments of a given length migrate in a single peak. The DNA fragments are illuminated with a laser beam. Then the fluorescent dyes are excited and emit light of specific wavelengths which is recorded by sequences that are read by determining the sequence of the colors emitted from specific peaks as they pass the detector. This information is directly seen on a computer ,which determines the sequence. A tracing electrogram of emitted light of the four dyes is generated by the computer. Color of each dye represents the different nucleotides.

Chimeric DNA

In this chapter, we will describe different types of restriction enzyme used regularly as a routine in all DNA cloning and recombinant DNA enzymes. The term gene cloning can be defined as the isolation and amplification of an individual gene sequence by insertion of that sequence into a bacterium where it can be replicated. Cloning a fragment of DNA allows indefinite amounts to be produced from even a single original molecule. Cloning technology involves the construction of novel DNA molecules by joining DNA sequences from different sources. The product is often described as recombinant DNA.

Gene cloning have been described in the following steps:

- Isolation gene of interest.
- A fragment of DNA (gene of interest) to be cloned is incorporated into a small replicating (usually circular) DNA molecule called a vector. This can be *E.coli* plasmid, a virus, a cosmid ,etc. The vector with an incorporated gene is called a recombinant vector or molecule.
- The recombinant vector is introduced into a host cell by transformation.
- Cells that have acquired the recombinant DNA molecule are selected.
- Recombinant DNA molecule is multiplied within the host to produce a number of identical copies of the cloned gene.

Restriction Enzymes for Cloning

For the cloning of DNA, often we need to cut DNA at specific sites, recognized and cleaved by specific enzymes. Also by locating the positions of the cleavage site of a number of restriction enzymes, restriction maps can be prepared. These restriction enzymes recognize short sequences of double-stranded DNA as targets for cleavage. Different enzymes recognize different but specific sequences, each ranging in length from 4 to 8 base pairs. The enzymes are named by a three letter that identifies their origin. Roman numerals are added to distinguish several enzymes with the same origin. For instance, *EcoRI* is derived from *E.coli* and *HpaI* is derived from *Haemophilus parainfluenzae.* Besides cleavage, modification in the form of methylation is also brought about by some enzymes called modification enzymes. This methylation distinguishes genes in different states of functioning. There are also some enzymes that perform the functions of both restriction and modification. Based on the attributes, restriction enzymes have been grouped into two classes: (i) Type II restriction enzymes systems which have separate enzymes for modification and restriction; and (ii) type I and type III enzymes systems, in which the same enzyme possesses both activities, although the restriction and modification sites differ in position.

The above two classes of restriction enzymes, type II enzymes are most important for cloning purposes. The target sites for these enzymes are 4-8 bp with symmetry. Most type II restriction enzymes cleave DNA at an unmethylated target site. Some enzymes introduce staggered cuts, others generate blunt ends. Enzymes with 4 bp target sites are used when frequent cuts are desired and those with 8 bp are used when rare cuts are desired to get long DNA segments. Otherwise the majority of enzymes used to have 6 bp target sites. Some of them can cleave both methylated as well as unmethylated targets, but the majority of them cleave only unmethylated targets.

Techniques of restriction mapping

The recombination frequencies, which are the function of distances between genes, are not completely independent of (i) the nature of mutants used, (ii) the position of these mutants on chromosomes, (iii) the genetic background, (iv) the environmental conditions and a variety of other factors. Consequently, the distances between genes on a genetic map may not correspond to the distances between them on the DNA molecule of which they are a part at the molecular level. Gaps may also be present on a genetic map due to the non-availability of mutants in that region. At the molecular level, the fine structure of a gene can be studied through the determination of the nucleotide sequence of the concerned DNA segment. Sometimes this is done through isolation of DNA segment corresponding to a gene followed by sequencing of the DNA segment, which needs time and energy. Instead, we can prepare a map of the DNA by its cleavage at specific sites with the help of restriction endonucleases, which recognize very short specific DNA sequences and cleave the DNA at these specific sites.

These sites of cleavage can be identified and mapped to give rise to a restriction map. In a restriction map, the nucleotide sequence of DNA between sites close to one another (300 bp or less) can be determined. These regions can then be connected into a sequence of the entire gone, whose nucleotide sequence, then can be compared with the amino acid sequence of the protein coded by it. This can be further extended to adjoining genes to saturate the physical map, an objective which is difficult but is being attempted for human genome. On a restriction map, is found a linear sequence of sites, each for a specific enzyme and the distances between them is measured in terms of the number of base pairs of DNA. This technique can be used both in prokaryotes and eukaryotes, although, all regions of chromosomes can not be easily mapped in eukaryotes. Different steps involved in restriction mapping include the following.

Restriction Cleavage and Gel Electrophoresis

If we take a particular DNA molecule or DNA sample, digest it with a specific restriction enzyme and then subject to gel electrophoresis we will notice a series of bands on the gel slab or cylinder. The bands can be observed under ultraviolet light after staining with ethidium bromide. The position of different bands will depend on the DNA fragment size, such that the smaller the fragment, the more rapidly it will move, and the longer the segment more slowly it will move. It will mean that the fragment away from the loading site will be smaller and that close to the loading site will be longer DNA fragments. The gel can be calibrated by using a mixture of DNA fragments of known length so that the position of the bands on this standard gel can be compared with the bands in the experimental DNA digest and the fragment length in each band of DNA digest can thus be determined sometimes using a computer device. The results of digestion of 5000 bp long hypothetical DNA molecule digested separately by two enzymes A and B. A cleaves the DNA into four fragments of lengths 2100, 1400, 1000 and 500 bp, while the enzyme B cleaves it into three fragments of length 2500, 1300, 1200 bp. These data with some additional experiments can be used to generate a restriction map.

Construction of a restriction map

DNA digestion by more than one endonucleases as we discussed above to be utilized to arrange the sites of cleavage. This can be done by several methods. This method involves successive digests with two individual enzymes, where we digest with either enzyme A or enzyme B and then cleaves it with the other enzyme. The original DNA sample is also digested by a mixture of both the enzymes to confirm the results of individual successive digests. When fragment A-2100 is obtained and digested with enzyme B, it is cut into two fragments of 1900 bp and 200 bp. Fragment B-2500 (obtained from individual digest with enzyme B) is similarly cut by enzyme A into two fragments of 1900 bp and 600 bp, suggesting overlap of A-2100 and B-2500 in the region of the fragment of 1900 bp that is obtained by one cut with enzyme A and the other by enzyme B. The DNA fragment of 1900 bp is also available, when DNA is digested with a mixture of both enzymes A and B (double digest). Using this information, we get the overlapping regions in A and B digests and find out the sites of cleavage by A and B. This will then allow us to prepare the restriction map.

Use partial digests, end Labeling Hybridization In Restriction Mapping

The above technique of individual and double digests can be supplemented with other techniques for the actual construction of maps. For instance, by permitting incomplete digestion, fragments longer than those obtained by complete digestion may be obtained. These will be called partial digests. In the example used in the last section, in partial digests with enzyme A, instead of getting A-1000, A-2100, A-1400, A-500, we may get A-3100 A-1400 and A-500, which will suggest that A-1000 and A-2100 lie in the adjacent regions. In another partial digest if we get A-1000, and A-500, this will suggest that A-2100 and A-1400 are adjoining, which will mean that A-1000 and A-1400 are found on the two sides of A-2100s. This technique will thus allow the arrangement of fragments in a linear order. Another useful technique involves labeling of the ends (end labeling) of DNA ends can be identified due to labeling, even alter digestion. For example, using end labeling, if A-1000 and A-500 are found to be radioactively labeled, these will be present at the two ends of the restriction map. Some of the features of a restriction map can also be confirmed by nucleic acid hybridization. A molecule before digestion, so that the fragments containing these 2100 and B-2500 have an overlapping region of 1900bp, they should hybridize with each other, which will confirm overlapping. The above technique will help in accurate completion of a restriction map, but will require that we have a complete set of restriction fragments which make the entire DNA region being mapped. Once the restriction map (which is a physical map) is ready, it can be compared with the genetic map. Although large changes, already located on the genetic map can be easily located on the restriction map, but point mutations cannot be always and easily located on the restriction map, since the restriction sites often do not change due to mutation. In such a situation, one may like to determine the nucleotide sequences on individual fragments and compare them in normal and mutant individuals. Since sequencing of nucleotides is more laborious and cannot be easily undertaken for the whole genome, another technique of molecular genetic markers in the form of restriction fragment length polymorphisms (RFLPs) has been utilized during the 1980s to generate genetic linkage maps in human, mouse, rat, pig, sheep, fruitfly, (Drosophila), nematode (*Caenorhabditis elegans),* yeast, maize, wheat, rice, barley, tomato, potato, lettuce, rice, pepper, pea, etc.

Construction of Chimeric DNA

DNA molecules are produced by inserting a foreign DNA segment into the DNA molecule of a vector. These chimeric DNA molecules are the means of cloning genes or other small DNA sequences. The construction of chimeric DNA, as we know it now, involves the joining of the broken ends of the vector with the two ends of the sequence to be cloned. There are at least three methods available for the construction of this chimeric DNA by palindromes and staggered cleavage. This is the most common meth palindromic sequence, a restriction enzyme causes staggered cuts producing short complementary single-stranded sticky ends (a palindromic sequence is one, which has complementary sequences at the two ends of a single strand, e.g. AT.GAT). This can be illustrated using the example of enzyme *EcoRI*, which cuts the sequences at specific positions. When another DNA molecule is similarly cut by the same order enzyme, similar sticky ends having the same sequences in the single-stranded ends will be produced, so that when it is mixed with the previous molecule similarly treated, the two will anneal producing a chimeric DNA or chi plasmid if plasmid DNA is involved. The enzyme DNA ligase helps in joining the bonds at the cut ends of two molecules. The above technique has the advantage of regenerating two restriction sites (eg; *EcoRI* site) in the chimeric DNA so that the foreign

DNA segment can be retrieved rather easily from the cloned copies of chimeric DNA by cleavage again with the same enzyme.

There are also the following disadvantages with this technique.

(i) The two cleaved ends of a vector or of a foreign DNA may join end to end before getting inserted. Therefore while isolating chimeric DNA; one will have to select chimeric molecules each having only a single insert. This can be achieved by separating molecules of different sizes by gel electrophoresis.

(ii) The recognition site, particularly in the sequence to be cloned may not lie at a convenient position so that sometimes only a part of the desired segment will be inserted.

Adding poly dA at the 3' ends of the vector and poly dT at the 3' ends of the DNA clone

In this method, DNA can be cut at the desired position both in the vector and in the clone, without a staggered cleavage. Using precursor dATP, poly dA is added at both the 3' cut ends of the vector with the help of enzyme terminal transferase. Similarly, using precursor dTTP, poly dT is added at both the 3' cut ends of the DNA sequence to be cloned. The vector and clone can then be joined by annealing the poly dA with poly dT tails and then ligating them using a DNA ligase enzyme. In this technique, there is no chance of reannealing between the two cut ends of the same DNA molecule or of the two similar DNA molecules so that one of the disadvantages of the first method has been overcome in this case. However, in this technique, it will not be easy to retrieve the cloned DNA, because the recognition site of the enzyme due to the insertion of poly dA and poly dT has been lost in this case. But if poly dG and poly dC are used instead (which serve the same purpose), the cloned sequence can be retrieved, since poly dG: poly dC regenerate the recognition for the enzyme *PstI.*

Blunt end ligation by T4 DNA ligase

In this technique, a restriction enzyme is used to cut a duplex DNA at the same place in both the DNA strands. The broken ends are then used to joining with the two ends of another DNA molecule irrespective of the sequences present at the broken ends of the two DNA molecules. T4 DNA ligase is used for this joining reaction. The disadvantage of this technique is that any two broken ends may join including those belonging to the same

DNA molecule. This leads to the production of a variety of products and one will have to select the desired product from a mixture of products. This blunt end ligation is used for developing a method in which the DNA to be cloned can be easily retrieved whenever required. This method makes use of short DNA duplexes (linkers), that contain *EcoRI* palindrome or some equivalent palindrome, which is small in size, can be synthesized chemically. These linkers can be linked to the blunt ends of vector DNA of an insert by blunt-end ligation. This will allow the creation of an *EcoRI* site in the linker region of the vector so that when a DNA segment is now cloned, the inserted DNA segment can be retrieved with *EcoRI.* However, in this method, there is no restriction on the original choice of sites to generate the cut ends. Therefore, with this method, it is now possible to insert a foreign DNA segment at a particular site in the linker region of the vector and then retrieve this foreign segment whenever necessary.

DNA Cloning in Bacteria and Eukaryotes

Cloning in bacteria

Bacteria are often used as hosts for the multiplication of chimeric vectors or recombinant DNA molecules. Therefore, a bacterial strain should be available, which should be the host for the vector and should be modified or selected for high transformation efficiency. Bacteria can be classified into.

(i) Gram-negative bacteria e.g. *Escherichia coli. Pseudomonas, Rhizobium* and

(ii) Gram-positive bacteria e.g. *Bacillus subtilis* and *Streptomycetes sp*. ordinarily same plasmid can not maintain and stable form in both classes of bacteria and, therefore, plasmid vectors for two classes usually differ. While gram-negative bacteria can be used for plasmids, phages and cosmids vectors, gram-positive bacteria are used as hosts for only plasmids. A chimeric vector is inserted bacterial cells and transformed bacterial colonies are selected and for multiplication in suspension cultures the chimeric DNA is retrieved whenever required.

Cloning in eukaryotes

Since chromosomes found in the nucleus of eukaryotes are separated from the rest of the cell through the nuclear membrane, and since many of the genes are split genes with exons and introns, genetic engineering with eukaryotes requires new methods and tools. When eukaryotic genes are cloned prokaryotes, the split genes cannot be correctly expressed, because prokaryotes do not have the equipment necessary for splicing out the RNA transcribed from the introns of a gene. In view of this, eukaryotic cells may sometimes be needed for cloning and particularly for the expression of cloned eukaryotic genes. Among eukaryotes, DNA cloning has been done in yeast, mouse and to some extent even in some higher plant species. In yeast, a plasmid called 2ug DNA (63 bp) is found, which is an appropriate cloning vehicle. An efficient transformation method is also available, which involves protoplast production followed by PEG directed introduction of DNA into protoplast. In animal cells like mouse cells, special animal viruses were used as cloning vehicles. *Simian virus* 40 (SV40) is one such virus, in which the globin gene could be integrated. This gene integrated into SV40 could be transcribed and translated in mouse kidney cells. A variety of human cell lines in culture is also used for cloning foreign DNA, using some animal viruses as vectors.

Molecular probes

Molecular probes are small DNA or RNA segments that complementary sequences in DNA or RNA molecules and thus allow identification and isolation of these specific DNA sequences from an organism. Antibodies are also occasionally used as probes to recognize specific protein sequences. Although, initially, these probes were developed and used for genetic engineering research but now frequently used for a variety of purposes including diagnosis of infectious diseases, identification of food contaminants, variety of microbiological tests, forensic (e.g. fingerprinting of murderers or rapists ,etc. Probes can also be used to identify strains of an organism e.g. varieties of a crop species (plant breeder likes to have a quick test to identify his variety to maintain his patent or breeder's right, so that a competitor may not use it in another name leading to infringement of rights). For basic studies in molecular biology laboratories, molecular probes are frequently used for the identification and isolation of genes or related sequences. In theory, any nucleic acid (or rarely protein, e.g. antibody) can be used as a probe, provided it can be labeled to permit identification and quantify the hybrid molecules formed between the probe and the sequence to be identified. In practice, double and single-stranded DNAs, mRNA and other RNA's synthesized *in vitro* are all used s probes. DNA/RNA probe assays are faster and sensitive, so that many conventional diagnostic tests for viruses and bacteria involving culturing of the organism, are being fast replaced by antibody and DNA probe assays. While culture tests can take days or even months, molecular probe assays can be performed within a few hours or minutes. Therefore, the use of DNA probes has become today's most sophisticated and sensitive technology for a variety of uses involving biological systems both in basic and applied studies including their commercial use.

Preparation of Probes

Genomic DNA probes

Simple, reproducible methods have been devised for the preparation of molecular probes. A random or specific DNA sequence (probe) can be isolated from the genomic DNA (DNA extracted from cell nuclei), using the following steps extract DNA from plant or animal tissue (i) Digest extracted DNA with a restriction enzyme (e.g. *EcoRI or HindIII*), which cuts DNA on specific sites or positions where a specific sequence recognized by the enzyme is found; (ii) Run the digested DNA on an agarose or polyacrylamide gel for electrophoresis to separate fragments of different sizes; (iii) Isolates DNA of specific fragment size from a particular band identified through Southern blots by hybridization with a specific labeled mRNA or c DNA molecules, if available. If random probes are needed, no hybridization of Southern blots is required (Southern blots are DNA bands transferred from gel to nitrocellulose membrane with the help of a buffer rising through capillary action; isolate DNA of specific (iv) Clone DNA in a vector allow the chimeric vector to infect bacteria for multiplication, where it can make billions of copies.

The steps (ii) to (vi) above are also used for preparing cDNA probes. From the transformed bacteria, the chimeric vector can be obtained and used in one of the following ways: (i) It may directly as the probe (the presence of vector DNA will not interfere, so that it is not really necessary to remove it) (ii) The segment may be separated (or retrieved), by using the same enzyme which was used for cloning. In the latter case, cleaved chimeric vector DNA will be again electrophoresed for separating the inserted segment on the gel. This inserted segment thus retrieved can now be used as a probe. (iii) The chimeric DNA may be used for PCR, using flanking sequences as primers; (the PCR probe) product can be separated and used as a probe

cDNA probes

A DNA sequence corresponding to a part of a gene(s) (an oligonucleotide) can be obtained by reverse transcription of mRNA. cDNA thus obtained can be cloned and used as a probe.

Synthetic oligonucleotides as probes

DNA probes with nucleotide sequence can also be synthesized chemically using automated DNA synthesizers now available commercially. These synthetic probes will be efficient only when they are not more than 20 to 40 nucleotides in length.

RNA Probes OR Riboprobes

High specific activity RNA probes or expression vector. This is achieved through RNA synthesized *in vitro* (on DNA inserted in vector) and labeled simultaneously employing specific RNA polymerase and labeled nucleotides which are supplied by commercial firms in a kit for riboprobe preparation. RNA probes off several advantages over DNA probes, since these are single stranded, and provide an improved signal or hybridization blots. However, the widespread presence of ribonucleases creates some problems in their preparation and riboprobes may also be synthesized from DNA templates cloned because these ribonucleases will cause degradation of RNA which is de synthesized. Therefore, extreme care is needed in the preparation of a riboprobe, keeping all glassware ribonuclease-free (no glassware should be touched with bare hands, because ribonucleases may be available on skin).

Labeling of probes

The detection of homologous sequences after hybridization with the probe is like finding a needle in the haystack. Therefore, for the success of DNA probe assay, it is necessary to develop simple, safe and sensitive techniques for their use. As probes transmit no signal of their own, they have to (b) either labeled with radioactive isotopes or (i) methods are used to couple nonradioactive signal molecules to the probes.Without impairing the hybridization ability of these probes. These signal molecules may include fluorescent antibodies and enzymes that produce color changes and chemiluminsccent catalysts.

Radiolabeled probes

Traditionally radioactively labeled probes are used for a variety of experiments. p is the most commonly used radioisotope, although other labels such as H, 35S and 1251 have also been used. Conventional labeling replaces a proportion of the nucleotides in a nucleic acid sequence, with ^{32}P derivatives (e.g. ^{32}PUCTP) or adds ^{32}P to be the end of the molecule. At least three methods are available for labeling the probe

- Nick translation involves the creation of nicks in the probe DNA followed by extension of the broken ends using a labeled deoxyribonucleotide with help of DNA polymerase I (or Klenow fragment of this enzyme)
- Oligonucleotide labeling involves the use of short random oligonucleotides, which are used as primers for copying the probe in presence of labeled deoxyribonucleotides;
- Riboprobe prepration this involves the synthesis of labeled RNA, using a DNA probe as a template, in presence of a labeled ribonucleotide.

After hybridization with the labeled probe, hybrids are detected by autoradiography.P^{32} has the advantage over other radioisotopes since it has a high specific activity. However, in general, radioisotopes have some disadvantages. They are difficult to handle and expensive to dispose of Detection by autoradiography, while sensitive, may take a long time if there are few counts in the hybrid. Furthermore, radioisotopes have a short half-life (half-life of 14.3 days) and therefore experiments should be completed preferably within one half life

Non-radioactive probes

(a)Biotin labeled probes

Recent advances in nucleic acid technology now offer alternatives to radioactively labeled probes. One procedure that is becoming increasingly popular is biotin labeling of nucleic acids. This system exploits the affinity, which the glycoprotein '**avidin**' has for biotin (vitamin H). Avidin is commonly found in egg white. Biotinylated probes are prepared through nick translation derivatives. Alter hybridization and washing, detection of hybrids is done by a series of cytochemical reactions, which finally give a blue color whose intensity is proportional to the amount of biotin in the hybrid. There are several advantages of using biotinylated probes. For example, these assays employ in advance in bulk and stored at –20°C for repeated uses. Detection technology is that very small probes non- toxic materials, whose half-life is longer. These probes can be prepared by hybrids is much faster than by radioactive probes. A limitation of this technology is that very small probes contain only a small number of biotinylated sites, limiting the intensity of the signal obtained. It has been solved by adding long tails' of biotinylated nucleotides to the probes through enzymatic methods. Sometimes, the probe does not need to be labeled with biotin, but only coupled with a tail.Another disadvantage of biotin labeled probes is that the cytochemical visualization reactions lead to precipitation of insoluble material which cannot be removed and therefore, the filter cannot be reused, while in radiolabeled probes, the filters can be used repeatedly for hybridization with a number of probes one at a time.

(b) Digoxigenin ***labeled probes***

Digoxigenin is another chemical derived from plants and used for non-radioactive labeling of probes. An antibody associated with an enzyme (anti-digoxigenin- alkaline phosphatase conjugate) is used for the detection of the presence of digoxigenin. The probe may be labeled with digoxigenin 11 dUTP supplied with a digoxigenin kit (these kits are available from any commercial firm, e.g. Boehringer Mannheim. The labeled and denatured probe may be used for hybridization with denatured DSA on Southern blots or for in-situ hybridization. After hybridization, the membrane or the slide may be washed and the membrane or the slide is transferred into a detection buffer containing 20 mg/ml of antidigoxigenin fluorescein and 5 % (w/v) BSA (bovine serum albumin). The system is incubated for 1h at 37°C, and then the membrane/slide is washed in detection buffer three times (8 min each at 37° C). Alkaline phosphatase activity can be detected using 0-17 mg/ml BCIP (5-bromo-4 loro-3 indolyl phosphate) and 0.33 mg/ml NBL (nitroblue tetrazolium) as dye substrate.

(c) Alternatives to biotin and digoxigenin labeling

The techniques of non-radioisotopic labeling have been further expanded and new methods have been devised for attaching other ligands (e.g hapten determinant 2, 4-dinitrophenol, arsenate derivatives, etc) to nucleotides without hampering their ability to be incorporated into DNA. These alternatives require binding of attached ligands to specific proteins can monitor many probes simultaneously by using several different ligands since each ligand would yield a different DNA. A chemiluminescent (light emitting) probe system has also been developed in which two different probes complementary to a continuous segment of DNA hybridize adjacent segments of a gene. The first label is a chemiluminescent complex that emits light at a specific wavelength; this emission excites the label molecule on the second probe to emit light at a different wave length which can be detected using a photomultiplier device. This process called non-radioactive energy transfer can occur only if the two probes hybridize correctly and the two labels are close to each other. This system has great fidelity and provides basic technology for a homogeneous assay. In addition, DNA does not need to be immobilized and no washing steps are necessary which may be an additional advantage for large-scale testing.

Amplification of DNA Probe Signals

In recent years, DNA probe assays and immunoassays (using antibodies) have been competing with each other for a variety of purposes. While DNA probes have an advantage of detecting the presence of gene rather than its product, the immunoassay has the advantage of getting amplified target (the gene product) firstly due to transcription (several mRNA molecules can be synthesized on a single copy of the gene) and secondly due to translation (several protein molecules can be produced from same mRNA molecule). The amplified target gives an amplified signal. Therefore, if DNA probes have to compete with monoclonal antibodies for immunoassay, DNA probe signals need to be amplified.

Following devices have been used for the signal amplification:

- Probes have been developed for tRNA since rRNA genes are present in thousands thus amplifying the target (as done by companies like Gene-Probe and Gene-Trak systems in the USA).
- Polymerase chain reaction (PCR) technology (utilizes a thermostable polymerase) is used to generate millions of copies of target DNA sequences, which can then be identified by standard methods.
- Signal generating capacity of the probe can also be increased by any one of the following methods: (i) concentrating more labels at the site of the target molecule by attaching multiple enzyme molecules to each of its DNA probes.

- Amplification probe technology uses multiple secondary probes (each having one or more enzyme molecules) that hybridize to multiple target-specific primary probes. The secondary probe is independent of the target but helps in amplifying the signal. These approaches have been multiple enzymes to each of the multiple probes.

Techniques used in Molecular Probing

Separation of DNA fragments using agarose or polyacrylamide gelelectrophoresis

When genomic DNA, extracted from any tissue of a plant or animal species, is digested with a restriction enzyme, it is cleaved into segments. The segments of different sizes can be separated through **gel electrophoresis** before a molecular probe is used to detect the segments which have sequences similar to those in the probe. Gel electrophoresis involves the movement of fragments or molecules under a high voltage electric current. The mixture of DNA fragments is loaded in a well created on one edge of the gel. The gel may be a cylinder or a slab (usually a slab for cloning expt.), about 10 cm long and 0-5 cm thick. The rate of movement of fragments is inversely correlated with the size of fragments molecules so that heavier fragments will remain closer to the site of loading and the lighter fragments will move away. Fragments of different sizes will appear as bands on the gel and can be examined or isolated for further study. More often agarose gels are used, but for the separation of fragments differing by few base pairs, polyacrylamide gels are used. Polyacrylamide gels are more commonly used for DNA sequencing experiments

Separation of large DNA molecule molecules using (PFGE)

The technique of gel electrophoresis is used for the separation of DNA molecules of different sizes. However, DNA molecules of large size not be handled in this technique. A new technique called separation of large DNA molecules (whole chromosomes) using PFGE. The technique of gel electrophoresis is used for separation of DNA pulsed-field gel electrophoresis (PFGE) has been used, for separation large-sized DNA molecules, sometimes representing whole chromosomes. Using this technique, the separation of DNA molecules belonging to each of the individual chromosomes of yeast has become a possible technique, short pulses of electricity are used in two different directions and DNA is embedded and used in the form of agarose plugs to avoid fragmentation of large DNA molecules. Using the technique: PFGE and a more refined technique CHEFE (countor clamped homogeneous electric field electrophoresis), the genome of several fungi could be resolved into chromosomal bands and used for the mapping of DNA.

Dots and slot blots

Another variety of Southern or Northern is dot blots or slot blots. In dot blots, cloned or pure extracted DNAs to be tested are spotted adjacent to each other on a nitrocellulose membrane. "DNA blots thus produced are immobilized and denatured so that they are bound on the membrane as single-stranded DNA blots. The membrane is then hybridized with a radioactively labeled probe (DNA, or RNA). The dot representing sequence related to the probe will light up autoradiography. The intensity of the dot will indicate the relative concentration of sequences related to the probe in the DNA sample used for a particular dot. No steps involving digestion with enzyme concentration of a sequence related to the probe in the DNA electrophoresis or transfer from the gel to the membrane are needed in this technique". Therefore, it is considered much more convenient, when o restriction sites are needed to be studied. Sometimes, an apparatus is used for placing spots on the membrane through slots made in this equipment. The spots made thus are in form of oblong slots rather than round blots. These slots are used just like dot blots and are described as slot blots.

Applications of Molecular Probes

Molecular probes can be used for both basic and applied studies in molecular biology and biotechnology.

Use of molecular probes in restriction fragment length polymorphisms RFLPs) and related analyses

(a) What are RFLP

When genomic DNAs from each of several individuals belonging to one or more species are digested separately with restriction enzymes, shown with electrophoresed, transfer on blotted a membrane, and tag with probed with a labeled DNA clone, polymorphism the hybridization pattern is sometimes revealed and attributed to sequence differences between the individuals of different species. Such variation has been termed as "restriction fragment length polymorphism" (RFLP).

(b) Detection of RFLPs

RFLP can be demonstrated using the following steps:

(i) Extract and purify DNA from several individuals which may differ in some respects among themselves.

(ii) Digest with a restriction endonuclease.

(iii) Subject each "DNA digest to gel electrophoresis separately but on the same gel slab.

(iv) The DNA fragments of different lengths resulting from digestion can be separated by gel electrophoresis, but the differences in the distribution patterns of fragments in several DNA samples due to digestion by several endonucleases cannot be detected directly. This is because the number of fragments is large and the range in size is rather continuous, such that they form a continuous smear on the gel. Electrophoresis is, therefore, followed b the following steps.

(v) Fragments are transferred from the gel to nitrocellulose filters using the technique of Southern blotting.

(vi) The filters with small fragments, fully or partially homologous to the probe, can be detected autoradiography after hybridization.

Since repetitive DNA may hybridize with many fragments, they will give a smear on autoradiography. Therefore, unique DNA sequences are generally used as probes for detecting RFLPs. The DNA fragments hybridizing with the probe can be visualized by autoradiography and will be called RFLPs at the phenotypic level. In results of a hypothetical experiment are shown where RFLPs are detected by using four plants (which differ by point mutation as well as by insertions) and three enzymes individually and in combination.

(c) RFLPs for evolutionary studies

- The restriction fragment length polymorphisms (RFLPs) can be studied in a set of related species, using a random or a specific DNA probe.

- The similarities and differences can be used to infer phylogenetic relationships. This has actually been done in a number of cases, both in plants and in animals.

(d) RFLP maps and linkage of RFLP with specific genes

- RFLP has been used to prepare chromosome maps in humans, mice, fruit fly and Mendelian markers for genetic mapping is sometimes limited due to the non-availability of mutants.
- The list of markers can be extended by using molecular markers, which are examined in the form of RFLP plants including maize, tomato, lettuce and rice.
- The use of a mendelian marker for genetic mapping is sometimes limited due to the non-availability of mutants.
- The list of markers can be extended by using molecular markers, which may be examined in the form of RFLP.
- "Once a large number of RFLP are available in a species, the parents, F, and F2's can be used to study their inheritance and linkage relationship and genetic linkage maps can be prepared.
- They have been assigned to specific chromosomes using monosomic in maize and efforts are being made to relate them with morphological and economic traits so that they can be used for practical plant breeding.
- A difference in restriction maps between two individuals or an RFLP can be used as a genetic marker in much the same way as any other phenotypic marker. The polymorphism at the molecular level may be due to the presence or absence of a restriction site.
- Shows how the presence of an additional site will produce two fragments and its absence will produce only one fragment.
- Although, there may be the linkage between a molecular marker and a phenotypic trait, the change in the molecular map (restriction map) may not affect the phenotype.
- In order to study linkage relation between a molecular marker and present in the DNA shown parents that differ morphologically as well as for molecular markers by we may examine the restriction fragment patterns and the phenotypes of making crosses between two such parents, we can estimate the recombination frequencies

between molecular markers and phenotypic markers using" a test cross.

- We used a hypothetical example showing the linkage of red and white eyes with different molecular markers in Drosophila.
- "It can be seen that as a result of recombination, in test cross, 15 % progeny shows an association of red-eye with a molecular marker associated the white eye in the parent, and vice versa.
- Therefore, it is concluded that the restriction marker is 15 map units away from the marker eye color although it has no causal relation with eye color.
- Such a linkage between the molecular marker and phenotypic marker allows the identification of genetic loci at the molecular level.
- A tight linkage of such a molecular n with discase resistance may also be recommended as a tool for identification of disease-resistant cultivars in many of our crop plants and will be increasingly utilized in the future Molecular markers are also being used for the selection of plants at the seedling stage for plant breeding purposes".

Use of molecular probes in molecular cytogenetics

(a) Isolation of gene or related sequence

This can be facilitated through the use of a probe carrying a known sequence.

For this purpose sequences may be hybridized following any one of the following techniques:

- Southern blots can be prepared (from genomic DNA or cDNA) to isolate DNA corresponding to a band that hybridizes with the probe.
- This DNA can then be used for cloning experiments to get a partial genomic library or cDNA library.
- The genomic or cDNA library can be screened through colony hybridization or plaque hybridization to identify and isolate the clone carrying the desired sequence.

- Sometimes it is not used for complete genomic/cDNA libraries can be prepared and used in the step.
- The identified clone should carry the gene or sequence of interest, which can be multiplied and the sequence or gene retrieved according to standard procedures.

In situ hybridization (ISH) for the location of sequences chromosomes

- A molecular probe can also be used for in situ hybridization "where chromosomes in mitotic or meiotic preparations are used for hybridization with the probe (without extracting DNA).
- In a number of cases, chromosomes in the wheat background have been identified using this technique.
- Sat- chromosomes with NORs (nucleolar -organizing regions) are recognized using probes for ribosomal DNA as done in many In wheat and can be identified using a D genome specific probe.
- A large (wheat, barley, etc). Telomeres have been identified in human eukaryotes using a telomeric sequence as a molecular probe, showing that all chromosomes carry the same sequence.
- In genome telomeric related species, chromosomes of genome-specific probe (pASI separated from Aegilops be cited) where the use of molecular probe hybridization served a very useful purpose.

Use of molecular probes in DNA fingerprinting

- With the help of probes prepared from human minisatellite DNA dispersed throughout the human genome, it has become possible to identify individual humans, in the possibility of any two humans (except identical twins) showing the same pattern of fingerprinting is remote". This aspect is already playing an important role in identifying criminals like murderers and rapists.

Construction and screening of genomic and cDNA libraries

- In order to isolate one or more related genes from a genome, we like to prepare a mixture of clones each carrying DNA derived either from the genomic DNA or from cDNA (derived from the mRNA isolated from a specific metabolically active tissue of an organism).

- This mixture may contain thousands of clones, which when derived directly from the genomic DNA are collectively called a genomic library.
- Similarly, when these clones are derived from cDNA, they are collectively called a cDNA library construction.

Genomic library by shotgun experiment

- Cloning an entire genome in the form of a library of random genomic clones (without identifying them) is often called a shotgun experiment.
- In this experiment, genomic DNA is extracted, broken into fragments of reasonable size by a restriction endonuclease and then inserted into a cloning vector to generate a population of chimeric vector molecules.
- A set of fragments cloned in this manner is called a genome library.
- Once such a library is available, then clones can be perpetuated indefinitely in a plasmid vector and retrieved whenever needed for a variety of poses, including identification and isolation of a gene, when a specific probe is available.
- Genomic libraries can be prepared by using a number of restriction endonucleases, one at a time, so that fragments of varying sizes having different places of the genome will be available.
- However, this may to cuts at inconvenient places, including sites within a gene, so that is having complete genes will be difficult to obtain.
- In order to come this difficulty, we use the following strategy in the shotgun experiment.
- We use restriction endonucleases, which have short (4 bp) recognition sequences, so that such a sequence may be frequently distributed.
- Conditions are used which give only partial digests, so that a particular restriction site is only occasionally cleaved, and long fragments without having any breaks on recognition sites available within a gene can be easily obtained.

- This technique of shotgun experiment leads to the construction of a random genomic library, in which all fragments have the same fragment ends thus helping retrieval of a fragment from the vector with the help of cuts of the same enzyme.
- The number of fragments representing every sequence of the genome increases with genome size.
- For instance, for a probability level of 99 that all sequences are present in our library of a species, we may need 1,500 cloned fragments for *E.coli*, 4,600 for yeast, 48,000 for *Drosophila melanogaster* and 8,00,000 for a mammal like human being, Libraries reaching these desired limits have been prepared in all these cases.

cDNA library from mRNAs

- Complementary DNA (cDNA) libraries can also be prepared by isolating. As from tissues which are actively synthesizing proteins, like roots in plants, ovaries or reticulocytes in mammals, etc.
- The mRNA for copying it into cDNA through the use of reverse transcriptase. The mRNA from leaves is used for copying it into cDNA through the use of re cDNA molecule can be made double-stranded, and cloned.
- However cDNA clones will differ from genomic clones in lacking the introns present in split genes, but the advantage of being capable to be expressed bacteria, which do not have the machinery to process HnRNA (obtained cDNA clones from split genes) mRNA.

Colony (plaque) hybridization for the screening of libraries

- Once a genomic library or cDNA library is available, we may use it for the isolation of gene sequences.
- This can be achieved by colony hybridization technique. In this technique, bacterial transformed chimeric vectors are grown into colonies, which are lysed on nitrocellulose filters.
- Their DNA is denatured in situ and fixed on the filter, which is hybridized with a radioactively labeled probe carrying sequence related to the gene to be isolated (usually a cloned cDNA for the screening of a genomic library).

- Colonies carrying this sequence will be identified by dark spots after autoradiography so that the original chimeric vector carrying the desired gene sequence can be recovered from one or more colonics in the original master plate and used for further experiments. This technique is described as colony hybridization.
- It is possible that the probe may identify more than one clone or that a gene is fragmented by the library.
- In such a case, one needs to reconstruct the desired sequence using several overlapping sequences available in the library.
- This is a very routine exercise whenever we like to isolate specific DNA sequences from the genome of a species, or from cDNA derived from mRNA of a specific tissue of species.
- Sometimes the library may be available not in the form of bacteria is formed with chimeric DNA molecules, but in the form of chimeric particles carrying the cloned segments.
- In such a situation, a bacterial is infected with a mixture of chimeric phage particles (i.e. the library) and a large number of plaques develop overnight.
- These plaques can be treated just like the colonies in colony hybridization to identify and isolate the chimeric phage particle carrying the gene of interest. This technique is then described as plaque hybridization.

Chromosome walking and characterization of chromosome segments

- When a probe is used for the identification of a gene sequence in a genomic library, the probe may hybridize with a number of clones, each carrying a part of a large gene fragmented during the preparation of the genome library.
- If we obtained partial digests (by digesting the DNA only partially) from the genome, different genomes (from the large number of cells) may give fragments that have overlapping sequences, because sites cleaved different genomes of the same organism, will differ being random.
- Since one of these fragments may have an entire sequence represented in the probe, overlapping, sequences may be used to construct the original genome sequence.

- Identification of fragments with overlapping sequences may be a key to the reconstruction or characterization of large chromosome regions.
- This is achieved by the technique popularly described as chromosome walking.
- The technique of chromosome walking involves the following
- From the genomic library select a clone of interest (identified by a probe) and subclone a small fragment from one end of the clone (there is a technique available to subclone a fragment form the end).
- The subcloned fragment of the selected clone may be hybridized with other clones in the library and a second clone hybridizing with the subclone of the first clone is identified due to the presence of overlapping region.
- the end of the second clone is then subcloned and used for hybridization with other clones to identify, a third clone having overlapping regions with the subcloned end of the second clone.
  - The third clone identified as above is also sub cloned and hybridized with clones in the same manner and the procedure may be continued.
- Restriction map of each selected clone may be prepared and compared to know the region of overlapping, so that identification of few overlapping restriction sites will amount to walking along the chromosome, or along a long in chromosome segment.
- Regions of chromosomes approaching 1000 kb have been mapped following the above technique.
- Restriction maps of entire chromosomes can be prepared in this manner following the technique of chromosome walking.

Reverse genetics and chromosome jumping (or hopping libraries)

In most cases, we identify the genes by their protein products, which help in the isolation and cloning of a specific gene. However, for many human disorders which are each controlled by a single gene (e.g. Huntington disease, cystic fibrosis, Duchene muscular dystrophy, etc.), the gene product is not known despite intensive investigations. Such genes are identified primarily by their map position. Once the gene is cloned, it can be utilized for getting information about the encoded protein, the gene mutation and the metabolic defect that may be associated with the disease. Since the information about the is coming in the reverse order, this is popularly described his reverse genetics approach involves the following two steps: (i) locate the gene to a particular chromosome through RFLP linkage analysis; (ii) map the position of the gene with respect to molecular marker, both through recombinational analysis (in terms of recombination in cM=centiMorgan units) and by physical mapping (in terms of kilobase pairs) with pulsed field gel electrophoresis (PFGE) and preparation of restriction map. Even if the distance is 1 cM, which is the limit of resolution, this will be equivalent to 1,000 kb. Therefore, if the segment carrying the gene is to be cloned using the linked molecular marker such a segment will be too large to be cloned.

In order to bring the molecular marker close to the gene of interest, 'chromosome jumping' approach has been utilized. The technique of 'chromosome jumping' is based on the following steps depending upon the distance between the gene and the marker, decide about the distance of jumps' or 'hopsize' (eg 100 kb or 200 kb): (i) genomic DNA molecules in the range of selected size (say 80 kb-130 kb in case of 100 kb 'hopsize are 16)-240 kb for 'hopsize' of 200 kb) are selected through pulsed-field gel electrophoresis (ii) for circularization of DNA segments, ligation been two ends of each long linear DNA molecule was allowed using T4 ligase in the presence of $supF^+$ (iv) DNA circles obtained in (ii) above are digested with EcoRI;(v) the vector iCh3A A lac, an amber mutated phage vector (supF"is also cut with *EcoRI* and used for cloning mal DNA fragments representing the junctions of the circularized genomic molecules and carrying sup F (vi) the cloned DNA fragments and (v) above represent the jumping library, which can be plated on the host and screened through the technique of plaque hybridization described earlier. The above technique of chromosome jumping will help to narrow the gene and available molecular markers. After several cycles the jumping followed by cloning the regions that are close to approaching very close to the desired gene and clone it.

It is thus obvious that in the field of 'reverse genetics' using the technique of chromosome jumping, it will characterize genes, whose gene products are unknown. The gene for cystic fibrosis has recently been isolated using this technique. It could also be shown that 'cystic fibrosis' disease is caused by a single base substitution in this gene. This technique of 'reverse genetics' will become a very active area of research, particularly in human genetics in the coming years.The term 'reverse genetics' has however been redefined recently. Classical genetics, sometimes described as forward genetics' involves a study that starts from the phenotype, and follows through the identification of the gene and finally concludes with the isolation and sequencing of the gene. In view of this, if 'reverse genetics' has to use steps in the reverse order for a genetic study, then it should start with DNA molecules or clones with unknown effects and should conclude with the determination of the phenotype which it controls. This can be achieved by preparing random genomic DNA clones, which can be subsequently utilized for a genetic study in the reverse order. This will then be described as 'reverse genetics'. This definition of 'reverse genetics' differs from the earlier conventional meaning of this term, where reverse genetics was used to describe cases of genes with unknown products, which are isolated (with the help of phenotype and its linkage with molecular markers).

27. ISOLATION, SEQUENCING AND SYNTHESIS OF GENE: DIFFERENT METHODS OF GENE ISOLATION

Introduction

Molecular manipulation of specific genes are of research in biotechnology. These genes are isolated or artificially synthesized before they are manipulated and used or transformation heading to the production of transgenic animals and plants that are discussed respectively, genes are also sometimes sequenced for a better understanding of its structure that needs to be manipulated. We therefore, discuss the gene as one of the very popular areas that should be either technology available for isolation, sequencing and synthesis of specific genes.

Isolation of Genes

The significant progress has been made in the techniques for (i) isolation of a variety of genes, including those for (ribosomal RNA; (ii) specific protein products; (iii) phenotypic traits with unknown product and those for (iv) regulatory functions e.g. promoter genes etc. Different techniques have been used for the isolation of the se different types of genes. Early attempts for isolation of ribosomal RNA and proteins As we know, ribosomes consist of ribosomal RNA (rRNA) and proteins ribosomal RNA (tRNA) can be of four sizes namely, 28S, 16S, 58S and 5S (S stands for Svedberg units, a measure of speed with which a molecule sediments on centrifugation). This ribosomal RNA makes 80 percent of cellular RNA and is synthesized on ribosomal genes, which could be isolated. Isolation of ribosomal genes was considered convenient due to their following three characteristic features (i) availability of homogeneous rRNA; (ii) differences between ribosomal RNA genes and other gene due to relatively high G + C content in rRNA (rRNA has 45-60 % G + C while the remaining RNA has only 40 % G + C); (iii) ribosomal genes are present in multiple copies, their number within several thousand. The ribosomal RNA genes were isolated for the first time (in 1965) in an amphibian named *Xenopus*, by Hugh Wallace and Max L. Birnstiel, working at the University of Edinburgh (UK). Following steps were involved in this isolation : (i) RNA was isolated from ribosomes of Xenopus and made radioactively labeled, due to its replication in a medium containing tritiated uridine; (ii) ribosomal DNA was isolated by density gradient centrifugation followed by its denaturation (G +C content of rDNA differs from that of bulk DNA and helps in separation by centrifugation); (ii) single-stranded DNA was fixed on filter paper; (iv) labeled rRNA was added on filter paper carrying single-stranded DNA;(v) DNA-RNA hybridization was allowed to take place (vi) excess labeled RNA was washed (vi) radioactivity was measured and duplex hybrids isolated, which on denaturation, gave single-stranded DNA, which could be made double-stranded. Isolation of ribosomal DNA allowed the first characterization of ribosomal genes, which was followed by the isolation of 5S genes by Donald Brown (August, 1973).

Isolation of Genes Coding for Specific Proteins

Isolation of genes for specific proteins became possible only after the discovery of the reverse transcriptase enzyme in 1970. This enzyme can be easily used for the synthesis of copy DNA or complementary DNA (cDNA) from mRNA. This complementary DNA can then be used for the isolation of the corresponding gene from genomic DNA. It is therefore, obvious that for the isolation of a specific gene, techniques should first be available for the isolation of specific mRNA. For this purpose, antibodies are produced against a specific protein for which the gene is to be isolated. Therefore, isolation of a gene coding for a specific protein involves the following basic steps are (i) purification of the protein product of gene; (ii) production of antibodies against this protein product by immunizing animals like rabbit or mouse; (iii) precipitation of polysomes engaged in synthesizing specific protein with the help of the antibody raised in step (i): (iv) isolation and purification of mRNA from the polysome fraction; (v) using mRNA for synthesizing cDNA with the help reverse transcriptase; (vi) cloning of cDNA thus synthesized for preparation of a cDNA library, (vi) immunological and electrophoretic analysis translation products of cDNA clones to identify specific cDNA meant for the specific protein in mind; (vii) use of the specific DNA probes selected in step (vii) the identification and isolation of the gene from genomic DNA through screening a complete or partial genomic library. In step (vi) above for the preparation of cDNA library, cDNA may be cloned in expression vectors (vectors in which gene can be expressed due to the presence of promoter sequences for RNA polymerase enzyme) like agtll, which can accept gene insertions into B galactosidase gene.

The chimeric vector will then produce hybrid protein is correctly expressed. The hybrid protein can be identified using an antibody as above. Plaques belonging to the chimeric vector carrying the desired gene can also be identified by the reaction of antibodies attached to a radioactive protein.This procedure was used to clone a number of genes including a gene regulating zein synthesis in maize. The cDNA clones may also be used directly for gene manipulation I transformation experiments. However, when a cDNA clone is used, it be described as a synthesis of the gene rather than as isolation of gene, cDNA has been artificially synthesized, and not isolated from the organism. The above technique for isolation of gene has now been successfully utilized both in plants and animals. However, the first genes isolated were those which existed in multiple copies and were expressed at high levels in specific tissues, e.g ovalbumin gene in chicks, globin and immunoglobulin genes in mouse, genes for storage proteins in cereals and legumes, amylase genes in barley, actin genes in some legumes, etc.

Isolation of Genes Which Are Tissue Specific in Expression

It is much easier to isolate genes that are expressed in specific tissues. For instance, genes for storage proteins are expressed only in developing seeds, oval albumin gene is expressed in oviduct or globin gene is expressed. Such genes can be easily isolated because mRNA extracted of an erythrocyte from these specific tissues will either exclusively belong to the gene of interest or it will be rich in this species of mRNA. Other mRNA molecules in minor quantities can be eliminated since these can be identified through their isolation from tissues where this gene is silent. This strategy which was actually followed by isolation and cloning carrot genes expressed during the development of the somatic embryos and also for the isolation of several genes for storage proteins in crops is outlined.

Isolation of genes using DNA or RNA Probes

Specific molecular probes (whether DNA or RNA probes) can be used for the isolation of specific genes. These probes may be available either from another species for the same gene or may be artificially synthesized using a part of the amino acid sequence of the protein product of the gene.

Use of Heterologous Probes

The probes obtained from one species are heterologous probes. These are effective in identifying gene heterologous probes have been found to be effective in identifying blots. Southern clones during colony hybridization or plaque hybridization or on Southern blots- synthase has been isolated from *Anuirhinum majus* and *Petunia hybrida* heterologous cDNA probe from parsley. Similarly, heterologous probes from maize were used for isolation of barley genes be used with cDNA library and not with the genomic library since vector (e.g pGEM-4 blue) can also be used for getting 89 waxy gene) and Al (aleurone gene). Heterologous probes should be the latter case, unrelated genes or pseudogenes (which do not isolated and cloned. These heterologous clones if available robes, which have been found to be more sensitive and efficient.

Use of cDNA or synthetic probes

We have already discussed cDNA probes above. However, if protein, purified using the technique of two -dimensional gel electrophoresis, is used for microgram of 5-15 consecutive amino acids, this information can be used for the synthesis of oligonucleotides (using automated DNA synthesizers).These oligonucleotides may then be directly utilized for the screening of cDNA genomic libraries for isolation of specific genes. These can also be used for designing primers for the amplification of a gene in DNA extracted from tissue. A 32 kilodalton (KD) glycoprotein (isolated from style) associated with the S2 incompatibility allele of *Nicotiana alaia*, was used for the synthesis of an oligonucleotide leading to the isolation of a related gene.

Isolation of Genes Coding for Unknown Products

In some cases, we are interested in isolation of a gene whose phenotypic effect is known, but the gene product has not been identified or cannot be isolated. Such genes include those for morphological traits like dormancy photoperiodicity, disease resistance, etc. This area of research, in which genetics is studied by isolating the gene first without knowing the gene product, is often described as reverse genetics. The method used for the isolation of these genes is different from those used for genes coding for known proteins.

Sequencing of the gene or a DNA segment

Once a gene or a DNA fragment has been cloned, its further study involves DNA sequencing. The technique of DNA sequencing was very laborious till 1975 when a breakthrough was made in DNA sequencing methods: two different methods are now routinely used for the determination of DNA sequences.

Maxam and Gilbert's chemical degradation method

In this method, which is illustrated, the following steps are involved:

(i) Label the 3'ends of DNA with ^{32}P.

(ii) Separate two strands, both labeled at 3'ends.

(iii) Divide the mixture into four samples, each treated with a different reagent having the property of destroying either only G, or only C, or 'A and. G' or T and C. The concentration of reagent is so adjusted - that 50 % of the target base is destroyed so that fragments of different sizes having ^{32}P are produced.

(iv) Electrophoresis each of the four samples in four different lanes of the gel.

(v) Autoradiograph the gel and determine the sequence from positions of bands in four lanes.

Sanger's Dideoxynucleotide Synthetic Method

Fred Sanger (who won Nobel Prize twice) had initially developed a method for DNA sequencing, which utilized DNA polymerase to extend DNA chain length. This was termed plus-minus method. Subsequently, he developed a more powerful method, utilizing single-stranded DNA as a template for DNA synthesis, in which 2',3' dideoxynucleotides were incorporated leading to termination of DNA synthesis. These dideoxynucleotides are used as triphosphates (ddNTP) and can be incorporated in a growing chain, but they terminate synthesis since they cannot form a phosphodiester bond with the next incoming deoxynucleotide triphosphate (dNTP). Following steps are involved in Sanger's dideoxy method for DNA containing single-stranded DNA sample (cloned in M13 phage) to be sequenced, all four dNTPs (radioactively labeled) and an enzyme for sequencing.

(i) Four reaction tubes are set up, each containing single-stranded DNA sample to be sequenced, all four dNTPs and an enzyme for DNA synthesis. Each tube also contains a small amount (much smaller amount relative to four dNTPs) of one of the four dNTP, so that four tubes have each a different ddNTP, bringing about at a specific base- adenine (A), cytosine (C), guanine (G) and thymine (T).

(ii) The fragments, generated by random incorporation of d-NTP leading to termination of the reaction, are then separated by electrophoresis on a high-resolution polyacrylamide gel. This is done for all the four reaction mixtures on adjoining lanes in the gel.

(iii) The gel is used for autoradiography so that the position of different bands in each lane can be visualized.

(iv) The bands on the autoradiogram can be used for getting the DNA sequence. A variant of the above dideoxy method was later developed, which has allowed the production of automatic sequencers.

In this new approach, a different fluorescent dye is tagged to the oligonucleotide primer in each of the four reaction tubes (blue for A, red for C, etc.). The four reaction mixtures are pooled and electrophoresed together in a single polyacrylamide gel tube. A high sensitivity fluorescence detector, placed near the bottom of the tube, measures the amount of each fluorophore as a function of time. The sequence is determined from the temporal order of peaks corresponding to four different dyes.

Direct DNA Sequencing using PCR (also Called Ligation-Mediated PCR LMPCR)

Polymerase chain reaction (PCR) has also been used for sequencing the amplified DNA produçt. This method of DNA sequencing is faster and more reliable and can utilize either the whole genomic DNA or cloned fragments for sequencing a particular DNA segment. The DNA sequencing using PCR involves two steps: (i) generation of sequencing template (double-stranded or single-stranded) using PCR and (ii) with the PCR products either with the thermolabile DNA polymerase or with the method or thermostable ***Tag***

DNA polymerase. Thus the DNA sequencing using PCR eliminates the need for cloning the DNA in single-stranded DNA phage vector i.e. M13. Although double-stranded DNA product difficulty due to can be utilized for sequencing, this may cause difficulty due to reassociation after denaturation thus preventing the sequencing from annealing to its complementary sequence to allow an extension. To reduce this problem, either a variant of the standard method for sequencing double-stranded DNA is employed or single-stranded DNA templates are produced following asymmetric PCR earlier. A number of thermolabile DNA polymerases have been used for sequencing of *in vitro* amplified DNA Alternatively, ***Tag*** polymerase (thermostable enzyme) used for PCR, can also be used for the sequencing reaction. In either case, Sanger's synthetic method involving incorporation of dideoxynucleotide (ddNTP) for chain termination, is used for sequencing. However, Maxam and Gilbert's chemical degradation method can also be used. In Sanger's method, as usual, four mixtures are prepared, each using one of the four ddNTPs. The sequencing primer is labeled with ^{32}P and the mixtures with amplified DNA, ***Taq*** polymerase and appropriate buffer are incubated at 70°C for 5 min. The reaction is stopped by the addition of formamide stop solution in each tube and mixtures are run on polyacrylamide sequencing gel to obtain the ladders, which can be read by a computer or manually. This method using PCR helped in the automation of DNA sequencing.

Synthesis of Genes

There are two approaches available for the synthesis of genes:

(i) When the detailed structure of a gene is available, this gene can be synthesized by a purely chemical method as done by HG Khorana for the synthesis of the gene for a tRNA (reported for the first time in 1970).

(ii) If the detailed nucleotide sequence of the gene is not available, one may utilize the 'RNA directed DNA polymerase' enzyme for the synthesis of the gene in question in the form of complementary DNA (cDNA) from the mRNA of the gene isolated in its pure form. As discussed above, cDNA can also be used for the isolation of a gene either

(i) From DNA extracted from living cells utilizing the technique of Southern blot hybridization.

(ii) A genomic library.

Chemical synthesis of tRNA genes.

As outlined above, before one can start the chemical synthesis of a gene, the structure of the gene should be known. The structure of a gene earlier could not be worked out by direct chemical analysis; since there were no means for isolating a gene. The structure of the gene could, therefore, be inferred only from its product. For instance, if a gene is responsible for giving rise to a polypeptide chain and the structure of this chain is known, then from the genetic code dictionary, the structure of the gene could be easily inferred. Such genes were initially considered to be too long to be synthesized because an average gene contains about 1,500 base pairs. On the other hand, since tRNA molecules are fairly small in size (about 80 nucleotides), a gene responsible for giving rise to a tRNA molecule was within the reach for synthesis.

Synthesis of the gene for yeast tRNA

As readers may know, structures of a large number of tRNA molecules are now known. The first RNA whose structure could be known was yeast alanyl tRNA. **R.W. Holley** who died in early 1993 and his coworkers (in 1965) gave the detailed structure of yeast alanyl tRNA. This information was immediately used by Khorana and his coworkers to deduce the structure of the gene for yeast alanyl tRNA. The structure of this gene would obviously be such that one of the two strands of DNA (gene) would be complementary to base sequence of yeast alanyl tRNA. The other strand would then automatically have the same sequence as in tRNA except that in place of C of tRNA there would be thymine in DNA. Synthesis of such a long chain (77 base pairs) of double-stranded DNA was rather difficult in 1965 (not in the 1990s) but Khorana and coworkers had extensive experience of synthesizing DNA of known bate sequences. It was decided that such a long chain could not be synthesized by adding a single base each time. Therefore, it was decided that small oligodeoxyribonucleotides ranging in length from 5 to 20 nucleotides should first be synthesized. These segments would be single-stranded and would cover the whole length of both strands of DNA. These would then be joined to form double-stranded DNA, 77 nucleotide pairs long in the actual synthesis of yeast alanyl tRNA gene.

(a) Synthesis of oligonucleotides

Fifteen oligonucleotides ranging from pentanucleotide (5 bases) to an icosanucleotide (20 bases) were synthesized. Synthesis was conducted through condensation between hydroxyl group a 3' position of one nucleotide and phosphate group at 5' position d the second nucleotide. In order to bring about condensation, all other functional groups, not taking part in condensation were protected using specific protective groups. After protecting the groups, the reaction between a nucleotide with a protected 5' end and another nucleotide with a protected end proceed. Then, condensation was consisting also brought about between groups of two, three or four nucleotides.

(b) Synthesis of three duplex fragments

With the help of 15 nucleotides (single-stranded), three duplex fragments, each with a stranded end, were prepared. These three fragments consisted of the following: The first 20 nucleotides, with 17-20 nucleotides single-stranded, this fragment was called 'A'. The nucleotide residue 17-50 with single-stranded regions 17- 20 and 46-50 segments called B. The nucleotide residues 45-77 with single-stranded 17-20 as region 46-50; this segment was called 'C'

(c) Synthesis of the gene from three duplex fragments

The three fragments 'A','B' ,and C" were joined to give complete gene in each of the two following ways; (i) In one scheme, 'A' was joined to 'B' taking advantage of the overlap in residues 17-20; C was then added, with the overlap in the region 46-50 residues. The complete double stranded DNA with 77 base pairs representing the gene was thus ready. (ii) In the other scheme, B was first added to C and to this A was added in the end to obtain the complete gene. This gene, which was synthesized in 1970, was hopefully used for subsequent work since this gene could replicate and make its own copies.

(d) Synthesis of gene for a true precursor tRNA.

Before **Khorana** could complete the synthesis of the gene for yeast alanyl tRNA in 1970, it became obvious that tRNA was not the direct product of transcription. In its place, a precursor molecule is first synthesized which subsequently, after losing segments of RNA by cleavage, gives rise to tRNA. Noticeably, so, the actual gene for yeast alanyl tRNA will be longer than the DNA duplex synthesized by Khorana. In view of this Khorana subsequently initiated the synthesis of a gene for *E. coli* tyrosine suppressor tRNA precursor. DNA duplex which will give rise to this tRNA precursor was synthesized in the form of 26 small oligonucleotide segments. These were then arranged into six DNA duplex fragments having sin stranded ends. These six fragments gave rise to the presumed gene for *E. coli* tyrosine suppressor tRNA precursor. This gene, however, still promoter region and other sequences essential for processing. Later, in 1979, Khorana reported completion of the total synthesis of a biologically functional tyrosine suppressor transfer RNA gene to carry all regulatory sequences. This gene was 207 base pairs long and include the following: (i) a 51 base pairs long DNA promoter region;(ii) a 126 base pairs long DNA corresponding to the precursor tRNA and (ii) a 25 base pairs long duplex DNA corresponding to 16 base pairs adjoining the CCA end of tRNA and the remainder, a modified sequence including EcoRI endonuclease specific sequence. The complete synthetic gene was cloned in the vector bacteriophage (lambda virus) by the gene cloning method discussed. On the transformation of *E. coli*, the phage could multiply with the cloned gene.

Gene Synthesis Machines

Methods for the synthesis of DNA molecules of known base sequences have improved dramatically during the 1980s and 1990s. Therefore, genes can now be synthesized rapidly and in high yields. For instance, a gene for RNA took 20 many years effort during 1965- 70 (described earlier in this chapter) but the same gene can now be synthesized within a matter of few days with the help of gene synthesis Machines developed in the 1980's. The key innovations which made the availability of gene synthesis described the following:

(i) Development of silica-based supports, which are e and provide support for solid-phase synthesis of DNA chains.

(ii) Development of stable protonated deoxyribonucleoside phoramidites as synthons, which are remarkably stable towards hydrolysis and thus are ideal for DNA synthesis.

These two procedures can be used to prepare DNA molecules more than 100 nucleotides long within a few days. Different steps are involved in solid-phase synthesis of a DNA chain. Several versions of gene machines are now available. These machines, under the control of a microprocessor, synthesize specific short sequences of single-stranded DNA automatically.

The desired sequence is entered on a keyboard and the microprocessor automatically opens the valves of the containers of successive nucleotides, reagents and solvents needed at each step, into a synthesizer column, which is packed with tiny silica beads. These beads provide support on which DNA molecules are assembled.

Synthesis of genes from mRNA

H.Temin and D. Baltimore, in 1970 discovered the presence of "RNA directed DNA polymerase enzyme" which has the ability to synthesize DNA on RNA templates. This enabled molecular biologists to synthesize complementary DNA (cDNA) using mRNA as a template. If mRNA transcribed from a specific gene is made available in purified form, the elementary DNA (cDNA) synthesized with its help will represent the synthesized gene. Synthesized gene is a faithful copy of the native gene will depend on the fidelity of copying mRNA and also on the stability of DNA thus synthesized. Furthermore, since the mRNA of a gene does not have the complete transcript of the gene *in vivo*, the synthesized gene will be smaller than the gene *in vivo*. By copying eukaryotic purified mRNA, several genes have been artificially synthesized. The most important of these genes are the genes for sea urchin histone proteins, ovalbumin gene in chicken and globin genes in mammals. The gene synthesized as cDNA from a B globin mRNA, has been inserted into a plasmid in order to study its behavior. As earlier pointed out, this gene will lack any additional regulatory sequences absent in mRNA but present in a globin gene *in vivo*.This type of gene will also lack intron sequences found in eukaryotic split genes. Another factor causing a difference between the synthesized gene and the gene *in vivo* will depend on the sincerity with which the enzyme will synthesize cDNA because in many cases cDNA was found to be smaller than the mRNA itself.

28. POLYMERASE CHAIN REACTION (PCR): BASIC PRINCIPLES, MODIFICATIONS, APPLICATIONS, MODIFYING GENES: SITE-DIRECTED MUTAGENESIS

Introduction

In polymerase chain reaction (PCR) provides a simple and ingenious method for exponentially amplification of specific DNA sequences by *in vitro* DNA synthesis. This technique was developed by **Kary Mullis** at Cetus Corporation in Emery Ville, California in 1985. **Kary Mullis** shared the Nobel Prize for chemistry in 1993. This technique has made it possible to synthesize large quantities of a DNA fragment without its cloning. It is ideally suited where the quantity of biological specimen available is very low such as a single fragment of hair or a tiny blood stain left at the site of a crime. The details of PCR techniques and its mechanism are described by Erlich (1989) in his edited book 'PCR Technology'. The PCR technique has now been automated and is carried out by a specially designed machine.

The PCR requires the following

- DNA Template: Any source that contains one or more target DNA molecules to be amplified can be taken as a template.
- Primers: A pair of oligonucleotides of about 180-30 nucleotides is with similar G+C contents act as primers. They direct DNA synthesis towards one another. The primers are designed to anneal on opposite strands of target sequence so that they will be extended towards each other by addition of nucleotides to their 3' ends.
- Enzyme: The most common enzyme used in PCR is a thermostable enzyme called Taq polymerase. It is isolated from a thermostable bacterium called *Thermus aquaticus*. It survives at 95°C for 1-2 minutes and has a half-life for more than 2 hours at this temperature. The other thermostable polymerases can also be used in PCR

DNA Amplification Fingerprinting (DAF)

- The DAF has been described by Caetano-Annoles and Gresshoff (1994) where a single arbitrary primer (about 5 bases long) is used to amplify DNA by using PCR. In this technique the parameters are carefully optimized.
- But for early and easy determination of amplified products, automation and fluorescent tagging of primers are extremely amenable.
- DAF differs from RAPD as (i) it requires higher concentration of shorter primers (5-8 nucleotides), (ii) two temperature cycles contrast to three cycles in RAPD, (ii) it produces a very complex banding pattern, (iv) DAF products are used on acrylamide gels and detected by using silver staining.

Application of PCR

- After the discovery of PCR, the modern biology has been revolutionized in each and every aspect. "Some of the areas of application of PCR have briefly been discussed herewith. (a) Diagnosis of Pathogens:
- There are several pathogens that grow slowly. Therefore the cells are found less in number in the infected cells/tissues.
- It is difficult to culture them on artificial medium. Hence, for their diagnosis, PCR-based assays have been developed.
- These detect the presence of certain specific sequences of the pathogens present in the infected cells/tissues.
- Besides, it is useful in detection of viral infection before they cause symptoms or serious diseases. (b) Diagnosis of Specific Mutation: in humans there are thousands of genetic diseases. Mutations are also related to genetic diseases.
- Presence of a faulty DNA sequence can be detected before the establishment of disease.
- By using PCR sickle cell anaemia, phenylketonuria and muscular distrophy can also be detected. The other diseases can also be diagnosed by using PCR. For example PCR based diagnostic for AIDS. Chlamydia, tuberculosis, hepatitis, human pappiloma virus, and other infectious agents and diseases are being developed.
- The tests are rapid, sensitive and specific for Prenatal Diagnosis: It is useful in prenatal diagnosis of several genetic diseases. If the genetic diseases are not curable, it is recommended to go for abortion

    (c) DNA Fingerprinting:

In recent years DNA fingerprinting is more successfully used in forensic science to search out criminals, rapists solving disputed parentage and uniting the lost children to their parents or relatives by confirming their identity.

- This is done through making lines between the DNA recovered from samples of blood, semen, hairs, etc. at the spot of crime and the DNA of suspected individuals or between child and his/her parents/relatives.

(d) In Research: In addition, DNA fingerprinting of new microorganisms isolated from various extreme environment (soil. water, sediments, air, extreme habitats, etc.) is also carried out to confim their identity by comparing with the DNA sequences of known microorganisms.

- Their DNA and RNA can be amplified. Besides, it is also useful to determine the orientation and location or restriction fragments relative to one another (In Molecular Archaeology (Paleontology): PCR has been used to clone the DNA fragments from the mummified remains of humans and extinct animals such as the woolly mammoth and dinosaurs from the remains of ancient animals as recently epitomized in Michel Crighton's Jurassic Park.
- DNA from buried humans has been amplified and used to trace the human migration that occurred in ancient time.

(e) Diagnosis of Plant Pathogens: It is also applied in diagnosis of plant diseases. A large number of plant pathogens in various hosts or environmental samples are detected by using PCR, example, viroids (associated with flower mosaic virus, bean yellow mosaic-virus, plum pox virus, potyviruses), mycoplasmas bacte (*Agrobacterium tumifaciens, Pseudomonas solanacearum, Rhizobium leguminosarum, Xanthomon Compestris,* etc), fungi (e.g., *collectotrichum gloeosporioides, Glomus spp., Laccaria sp Phytophthora spp.. Verticillium spp*)", and nematodes (eg. *Meloidogyne incoginta M. javanic*, pear, grape, citrus, etc), viruses (such as TMV) (Henson and French, 1993; Chawla, 1998).

Site-directed mutagenesis

Now, techniques have been developed the specific portion in the genome in a view to get some novel products of enormous value. Mutations can be done by directing insertions or deletions to a site on the DNA but they have minor changes may have some single amino acid is altered to improve its properties. For such modifications point- mutations i.e. changing of single nucleotide, is done on specific portion in the gene. Thus, the most suitable method that could bring about point-mutation in a gene to mutate specific are unlikely to be of use. Importance. If a protein in order to oligonucleotide-directed mutagenesis. This is known as technique has potential for protein engineering, the engineered enzymes would be better than the wild-type ones. Change in a single nucleotide base pair is called 'point mutation'. Some useful properties in proteins (like stability in subtilisin) that results in substitution of selected amino acids. Using this technique, a specific point of a gene can be mutated. Therefore, this method has been used to understand the function of many genes. Moreover, this technique can only be used when the nucleotide sequence of a gene is known.

CHAPTER 5

MOLECULAR BIOLOGY

29. CHROMOSOME STRUCTURE: BASIC CHEMICAL ASPECTS – DNA, HISTONES AND NON-HISTONES; BASIC STRUCTURAL ASPECTS – THE NUCLEOSOMES, EUCHROMATIN AND HETEROCHROMATIN.

Chromosomes

It contains the genetic material such as DNA and RNA.

Chromatin

The chromosomal material in its decondensed and threadlike state.

Mitosis

It forms asexual reproduction. It occurs when an organism grows or replaces damaged cells. Prior to mitosis, a cell undergoes replication. The process in which chromatin is copied. It produces diploid cells.

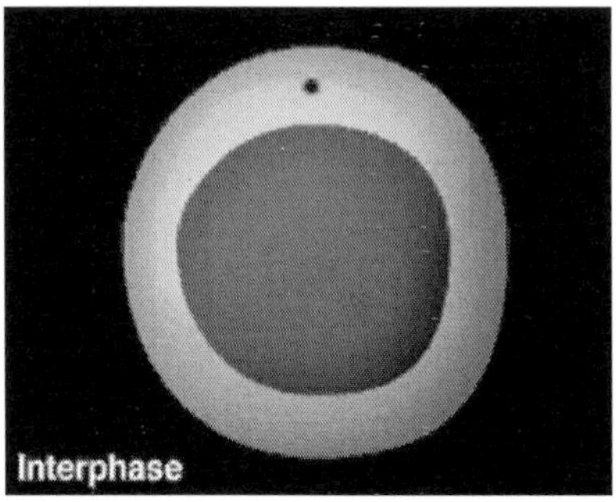

Prophase

Start of mitosis. The chromatin condenses into rod-like chromosomes. Each chromosome is consists of sister chromatids, connected at the centromere and nuclear membrane disappears.

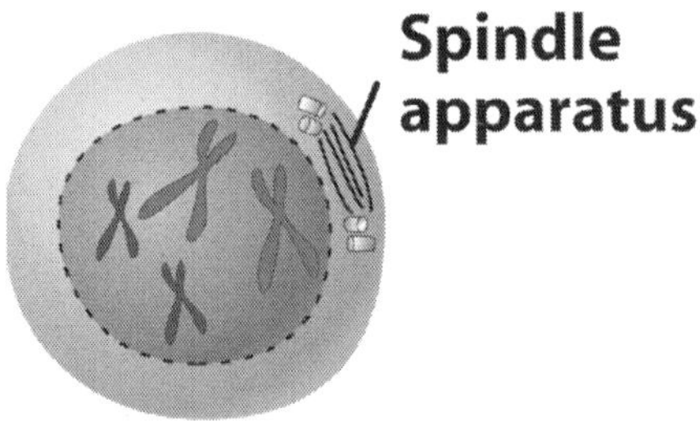

Prophase¶

Metaphase

Chromosomes align themselves at the flat plane at the cell equator in metaphase.

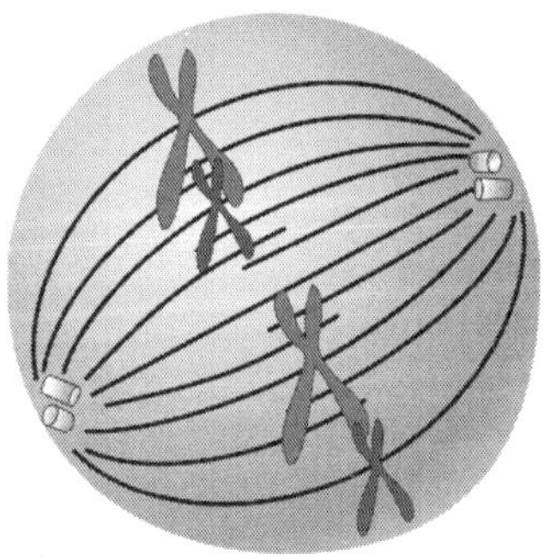

Metaphase

Anaphase

Centromeres are split. Sister chromatids-now chromosomes- are pulled to the opposite poles of the cell.

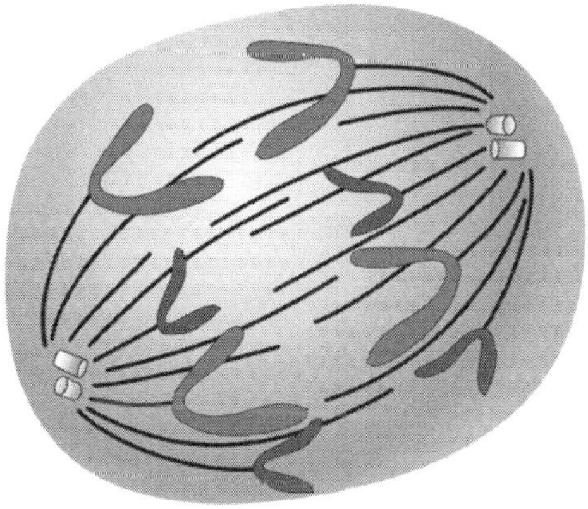

Anaphase

Telophase

Chromosomes are unravel, returning the chromatin to its non-dividing thread-like state. Nuclear membrane is assembles.

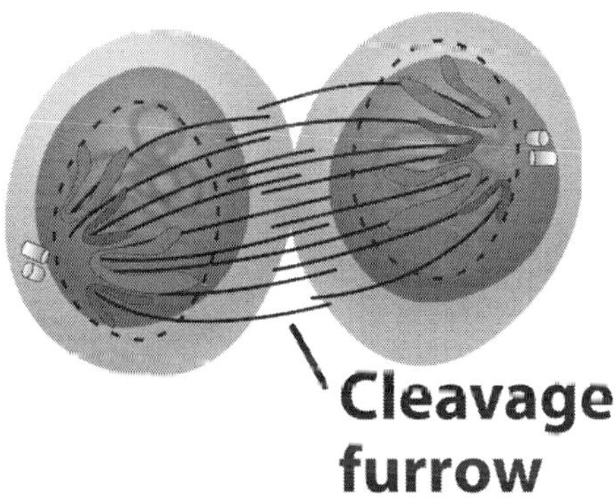

Telophase

Cytokinesis

Division of the cytoplasm in cytokinesis and it initiates during anaphase and telophase.

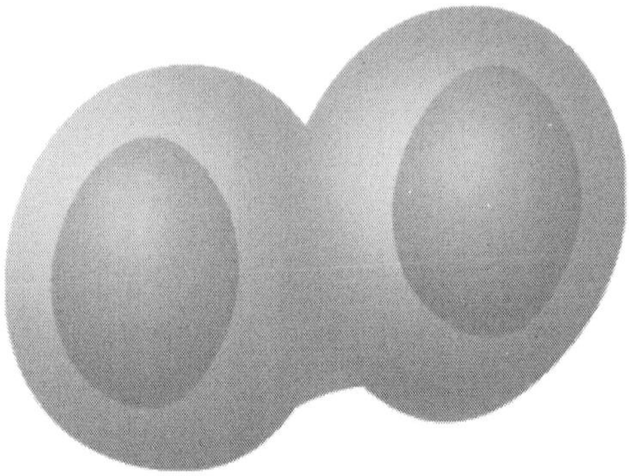

Cytokinesis

Differs in animals and plant cells. Plant cells form a cell plate. Membranous vesicles congregate at the center of the cell.Vesicles are contain cell wall material. Animal cells form a cleavage furrow. It forms around the periphery of the dividing cell. It furrow becomes deeper and deeper until membrane pinches off forming two cells.

Chromosomes come in Matched Pairs

Homologous pairs: chromosomes that are closely matched in size and shape. It determine the same traits. Sex chromosomes: Those that determine the gender of the organism.

Chromosomal theory of inheritance

The two members of each pair of a homologous pair of chromosomes carry alleles for the same genes and, therefore, affect the same traits. It was proposed in 1903.

Mapping genes

The location of a gene on a chromosome. It is currently trying to map all human genes to the appropriate chromosome. **Example:** gene responsible for sickle cell anemia is located on chromosome 11.

Sexual reproduction

The type of reproduction in which genetic information from females combines with males. It requires fertilization. The fusion of gametes (egg and sperm). These cells need to have half the amount of genetic information.

Meiosis

It produces haploid cells (gametes-eggs and sperm). It posses only one member of each pair of homologous chromosomes. The chromosomes are replicate before meiosis. It requires two rounds of division.

Prophase I

The chromatin is condenses into compact chromosomes. The nuclear envelope disappears. The synapsis occurs. The homologous pairs of chromosomes closely align allowing the exchange of chromosome segments

Metaphase I

The aligned pairs of replicated chromosomes are move to the equator of the dividing cell.

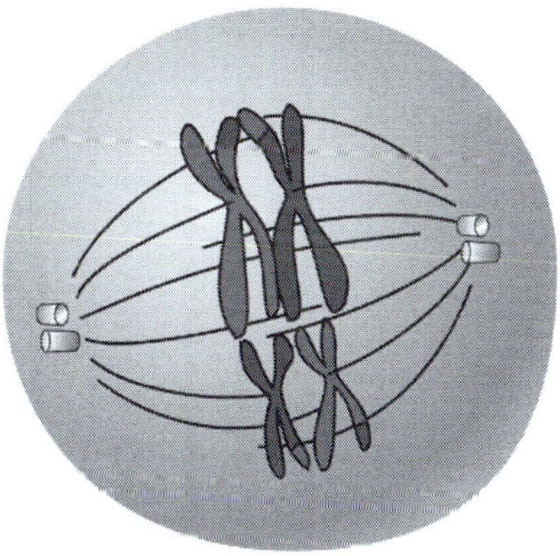

Metaphase I

Anaphase I

The members of homologous pairs of chromosomes are separate from each other. They move to the opposite poles of the cell. The chromosomes are experience independent assortment.

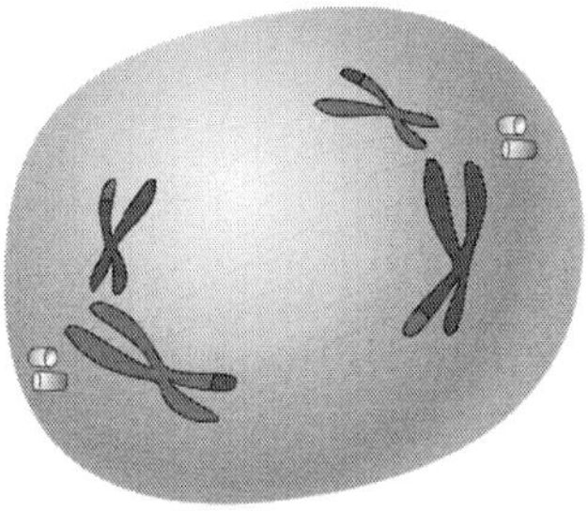

Anaphase I

Telophase I

The chromosomes are cluster at opposite poles of cell and begin to decondense. The nuclear envelope may reform. The cytokinesis occurs Interkinesis: phase between meiosis I and II. The daughter cells are now haploid.

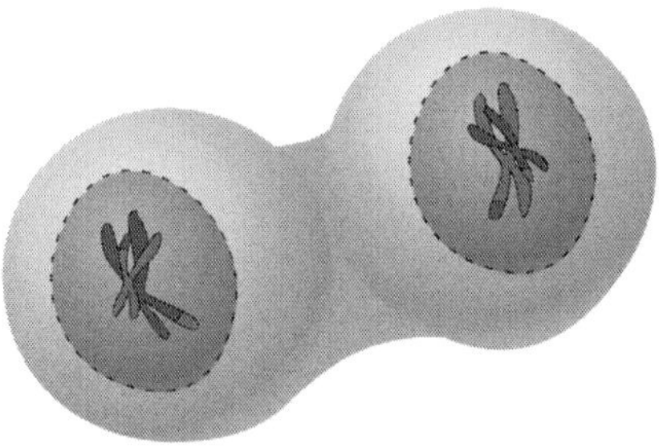

Telophase I

Prophase II

It is partially unraveled chromosomes condense again.

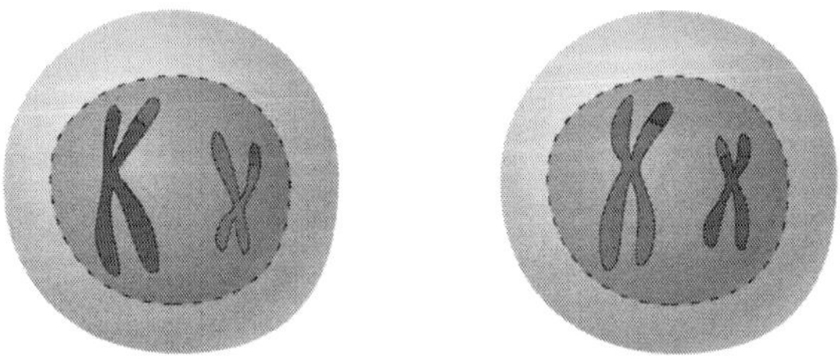

Prophase II

Metaphase II

The chromosomes are move to the cell equator. It has no longer homologous pairs, so chromosomes line up singly in the middle of the cell.

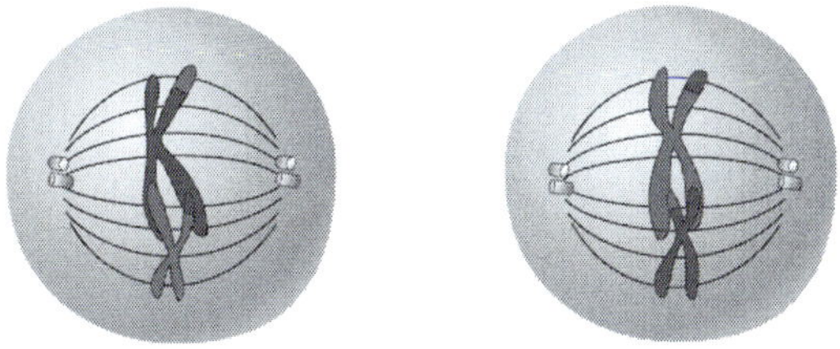

Metaphase II

Anaphase II

The centromeres are divide and chromosomes separate. It move to the opposite poles of the cell.

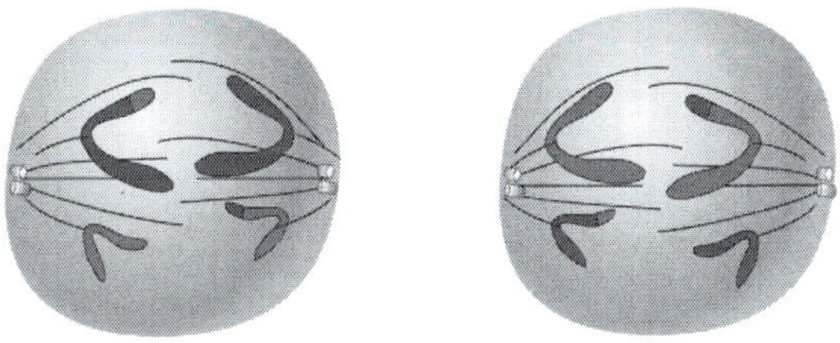

Anaphase II

Telophase II

The clustered chromosomes at cell pole begin to decondense. The nuclear membrane develops and cytokinesis occurs. It produces four non-identical haploid cells.

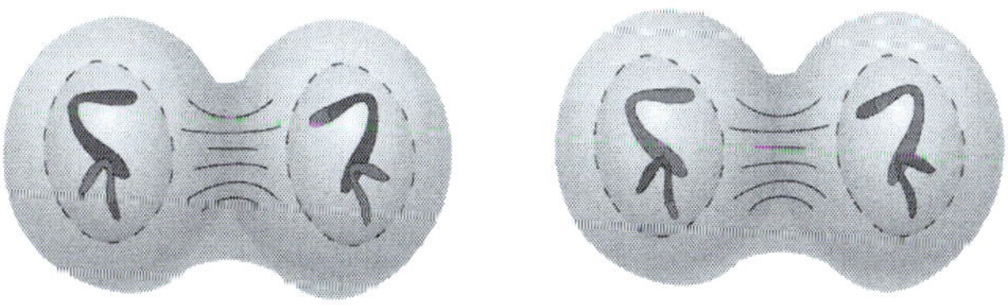

Telophase II

Eukaryotic Chromosome Structure

Chromatin: It is complex of DNA and chomosomal proteins and it is twice as much protein as DNA.

Two Major Types of Proteins

- *Histones*: It is abundant and basic proteins with a positive charge that bind to DNA 5 main types: H1, H2A, H2B, H3, H4, equal in mass to DNA, evolutionarily conserved.
- *Non-histones*: All the other proteins associated with DNA differ markedly in type and structure amounts vary widely, >> 100% DNA mass and **<< 50% DNA mass.**

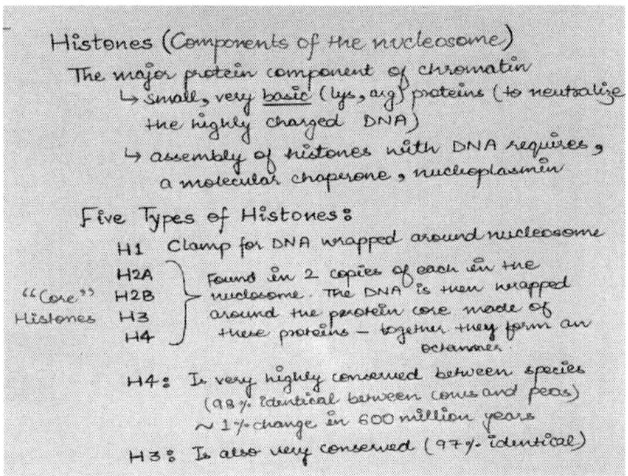

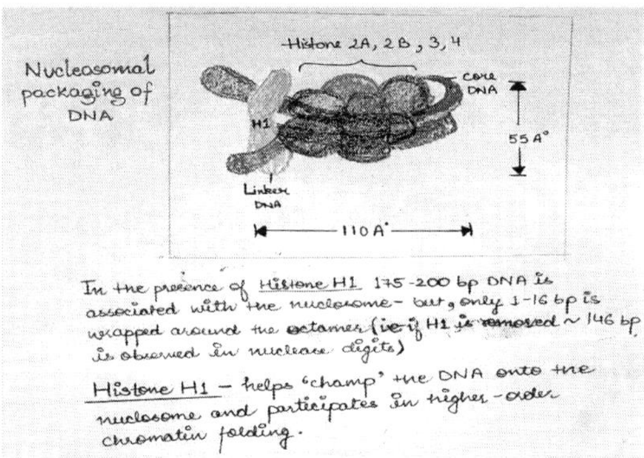

Histone modifications

Histones are also modified in order to regulate

1. chromosome structure; compaction
2. gene activity

The main types of modifications are

a) Phosphorylation of serines

b) Methylation of lysines

c) Acetylation of lysines: neutralizes + charges

$$--NH_3^+ \rightarrow --NH-\underset{\underset{O}{\|}}{C}-CH_3$$

11nm fiber is coiled into 30nm
Solenoid - requires histone H1

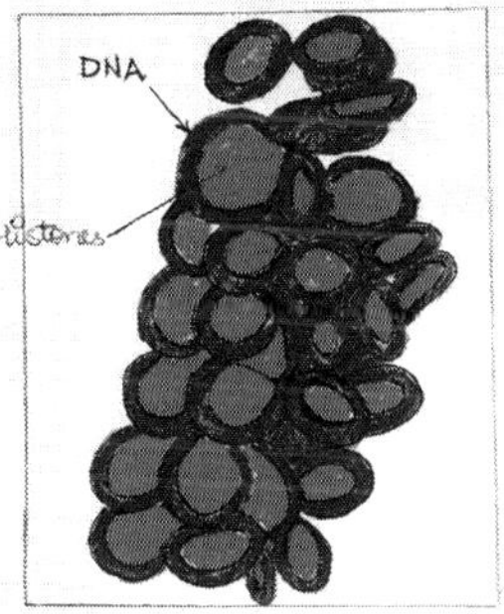

Schematic of formation of 30nm chromatin fiber

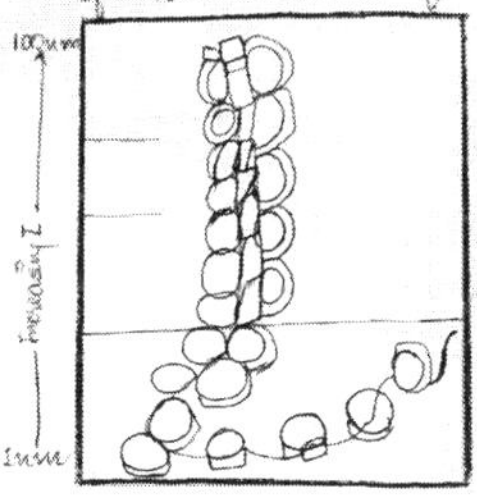

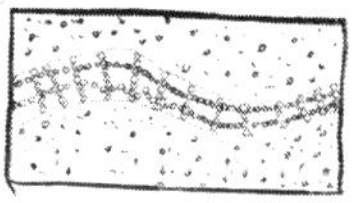

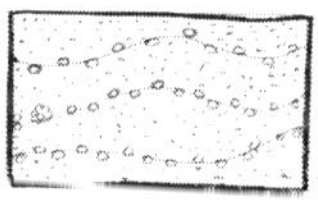

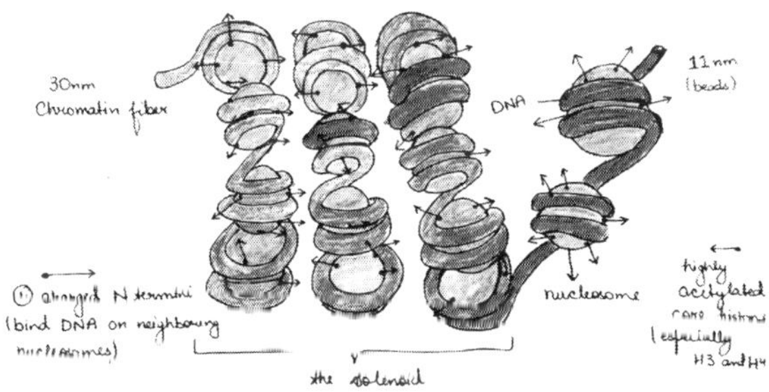

Solenoid:

- HIGH level of histone H1 "pull" nucleosomes into solenoid
- NO gene transcription in solenoid

Short extended regions can exist between solenoids:

- Reduced level of histone H1
- Gene transcription possible

Packing of DNA into chromosomes:

1. Level 1 Winding of DNA around histones to create a nucleosome structure.
2. Level 2 Nucleosomes connected by strands of linker DNA like beads on a string.
3. Level 3 Packaging of nucleosomes into 30-nm chromatin fibre.
4. Level 4 Formation of looped domains.

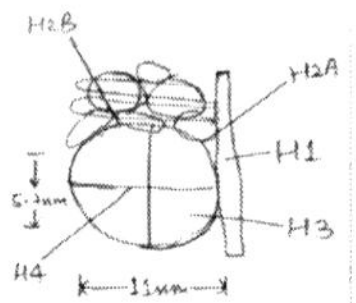

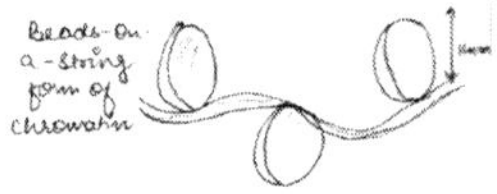

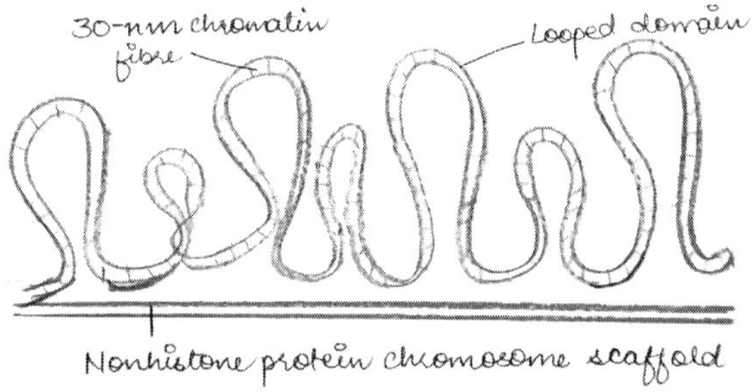

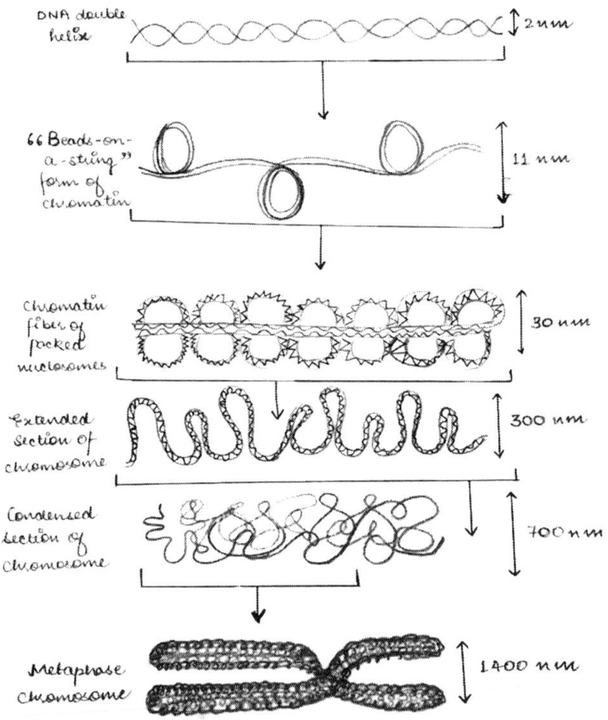

Chromatin is Divided Into Euchromatin And Heterochromatin

Individual chromosomes can be seen only during mitosis. During interphase, the general mass of chromatin is in the form of euchromatin. Euchromatin is less tightly packed than mitotic chromosomes. Regions of heterochromatin remain densely packed throughout interphase.

Chromosomes have banding patterns

Certain staining techniques cause the chromosomes to have the appearance of a series of striations called G-bands. The bands are lower in G • C content than the interbands. Genes are concentrated in the G • C-rich interbands.

Eukaryotic DNA has loops and Domains Attached to A Scaffold

DNA of interphase chromatin is negatively supercoiled into independent domains of ~85 kb. Metaphase chromosomes have a protein scaffold to which the loops of supercoiled DNA are attached. DNA is attached to the nuclear matrix at specific sequences called MARs or SARs. The MARs are A • T-rich but do not have any specific consensus sequence. The centromere is essential for segregation. A eukaryotic chromosome is held on the mitotic spindle by the attachment of microtubules to the kinetochore that forms in its centromeric region. The centromeres often have heterochromatin that is rich in satellite DNA sequences.

Telomeres

It seal the chromosome ends. The protein TRF2 catalyzes a reaction in which the 3 repeating unit of the G+T-rich strand forms a loop by displacing its homologue in an upstream region of the telomere.

Lampbrush chromosomes

Sites of gene expression on lampbrush chromosomes show loops that are extended from the chromosomal axis.

Polytene chromosomes

Polytene chromosomes of dipterans have a series of bands that can be used as a cytological map.

Bands that are sites of gene expression on polytene chromosomes expand to give "puffs." The nucleosome is the subunit of all chromatin. Micrococcal nuclease releases individual nucleosomes from chromatin as 11S particles. A nucleosome contains:

- ~200 bp of DNA
- It has two copies of each core histone (H2A, H2B, H3, and H4).
- One copy of H1.
- DNA is wrapped around the outside surface of the protein octamer.

DNA is coiled in arrays of nucleosomes. Greater than 95% of the DNA is recovered in nucleosomes or multimers when micrococcal nuclease cleaves DNA of chromatin. The length of DNA per nucleosome varies for individual tissues in a range from 154-260 bp.

Nucleosomes

Nucleosomal DNA is divided into the core DNA and linker DNA depending on its susceptibility to micrococcal nuclease. The core DNA is the length of 146 bp that is found on the core particles produced by prolonged digestion with micrococcal nuclease. Linker DNA is the region of 8-114 bp that is susceptible to early cleavage by the enzyme. Changes in the length of linker DNA account for the variation in the total length of nucleosomal DNA. H1 is associated with linker DNA and may lie at the point where DNA enters and leaves the nucleosome.

Polytene Chromosome and Lambrush Chromosome

Giant chromosomes were first observed by E. G. Balbiani in 1881 in the salivary gland cells of Dipteran species. Polytene chromosomes are giant chromosomes found in the salivary glands, malpighian tubules, gut lining and adipose tissue cells of some Dipteran larvae.These chromosomes may reach a size up to 200 times (or more) the size of corresponding chromosomes at meiosis or in nuclei of normal mitotic cells.These giant chromosomes are formed by repeated cycles of endo-reduplication of single chromatids. This means that the chromatids replicate without cell division, as a result, the number of chromatids keeps increasing. This process is called Polyteny and therefore, these chromosomes are called polytene chromosomes.

Polytene chromosomes show another characteristic feature i.e., somatic pairing. As a result of this, chromosomes always appear to be half of that in normal somatic cells. Lampbrush chromosomes were first seen in salamander oocytes by Fleming in 1882.The name lampbrush was given by Ruckert in 1892.The nomenclature comes from its similarity in appearance to the brushes used to clean lamp chimneys in old days. These chromosomes are transitory structures that exist during the extended Diplotene in oocytes of most invertebrates and vertebrates except mammals. The loops in these chromosomes are sites of RNA transcriptions.The condensed chromatin in the chromomeres is generally not transcribed.The position of each loop is fixed. A given loop always contains the same DNA sequence. They have been widely used in studies of chromosome organization, gene expression, molecular morphology of RNA transcription, etc.

30. DNA REPLICATION: DNA REPLICATION AND ITS OVERVIEW: SEMICONSERVATIVE MECHANISM OF DNA REPLICATION, REPLICONS, ORIGINS AND TERMINII, SEMI-DISCONTINUOUS REPLICATION, RNA PRIMING, REPLICATION IN PROKARYOTES AND EUKARYOTES

Introduction

The DNA is found in all plants, animals, prokaryotes and some viruses. DNA is a double-helical structure that consists of genetic material. It is joined by a phosphodiester bond. It is acidic in nature. The entire genetic message that controls the chemistry of every cell of the body acting in a specific way is actually written in the language of four nitrogenous bases of DNA i.e., purines and pyrimidines.

Composition of DNA

It is a polymeric compound containing four monomers known as deoxyribonucleotide monomers or deoxyribotides. Each DNA consists of a pentose sugar, a phosphate group and nitrogenous bases: **Figure 30.1**. Purines bases (adenine and guanine) are heterocyclic and two ringed bases and the pyrimidines (thymine and cytosine) are one ringed bases.

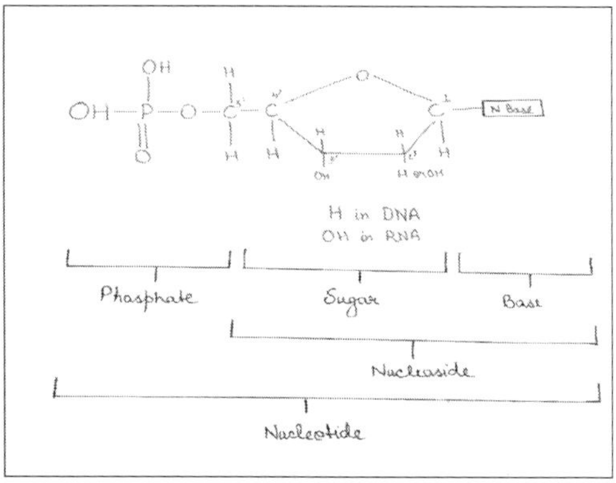

Figure 30. 1. Nucleotides

Nucleosides

The nitrogenous bases are combined with a pentose sugar are called nucleosides.

Nucleotides

A nucleoside linked with phosphate forms a nucleotide. Nucleoside has pentose sugar + nitrogenous base nucleoside + phosphate Nucleotide. On the basis of different nitrogenous bases the deoxynucleotides are of following types: (i) Adenine (A) deoxyadenosine-3'/5'-monophosphate (3'75'-d AMP) deoxyguanosine -5'-monophosphate (5'-d GMP) deoxythymidine -5-monophosphate (5'-d TMP) (ii) Guanine (G) (iii) Thymine (T) (iv) Cytosine (C) deoxyeytidine -5-monophosphate (5'-d CMP). In addition to the presence of nucleosides in the DNA helix, these are also present in nucleoplasm and cytoplasm in the form of deoxyribonucleotide phosphates eg. deoxyadenosine triphosphate (dATP), deoxyguanosine triphosphate (dGTP), deoxycytidine triphosphate (dCTP), deoxythymidine triphosphate (dTTP) **figure 30.2**. The advantage of these four deoxyribonucleotides in triphosphate forms is that the DNA polymerase acts only on triphosphates of nucleotides during DNA replication. Similarly, the ribonucleotides contain ribose sugar, nitrogenous bases and phosphate. Except for sugar, the other components are similar. However, uracil (U) is found in RNA instead of thymine. Generally, an RNA molecule is single-stranded besides some exceptions.

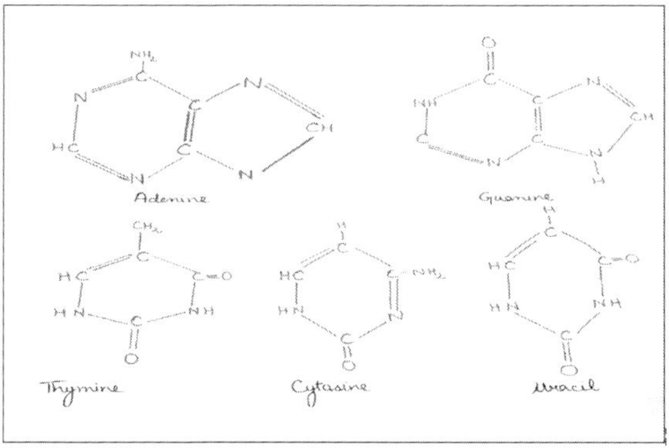

Figure 30.2 Nitrogenous base.

Polynucleotide

The nucleotides undergo the process of polymerization to form a long chain of polynucleotides. The nucleotides are designated by prefixing 'poly' to each repeating unit such as poly A (polyadenylic acid), poly T (polythymidilic acid), poly G (polyguanidylic acid), poly C (polycytidilic acid) and poly U (poly uridylic acid). The polynucleotides that consist of the same are called homopolynucleotides such as poly A, poly T, poly G, poly C and poly U.

Chargaff-Equivalence Rule

By 1948, chemist **Erwin Chargaff** started using paper chromatography to analyze the base composition of DNA from a number of studies. In 1950, **Chargaff** discovered that in the DNA of different types of organisms: the total amount of purines is equal to the total amount of pyrimidines i.e. The total number of A is equal to the total number of T (A-T), and the total number of G is equal to the total number of C (G-C). It means that A/T = G/C i.e. A+T/G+C 1. In the DNA molecules isolated from several organisms, regularity exists in the base composition.

Physical nature of DNA

Watson and crick's modelof DNA

J.D.Watson and F.H.C. Crick (1953) combined the physical and chemical data generated by earlier workers and proposed a double-helix model for DNA molecule. This model is widely accepted. According to this model, the DNA molecule consists of two strands that are connected together by hydrogen bonds and helically twisted. Each step on one strand consists of a nucleotide of purine base that alternates with that of a pyrimidine base. Thus, a strand of DNA molecule is a polymer of four nucleotides i.e. A. G. T, C. The two strands join together to form a double helix. Bases of two nucleotides form hydrogen bonds i.e A combines with T by two hydrogen bonds (A T) and G combines with C by three hydrogen bonds (G C). However the sequence of bonding is such that for every A.T.G.C. on one strand there would be T, A, C, G on the other strand. Therefore, the two chains are nucleotides on one chain are the photocopy of the sequence of nucleotides on the other chain.

The two strands of double helix run in antiparallel direction i.e. they have opposite polarity. The left-hand strand has 5'-3' polarity, whereas the right hand has 3'-5' polarity to the first one. The polarity is due to the direction of phosphodiester linkage complementary to each other i.e. sequence of as compared to the first one. The DNA model also suggested a copying mechanism of the genetic material. DNA replication is the fundamental and unique event underlying growth and reproduction in all living organisms ranging from the smallest viruses to the most complex of all creatures including man. DNA replicates by a semiconservative mechanism that was experimentally proved by Mathew, Meselson and Frank W. Stahlin in 1958. If changes occur in the sequence or composition of base pairs of DNA, mutation takes place. Though the presence of adenine, guanine, thymine and cytosine is a universal phenomenon, yet unusual bases in DNA molecule also occur. In some bacteriophages, 5- hydroxymethyl- cytosine (HMC) replaces cytosine of the DNA molecule when methylation of adenine, guanine and cytosine occurs. This results in changes in these bases: **figure 30.3.**

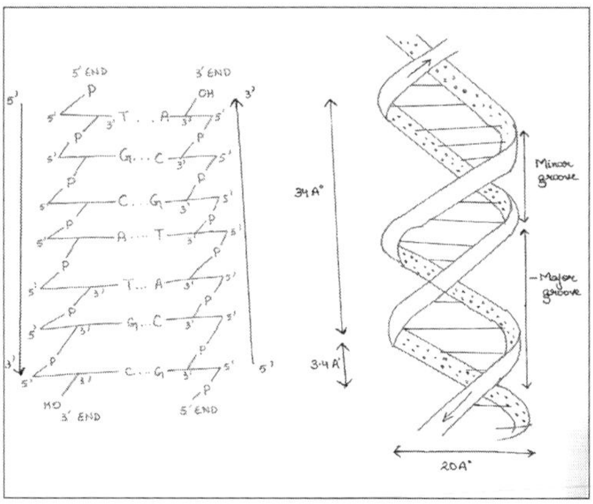

Figure 30.3 Watson Crick's Model of DNA

Three major forms

- A-DNA
- B-DNA
- Z-DNA

B-DNA is biologically the most common

- Right-handed (20 angstrom (a) diameter)
- Complementary base-pairing (Watson-crick)
  - A-T = G-C
- Each base pair has the same width
- 10.85 a from c1' to c1'
- A-T and G-C pairs are interchangeable.

"Pseudo-dyad" axis of symmetry

Ideal **B-DNA** has 10 base pairs per turn.

Base thickness. Aromatic rings with 3.4 A° thickness to rings.

- Pitch = 10 x 3.4 = 34 A° per complete turn.
- Axis passes through the middle of each bp
- Minor groove is narrow.
- Major groove is wide
- Class exercise: explore the structure of B-DNA. Pay special attention to the major, minor grooves.
- Minor: exposes edge from which c1' atoms extend.
- Major: exposes the opposite edge of the base pair.
- The pattern of H-bond possibilities is more specific and more discriminating in the major groove.

A-DNA

It has right-handed helix. It is wider and flatter than B-DNA.11.6 bp per turn. The Pitch of 34 A°. An axial hole. The base planes are tilted 20 degrees with respect to the helical axis. The helix axis passes "above" major groove. The deep major and shallow minor groove. It observed under dehydrating conditions. When relative humidity is ~ 75%. B-DNA - A-DNA (reversible). Most self-complementary oligonucleotides of < 10 bp crystallize in A-dna conf. A-DNA has been observed in 2 contexts:

- At active site of DNA polymerase (~ 3 bp)
- Gram (+) bacteria undergoing sporulation

It induce induce B-DNA to → A-DNA. The resistant to UV-induced damage and cross-linking of pyrimidine bases

Z-DNA

A left-handed helix. It is seen in conditions of high salt concentrations. It reduces repulsions between closest phosphate groups on opposite strands (B-DNA) in complementary polynucleotides with alternating purines and pyrimidines.

- poly d(GC) → poly d(GC)
- poly d(AC) → poly d(GT)

It might also be seen in DNA segments with the above characteristics.12 base pairs per turn. A pitch of 44 degrees. A deep minor groove. No discernible major groove. The reversible change from B-DNA to Z-DNA in localized regions may act as a "switch" to regulate gene expression. Transient formation behind actively transcribing RNA polymerase.

Palindromes in DNA Sequences

The genetic palindromes are similar to verbal palindromes. A palindromic sequence in DNA is one in which the 5' to 3' base pair sequence is identical on both strands (the 5' and 3' ends refers to the chemical structure of the DNA).

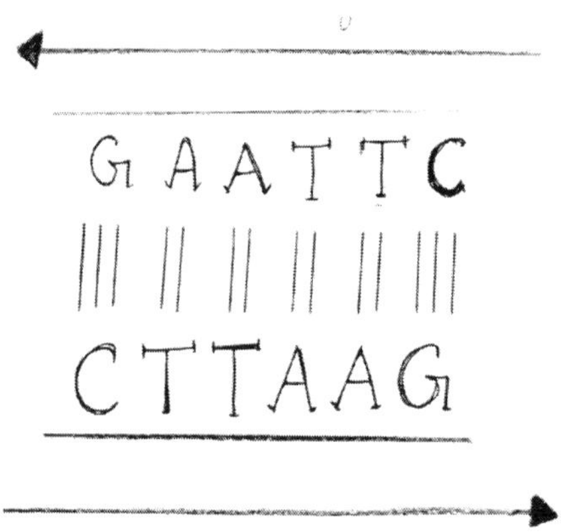

RNA is single-stranded and does not have to stay in the nucleus. RNA is not found in chromosomes because it does not carry the genetic code; however, it can read the DNA code and take the information out of the nucleus.

- RNA's main work is to build proteins.
- The building blocks of RNA are Nucleotides, just like DNA.

A Nucleotide in RNA is still made of 3 important things:

1. 5-Carbon Sugar - Ribose (instead of Deoxyribose)
2. Phosphate
3. Nitrogen base

There are 4 nitrogen bases in RNA, A,G,C, and **U** that pair together)

A → **U** C →G

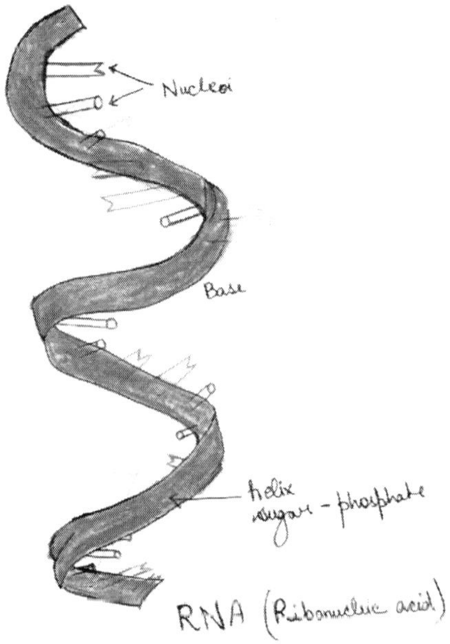

Three Main Types of RNA

- Messenger RNA (mRNA): It carries copies of instructions for the assembly of amino acids into proteins from DNA to the rest of the cell (serve as a "messenger").
- Ribosomal RNA (rRNA): It makes up the major part of ribosomes, which is where proteins are made.

- Transfer RNA (tRNA): The transfers amino acids to ribosomes during protein synthesis.

DNA denaturation (melting)

Strands are held together by weak, noncovalent bonds. The strands start to separate at a specific temperature.

- Within a few degrees, the process is complete.
- The solution contains single-stranded molecules.
- Higher single strand absorbance @ 260 nm.
- Hydrophobic base interactions reduced.
- Bases absorb photons more efficiently.

Melting temperature (T_m) is half denatured. T_m increases with G-C content (%G + %C). A-T-rich sections melt before G-C-rich segments.

DNA Renaturation

J. Marmur (Harvard, 1960) – first described Slowly cool heat-denatured DNA. The temperature quickly is brought to ~25°C below T_m & incubate a while complementary single-stranded DNAs can reassociate or reanneal

Renaturation very useful genome complexity: variety & copy number and Hybridization: molecular identification. Renaturation governed by ionic strength of the solution, Temperature, Time, DNA concentration and Size of the interacting molecules

Smaller Genome - Faster Renaturation

More copies of small genomes and increases chance of collision between complementary fragments and Viral/bacterial genomes are symmetrical curve.

Circular and superhelical DNA

Almost in all the prokaryotes and a few viruses, the DNA is organized in the form of closed circles. The two ends of the double helix get covalently sealed to form a closed circle. Thus, a closed circle contains two unbroken complementary strands. Some times one or more nicks or breaks may be present on one or both strands, for example, DNA of phage PM, Besides some exceptions, the covalently closed circles are twisted into super helix or super coils and is associated with basic proteins but not with histones found complexed with all eukaryotic DNA these histones like proteins appears to help the organization of bacterial chromatin structure with the result of nucleosome like structure, folding and supercoiling of DNA and association of DNA polymerase with nucleoids. These nucleoid-associated proteins include HU proteins, IHF, proteins HI, Fir A, H-NS and Fis. In archaeobacteria (e.g. Archaea) the chromosomal DNA exists in protein associated form. Histone-like proteins have been isolated from nucleoprotein complexes in *Thermoplasma acidophilum* and *Halobacterium salinarum.*

Thus, the protein associated DNA and nucleosome like structures are defected in a helix coils clockwise from the axis the coiling is termed as positive contrast, if the path of coiling is anticlockwise, the coil is called left-handed or negative coil. The two ends of a linear DNA helix can be joined to makeeach strand continuous. However if one end rotates at 360° with respect to the other.These histone like proteins are appearing to help the organization of bacterial DNA into a coiled variety of bacteria. The unwinding of the double joined resulting in the formation of a twisted circle in opposite sense i.e. opposite helix; the ends are to unwinding direction.

Such a twisted circle appears as an 8 it has one node or crossing over point. If it is twisted at 720° before joining, the resulting super helix will contain two nodes. The enzyme topoisomerases alter the topological form i.e. super coiling of a circular DNA molecule. Type I topoisomerases (e.g. *E.coli* top A) relax the negatively super coiled DNA by breaking the 5'-phosphoryl end and then resealing the nicked phosphodiester backbone (Moat and Foster, 1995). Type II Topoisomerases need energy to unwind the DNA molecules resulting in the introduction of super coils. One type of type II isomerases, DNA gyrase is apparently responsible for the negatively super coiled state of the bacterial chromosome. Super coiling is essential for efficient replication and transcription of prokaryotic DNA. The bacterial chromosomes is believed to contain about 50 negatively super coiled loops topological unit, the boundaries of which may be defined by the sites on DNA that limit its rotation (Wang, 1982; Mout and Foster, 1995). One of the phosphodiester bonds in dsDNA allow the 3'-OH end to swivel around or domains.

Organization of DNA

In addition to the organization of DNA in prokaryotes and lower eukaryotes as discussed earlier in eukaryotes, the DNA helix is highly organized into the well-defined DNA-protein complex termed as nucleosomes. Among the proteins, the most prominent are histones. The histones are small and basic proteins rich in amino acids such as cells there are five types of histones e.g. H,, H,A, H,B, H, and H. Eight histone molecules (two each of H, A, H, B, H, and H) form an octamer ellipsoidal structure of about 1 nm to 7 nm in diameter. DNA coils around the surface of the ellipsoidal structure of histones 166 base pairs (about 7/4 turns) before proceeding onto the next and form. Thus, a nucleosome is an octamer or four histone proteins complexed with DNA lysine and/or arginine. Almost in all eukaryotic long and 6.5- complex structure, the nucleosome.The histones play important role in determining eukaryotic chromosomes by determining the conformation known as chromatin. The nucleosomes are the repeating units of DNA organization which are often termed as beads. The DNA isolated from chromatin looks like a string of 146 base pairs of DNA lie in the helical path and the histone. DNA assembly is known as the nucleosome core. The stretch of DNA between the nucleosome is known as linker' which varies in length from 14 to over 100 base pairs. The H_1 is associated with the linker region and helps the folding of DNA into a complex structure called chromatin or beads. As a result of m chromatin becomes visible as chromosomes during cell division.

Structure of Ribonucleic acid (RNA)

The RN TMV, yellow mosaic virus, influenza virus, foot and mouth disease is usually single-stranded except one virus such as virus, etc. which have dsDNA. The retrovirus, is entirely single strand of the RNA which folded either at certain regions or to form a hairpin-shaped structure. In the hairpin-shaped structures, the complementary bases are linked by hydrogen bonds which give stability to the molecules. However, no complementary bases are found in the unfolded region. The RNA does not possess an equal purine-pyrimidine ratio, as it is found in the DNA. Like DNA, the RNA is also the polymer of four nucleotides each one contains D-ribose, phosphoric acid and a nitrogenous base. The bases are two purines (A, G) and two are not found in RNA. A pairing between bases occurs as A-U and G-C pyrimidines (C, U). Thymine is not found in RNA. The pairing is between bases occurs as A-U and G-C. The nucleotides formed by the four bases are adenosine monophosphate (AMP), guanosine monophosphate (GMP), cytosine monophosphate (CMP), and uridine monophosphate (UMP). These are found freely in nucleoplasm but in the form of triphosphates e.g. ATP, GTP, UTP and CTP. As a result of polymerization, the ribonucleotides form a polynucleotide chain of RNA. If the RNA is involved in genetic mechanisms, it is called genetic RNA as found in plants, animal and bacterial viruses. The DNA acts as genetic material and RNA follows the order of DNA. In such cells, the RNA does not have a genetic role. Therefore, it is called non-genetic RNA. The non-genetic RNA is of three types: (i) ribosomal RNA (rRNA), (ii) transfer RNA (tRNA) or soluble RNA (sRNA), and (iii) messenger RNA (mRNA) or template RNA. These three types of RNA differ from each other in structure, site of synthesis in eukaryotic cell and function.

DNA Replication

Transmission of chromosomal DNA from generation to generation is crucial to cell propagation. This can only be achieved when chromosomal DNA is accurately replicated, providing two copies of the entire genome for faithful distribution into each daughter cell. Cells need to make a copy of DNA before dividing so each daughter cell has a complete copy of genetic information.

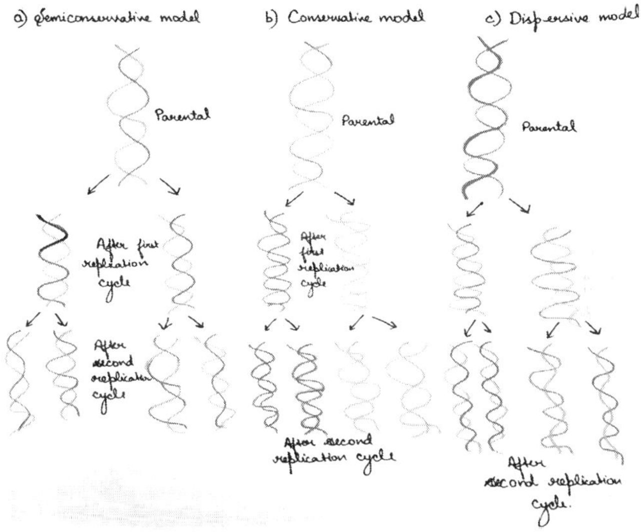

Semi-Conservative Model

Replication of DNA

Base pairing allows each strand to serve as a template for a new strand. The new strand is 1/2 parent template & 1/2 new DNA. It has anti-parallel strands. The nucleotides in the DNA ba ckbone are bonded from phosphate to sugar between 3′& 5′ carbons.

- DNA molecule has "direction"
- Complementary strand runs in the opposite direction

Meselson and Stahl's experiment

Two isotopes of nitrogen:

- ^{14}N common form; ^{15}N rare heavy form.
- *E. coli* were grown in a ^{15}N media first, and then transferred to ^{14}N media.
- Cultured *E. coli* were subjected to equilibrium density gradient centrifugation.

Modes of replication

Replicons: units of replication. Theta replication: circular DNA, *E. coli*; the single origin of replication forming a replication fork, usually a bidirectional replication. Rolling-circle replication: virus, F factor of *E. coli* and a single origin of replication.

Linear eukaryotic replication

Eukaryotic cells; thousands of origins; a typical replicon: 200,000 ~ 300,000 bp in length. Linear eukaryotic replication.

Direction of replication

DNA polymerase adds nucleotides only to the 3′ end of a growing strand. The replication can only go 5′→3′.

Bonding in DNA

Hydrogen bonds and covalent phosphodiester bonds.

Replication: 1st step

- Initiation: 245 bp in the *oriC* (single origin replicon); an initiation protein. Unwinding of DNA is performed by Helicase. Gyrase removes supercoiling ahead of the replication fork. Single-stranded DNA is prevented from annealing by single-stranded binding proteins. Primers: an existing group of RNA nucleotides with a 3′-OH group to which a new nucleotide can be added; usually 10 ~ 12 nucleotides long.

- Primase: RNA polymerase

Unwind DNA

- Helicase enzyme is unwinds part of DNA helix. Stabilized by single-stranded binding proteins.
- Prevents DNA molecule from closing
- DNA gyrase is an enzyme that prevents tangling upstream from the replication fork.
- Replication: 2nd step
- RNA Primase. Adds small section of RNA (RNA primer) to the 3’ end of template DNA. DNA polymerase 3 (the enzyme that builds a new DNA strand) can only add nucleotides to existing strands of DNA.

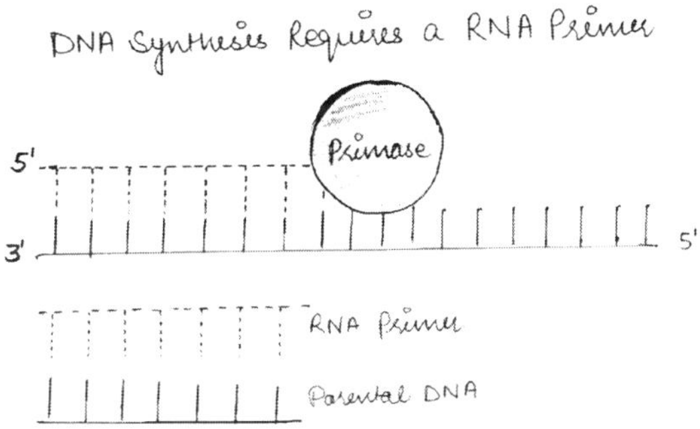

- Elongation: It is carried out by DNA polymerase III. The removing RNA primer: DNA polymerase I. DNA ligase: connecting nicks after RNA primers are removed.
- Termination: when a replication fork meets or by a termination protein.

Replication: 3rd step

Build the daughter DNA strand. Add new complementary bases. With the help of the enzyme DNA polymerase III.

Replication: 4th step

Replacement of RNA primer by DNA and it has done by DNA polymerase I.

Direction of Replication

- Leading strand: It undergoes continuous replication.
- Lagging strand: It undergoes discontinuous replication.
- Okazaki fragments: The discontinuously synthesized short DNA fragments forming the lagging strand.

Leading &lagging strands

Limits of DNA polymerase III. It can only build onto 3′ end of an existing DNA strand.

Lagging strand

Okazaki fragments. It is joined by ligase and "spot welder" enzyme.

Leading strand

It is produced by continuous synthesis.

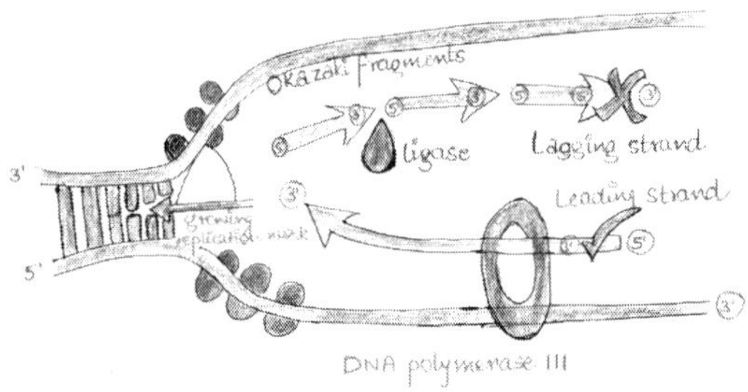

DNA replication on the lagging strand

RNA primer is added. It built by primase. It serves as a starter sequence for DNA polymerase III. However, short segments called Okazaki fragments are made because it can only go in a 5' 3' direction. Replacing RNA primers with DNA

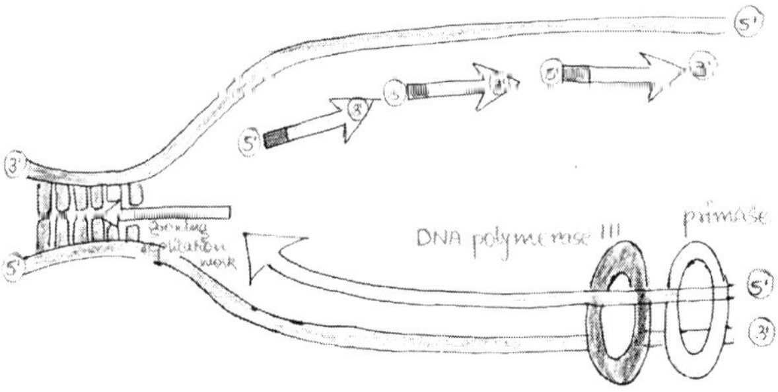

DNA polymerase I

It removes sections of the RNA primer and replaces them with DNA nucleotides.

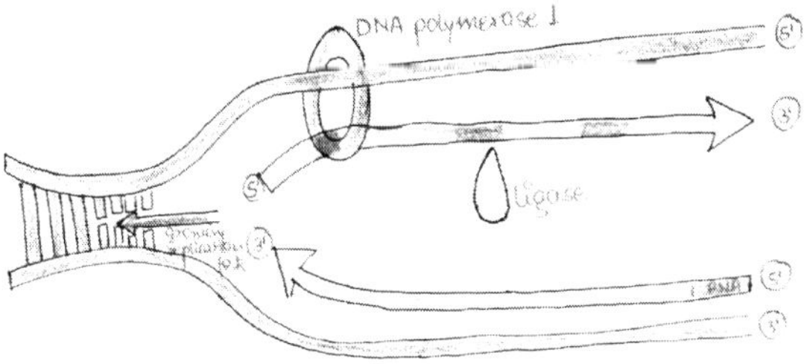

Replication fork

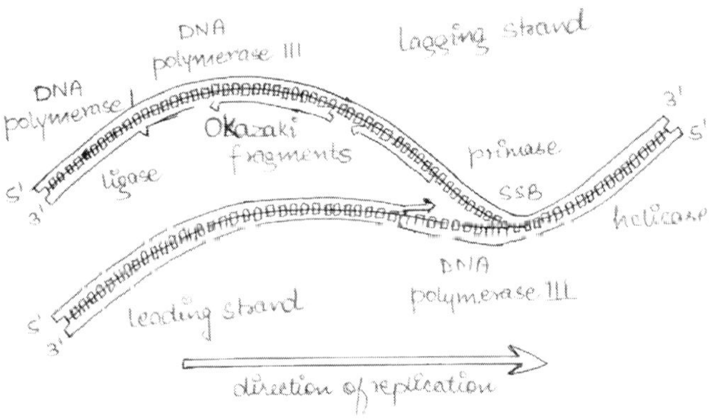

SSB – single-stranded binding proteins

Stabilization of the Replication Machine

DNA polymerase delta and epsilon are attached to the DNA and held in place by other proteins. Sliding-clamp proteins (proliferating cell nuclear antigen - PCNA), allow the stable binding of DNA polymerase and strand synthesis. Clamp-loading proteins (replication factor C - RFC), aid in attaching the sliding-clamp proteins. The DNA is copied by a replication machine that travels along each replication fork. One of the most important members of this complex is DNA polymerase. It is able to add DNA subunits and make new DNA.

DNA polymerase

It was discovered in 1956 by **Kornberg** first time from *E.coli* bacteria.

Bacteria have 3 types

DNA Pol I, II, and III. DNA Pol III involved in replication of DNA. DNA Pol I involved in repair. Humans have 4 types polymerase. DNA Pol alpha, beta, delta - nuclear DNA. DNA Pol gamma - mitochondrial DNA. All DNA Pol's have 2 properties. Only synthesize DNA in one direction 5' to 3'. Only add to the end of existing double-stranded DNA. Therefore, they cannot start the synthesis of DNA from scratch. RNA polymerases can, but not DNA polymerases.

DNA polymerase

It has a proofreading mechanism built. Checks for base matching. Removes mismatched bases by going backward.The reason why it is not able to build DNA in the 3' to 5' direction. No energy from ATP hydrolysis. Makes just one error in 10E8 or 10E9 (billion) bases added.Unwinding of DNA strands by helicases. Require ATP. Single-stranded DNA-binding proteins. Bind to single-stranded DNA to stabilize the structure. RPA (replication protein A - in eukaryotes). Topoisomerase - helps with the prevention of DNA strand twisting - 'swivels'.

Two types

- Type I - Break one strand only and then rejoin.
- Type II - Break both strands and then rejoin.

Telomerase

It is enzyme and role in reverse transcriptase. It acts without a DNA template, as it has its own RNA template stored inside.

The Fidelity of DNA Replication

- ***Proofreading***: DNA polymerase I: 3′→5′ exonuclease activity removes the incorrectly paired nucleotide.
- ***Mismatch repair***: correcting errors after replication is complete.

Eukaryotic DNA Replication

Eukaryotic DNA polymerase. DNA polymerase α- acts like Primase to initiate. DNA polymerase σ- replicates the lagging strand. DNA polymerase e- replicates the leading strand.

Replication at the Ends of Chromosomes

Telomeres and telomerase. DNA repair mechanism. Damaged DNA must be repaired. If the damage is passed on to subsequent generations, then we use the evolutionary term - mutation. It must take place in the germ cells - the gametes -eggs and sperm. If the damage is to somatic cells (all other cells of the body bar germ cells) then just that one individual is affected.

Damage from where Consequences of DNA replication errors.Chemical agents acting on the DNA.UV light imparting energy into a DNA molecule. Spontaneous changes to the DNA.

Repair mechanisms

1. Reversal of damage
2. Excision repair
3. Mismatch repair
4. Recombination repair
5. Error-prone repair
6. Restriction-modification systems

1. Reversal of damage

Enzymatically undo the damage

- Photoreactivation
- Removal of methyl groups

Photolyase breaks apart pyrimidine dimers

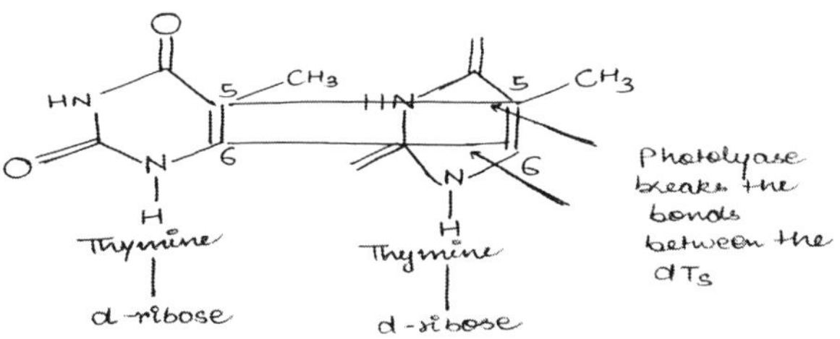

1a. Photoreactivation

Photolyase: It binds pyrimidine dimers and catalyzes a photochemical reaction. It breaks the cyclobutane ring and reforms two adjacent T's. 2 subunits, encoded by *phrA* and *phrB*.

Conversion of 6-O-methyl-dG back to dG

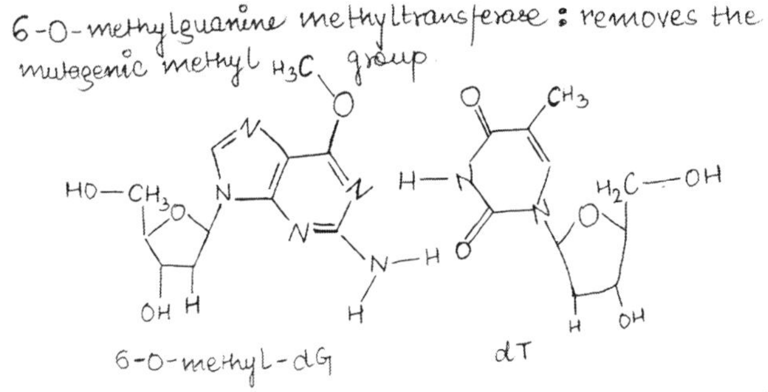

1b. Removal of methyl groups

6-O-methylguanine methyltransferase. It recognizes 6-O-methylguanine in DNA, removes the methyl group. Transfers the methyl group to an amino acid of the enzyme. "Suicide" mechanism. It is encoded by the ada gene in *E. coli*.

Excision repair

It has remove damage (base or DNA backbone) and ss nick/gap provides the 3'OH for DNA Pol I initiation. DNA ligase seals nick.

Nucleotide excision repair

It cut out a segment of DNA around a damaged base.

Base excision repair

It cut out the base, then cut next to the apurinic/apyrimidinic site, and let DNA Pol I repair.

Mismatch repair

The action of DNA polymerase III (including proofreading exonuclease) results in 1 misincorporation per 10^8 bases synthesized. Mismatch repair reduces this rate to 1 change in every 10^{10} or 10^{11} bases. Recognize mispaired bases in DNA, e.g. G-T or A-C base pairs. These do not cause large distortions in the helix: the mismatch repair system apparently reads the sequence of bases in the DNA.

Role of Methylation in Discriminating Parental and Progeny Strands

dam methylase acts on the A of GATC (note that this sequence is symmetrical or pseudopalindromic). Methylation is delayed for several minutes after replication. Mismatch repair works on the un-methylated strand (which is newly replicated) so that replication errors are removed preferentially.

Action of MutS, MutL, MutH

MutS: recognizes the mismatch (heteroduplex). MutL: a dimer; in presence of ATP, binds to MutS-heteroduplex complex to activate MutH. MutH: endonuclease that cleaves 5' to the G in an unmethylated GATC, leaves a nick.

Eukaryotic homologs in mismatch repair

Human homologs to *mutL* (hMLH1) and *mutS* (hMSH2, hMSH1) have been discovered. Mutations in them can cause one of the most common hereditary cancers, hereditary nonpolyposis colon cancer (HNPCC).

Role of umuC and umuD Genes in Error-Prone Repair

It named for the UV **non**-mutable phenotype of mutants with defects in these genes. It is needed for bypass synthesis; the mechanism is under investigation e.g. these proteins may reduce the template required for the polymerase. UmuD protein is proteolytically activated by LexA.

Evolution acts on mutations

If we did not have a mutation then we would all be the same. Any changes in the environment would be deleterious to all members of the population equally. There would be no evolution. But mutations does exist and are supported by comparison of related organisms.

31. DNA DAMAGE AND REPAIR: BIOLOGICAL INDICATIONS OF DAMAGE TO DNA, EVIDENCE FOR REPAIR SYSTEMS, REPAIR OF THYMINE DIMMERS

Introduction

DNA is a relatively stable molecule, but earth's natural environment is quite toxic, and damage to DNA is inevitable. DNA can also be altered by mistakes made during its own replication or recombination. Damage and sequence alterations to DNA are often quickly repaired, but when they are not, the DNA becomes permanently altered and harbors a mutation. Mutations are changes in the DNA sequence, and when mutations occur in germ-line cells, these changes are inheritable. A cancer cell has mutations that prevent cell death, resulting in loss of cell cycle control and unregulated cell division, which leads to malignant tumors that can end the life of the entire organism. The cell has a limited amount of time to fix the initial alteration and restore the DNA to its normal sequence before replication converts the alteration into a mutation that will be passed on to the next generation.

Mutation

- ***Somatic mutations***: It occurs in somatic cells and only affects the individual in which the mutation arises.
- ***Germ-line mutations***: They alter gametes and passed to the next generation.

Mutations are Quantified in Two Ways

- Mutation rate = probability of a particular type of mutation per unit time (or generation).
- Mutation frequency = number of times a particular mutation occurs in a population of cells or individuals.

Types of DNA Mutations

A mutation is a change in a DNA sequence that is propagated through cellular generations. Mutations can be as small as a single base pair or can range from a few base pairs to thousands. Mutations of one or a few base pairs usually result from errors in replication or damaged nucleotides. Mutation can have different effects on gene function.

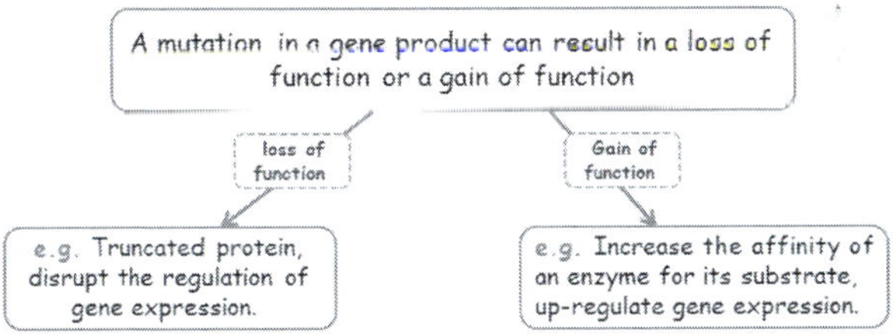

Studies in *Drosophila melanogaster* suggest that mutations that alter the protein sequence are most likely to be harmful; about 70% have a negative effect, and the rest either are neutral or have a weak beneficial effect.

A point mutation can alter one amino acid

A change in a single base pair is often referred to as a point mutation. Point mutations fall into two categories. A transition mutation is the exchange of a purine-pyrimidine base pair for the other purine-pyrimidine base pair: C=G becomes T=A, or T=A becomes C≡G. A transversion mutation is the replacement of a purine-pyrimidine base pair with a pyrimidine-purine base pair or vice versa.

For example: C≡G becomes either G≡C or A=T.

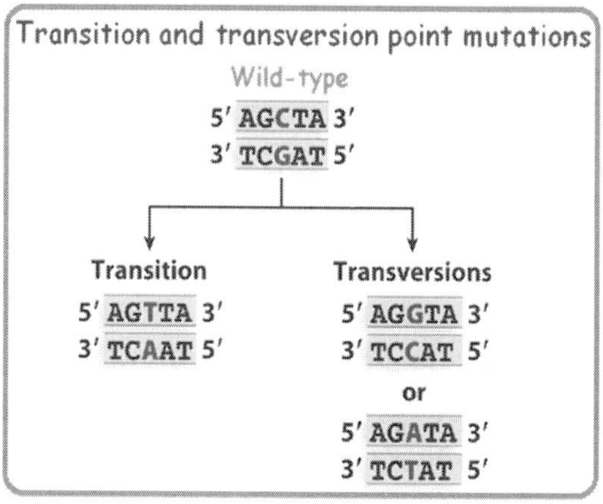

Transition mutations are nearly 10 times more frequent than transversions. A point mutation in the protein-coding region of a gene can result in an altered protein with partial or complete loss of function.If the protein is central to cell viability, the cell could die. Point mutations are known to cause a wide variety of human diseases. One example is sickle-cell anemia, resulting from a transversion that produces an amino acid change in hemoglobin: a GAG encoding glutamate (E) at residue 6 changes to GTG, encoding valine (V) (E6V).The most harmful mutations are those occurring in the genes involved in DNA repair because these often result in cancer. Many mutations in one cell can result in cancer because a mutation will occur in a gene (or genes) that encodes a protein needed to control cell division. In normal cells, oncogenes encode proteins that drive the cell division cycle forward, and tumor suppressor genes encode proteins that suppress cell division. Many tumor suppressors are transcription factors that regulate the expression of genes that drive the cell cycle. The transcription factor p53 and the retinoblastoma protein are examples of tumor suppressors that are mutated in many types of cancer. The DNA sequence encoding a protein is read in codons. Each codon corresponds to an amino acid. A silent mutation is a nucleotide change that produces a codon for the same amino acid. For example, GAA and GAG both code for glutamate. A missense mutation is a nucleotide change that results in a different amino acid, such as a change from glutamate (GAA) to glutamine (CAA). A nonsense mutation changes the nucleotide sequence so that instead of encoding an amino acid, the triplet functions as a stop codon, terminating the protein. Another type of mutation is the gain or loss of one or more base pairs. Insertion mutations occur when one or more base pairs are added to the wild-type sequence. Deletion mutations are due to the loss of one or more base pairs. Insertion and deletion mutations are collectively referred to as indels. The DNA sequence from the start codon to the stop codon is referred to as a reading frame. Because nucleotides are decoded in triplets, an indel mutation of only one or two base pairs in the coding sequence of a protein throws off the reading frame after the mutation, resulting in a frameshift mutation.

Mechanisms of DNA Repair

The most prevalent means that cells use to repair damaged DNA is excision repair, of which there are two types: base excision and nucleotide excision repair. Base excision repair (BER) functions at the level of a single damaged nucleotide that distorts DNA very little. It is also the main pathway for the repair of single-strand DNA breaks that lack a ligatable junction and therefore require "cleaning" of the 3′ or 5′ terminus for ligation. Nucleotide excision repair (NER) targets large, bulky lesions and removes DNA on either side of them. In contrast to base excision repair, NER does not require specific recognition of a damaged nucleotide and thus it can remove DNA lesions.

Base excision repair

(a) In bacteria, a glycosylase excises a damaged nucleotide base, and then an AP endonuclease nicks the backbone at the abasic site. Nick translation by Pol I excise the 5′ deoxyribose phosphate (5′-dRP) and some dNMPs, and synthesize a new strand. Ligase seals the gap.

(b) Eukaryotic BER, after the first two steps, can take either of two paths. In long patch repair, a DNA polymerase extends the DNA strand from the 3′ terminus, displacing the 5′ single-stranded DNA; this is followed by cleavage by a flap endonuclease and ligation. In short patch repair, only one nucleotide is inserted (by Pol β) prior to ligation.

The NER pathway uses several proteins, including UvrA, UvrB and UvrC, which recognize the lesion and make incisions on either side, allowing UvrD (helicase II) to displace a section of lesion-containing DNA. The single-strand gap is filled in by Pol I, and the DNA is sealed by ligase. A transcription-coupled repair (TCR) path can also be taken in which

RNA polymerase stalls at the lesion on the coding strand. After the RNA polymerase is displaced, the reaction proceeds as shown here, using UvrA through UvrD, Pol I, and ligase.

32. TRANSCRIPTION IN PROKARYOTES. TRANSCRIPTION IN EUKARYOTES, ITS REGULATION

Introduction

It is the process of formation of the RNA. It takes place through the usual process of complementary base pairing, catalyzed and scrutinized by the enzyme RNA polymerase. It occurs unidirectionally in which RNA chain (transcript) is synthesized from 5' to 3 directions. It is chemically and enzymatically very similar to DNA replication. However, there are the following important that differ from that of replication. The new strand is made from ribonucleotides (NTPs) rather than deoxyribonucleotides (NTPs). It is primer independent. RNA polymerase can initiate a new RNA chain on a DNA template without the need for a primer. It is a selective process i.e transcription selectively copies only certain parts of the genome called a transcription unit. In contrast, during replication entire genome replicates. Only one of the DNA strands acts as a template on which the RNA strand is built. It is unidirectional.

Prokaryotes

Promoter

Pribnow Box (also called the -10 element) – TATAAT. 35 element- TTGACA.

Eukaryotes

Promoter – (asymmetrical sequence). Basic core promoter –TATA box (TATAAA(A)); within 50bp upstream of the start site; found in unicellular eukaryotes. The core promoter plus have downstream promoters and proximal promoter elements.

Initiator sequences

Regulatory Elements/Response Element: Response elements are the recognition sites of certain transcription factors. Most of them are located within 1 kb from the transcriptional start site. Enhancer elements -upon binding with transcription factors (activators), can enhance transcription; located either upstream or downstream of the transcriptional initiation site.

- Upstream enhancer elements
- Downstream enhancers
- Distal enhancer elements

Silencers - upon binding with transcription factors (repressors), can repress transcription.

Proteins Involved in Transcription

RNA polymerase, general (or basal) transcription factors are tfiia, tfiib, tfiid, tfiie, tfiif, tfiih. transcription factors that bind to regulatory elements are holoenzyme or initiation complex and transcription factors have common dna binding motifs

- Zinc finger
- Helix-turn-helix
- Leucine zipper

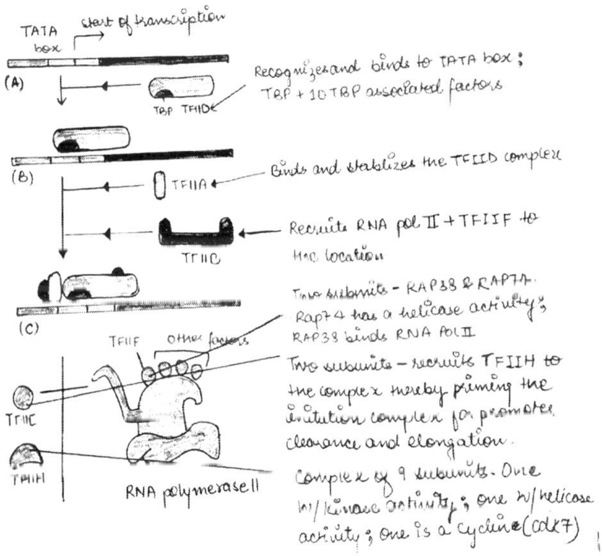

Synthesis of enzymes comes into the control of genetic material i.e., DNA in living organisms. Enzymes synthesis occurs by two processes such as translation and transcription. Translation refers to the synthesis of mRNA. It is controlled by the promoter region of the gene.

Pathway for gene expression

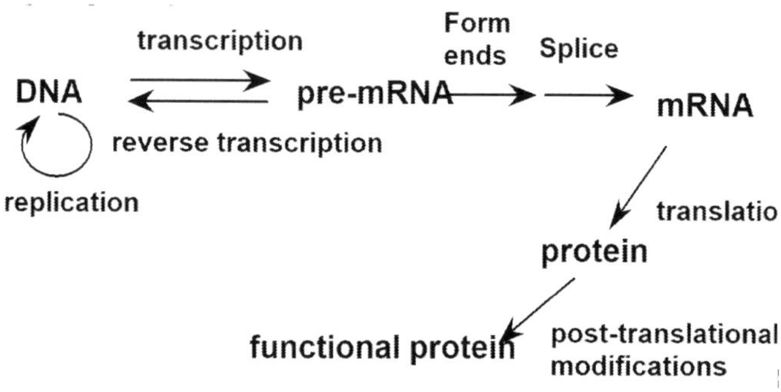

The Reaction Catalyzed by RNA Polymerase

It catalyzes the synthesis of RNA directed by DNA as a template = transcription. Makes an RNA chain with a sequence complementary to the template strand of DNA. Does not require a primer; can start RNA synthesis at a site on the DNA template *E. coli* RNA polymerase. Synthesizes all classes of RNA.

- mRNA
- rRNA
- tRNA

Core= $\alpha_2\beta\beta'$ catalyzes the elongation of an RNA chain. Holoenzyme = $\alpha_2\beta\beta'$**s** catalyzes initiation or RNA synthesis specifically at a promoter.

Promoter

The sequence of DNA is required for accurate, specific initiation of transcription. The sequence of DNA to which RNA polymerase binds to accurately initiate transcription. In most cases, RNA polymerase binds to a DNA sequence including the initiation site, but it can be directed there by sequences flanking the initiation site.

RNA polymerase at a promoter

- **α: assembly and binds to UP.**
- **ß+ß'**: form catalytic center.
- **σ: binds -10 and -35 of the** promoter to confer specificity during initiation.

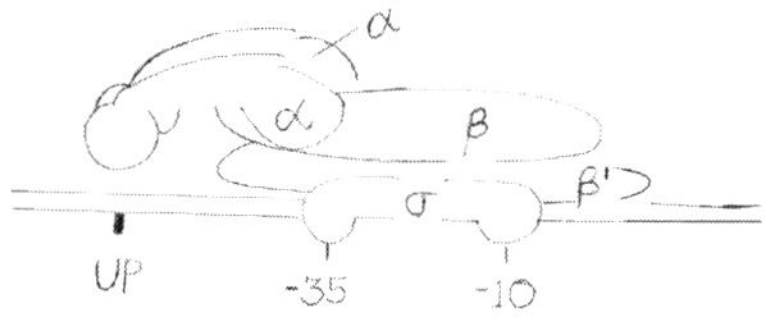

Mode of action of σ factors

The σ factor causes RNA polymerase to be selective in its choice of initiation sites by affecting the dissociation rate of polymerase from DNA. Core dissociates from general DNA with a half-time of 60 min; use in elongation. Holoenzyme dissociates from general DNA with a half-time of 1 sec. Holoenzyme dissociates from promoter DNA with a half-time of hours (in absence of rNTPs).

Transcription Cycle

- *Initiation*: Holoenzyme binds to the promoter, unwinds DNA, and forms phosphodiester bonds between 7 to 12 nucleotides.
- *Elongation*: Core elongates RNA with high processivity. May use NusA.
- *Termination*: Polymerase dissociates from template DNA and releases new RNA.

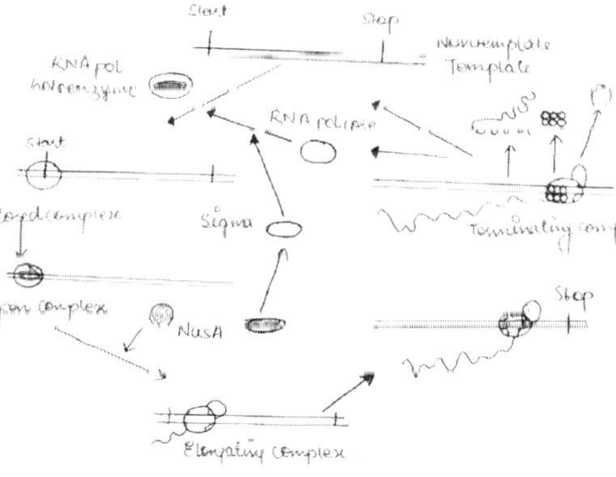

Sites on RNA Polymerase Core

Enzyme covers about 60 bp of DNA, with about 17 bp unwound = transcription bubble. The bubble must contact the active site for polymerization. At the beginning of the bubble, the DNA is unwound, implicating a helicase activity. At the end of the bubble, the DNA is rewound. Effects of transcription on supercoiling of the template. The unwinding (untwisting) and rewinding of the DNA template introduce positive supercoils ahead of the polymerase and negative supercoils behind it. Unwinding will decrease T by 1 for every 10 bp unwound. Since DL=0, DW=-DT, and DW=+1 for every 10 bp unwound. The opposite occurs upon rewinding. DT=+1 for every 10 bp rewound, and thus DW=-1 for every 10 bp rewound. Topoisomerases aid transcription in-vivo.

RNA polymerases in eukaryotes

Name	Makes	α-amanitin
RNA Polymerase I	pre-rRNA	insensitive
RNA Polymerase II	pre-mRNA some snRNAs	very sensitive
RNA Polymerase III	pre-tRNA other small RNAs some snRNAs	less sensitive

Subunit structure of eukaryotic RNA polymerases

All 3 have multiple subunits (8 to 14). The molecular weight of each polymerase is about 500,000. Some subunits are common to all 3 RNA polymerases. All 3 RNA polymerases have subunits that are homologous to the bacterial ß, ß' and α subunits.

33. RNA PROCESSING AND RNPS. MULTIPLE RNA POLYMERASES, SIGMA LIKE FACTORS IN EUKARYOTES; HETEROGENEOUS NUCLEAR RNA; MESSENGER RNA – STRUCTURE AND COMPLEXITY; INTERRUPTED GENES AND RNA SPLICING; EXPRESSION OF SPECIFIC GENES; GENES FOR RIBOSOMAL RNA; HISTONE GENES; GLOBIN GENES; HEAT-SHOCK GENES

Introduction

The transcription and translation take place in different cellular comparts: transcription takes place in the nucleus and translation takes place in the cytosol. In prokaryotes, transcription of Mrna and translation occur simultaneously. Thus, Mrna molecules undergo little or no modification after synthesis by RNA polymerase in prokaryotes. In contrast, pre-Trna and pre-Rrna undergo processing like cleavage, the addition of nucleotides and chemical modification after synthesis. Although both prokaryotes and eukaryotes modify pre-Trna and pre-Rrna, eukaryotes very extensively process pre-Mrna destined to become Mrna. The primary transcript of an RNA polymerase is referred to as pre-RNA. Processing of eukaryotic pre-Mrna involves 5' capping, 3' polyadenylation, splicing and RNA editing before being transported to the cytosol, where they are translated by ribosomes.The final product of gene expression is a polypeptide chain of amino acids whose sequence was prescribed by the genetic code.

- Mrna transcribed from genomic DNA.
- Transfer RNA (Trna) to transport amino acids.
- Ribosome to "read" Mrna, align amino acids attached to Trna and create the peptide bonds between adjacent amino acids.

Ribosome

Two subunits composed of protein and ribosomal RNA (Rrna). The Rrna is a structural component of the ribosome subunits.

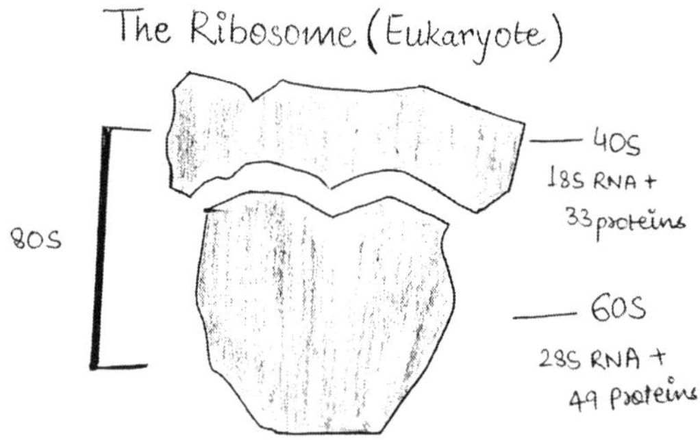

The Genetic Code

tRNA

tRNAs are small, extremely stable RNA structures shaped like a cloverleaf due to internal base pairing. They are almost identical in both prokaryotes and eukaryotes. They contain a number of post-transcriptional modifications, including non-traditional bases (other than (A, U, G and C). The amino acid attaches to the 3' end of the tRNA. tRNA can base pair with the codon in mRNA at the **anticodon** on the tRNA.

Charging tRNA with an amino acid

An individual tRNA is specific for a single amino acid. (There are at least 20 different tRNAs). **Charging** is the process of covalently attaching an amino acid to the tRNA. Charging is accomplished by aminoacyl tRNA synthetases, each of which is specific for an amino acid.

- Initiation—the two subunits of the ribosome come together and the start codon on the mRNA in the ribosome is aligned to set the reading frame.
- Elongation—charged tRNAs attach and peptide bonds form between the amino acids.
- Termination

Prokaryotes

Translation takes place directly after transcription. mRNA is not modified. Transcription and translation take place in the same area.

Eukaryotes

The transcript is modified before leaving the nucleus (5' cap and 3' poly-A tail).

- Modifications increase translation efficiency and lifespan of the mRNA.
- Translation takes place on ribosomes located in the rough endoplasmic reticulum (translation is physically separated from transcription).
- Influence on gene expression forms the basis of function in many antibiotics.
- The purpose is to impair function in the prokaryote without disrupting function in eukaryotes.

Antibiotics

- Tetracycline: Blocks attachment of the AA-tRNA to the A site in the ribosome.
- Erythromycin: Blocks the movement of the ribosome down the mRNA.
- Streptomycin: Blocks the formation of the initiation complex and causes miscoding.
- Rifamycin: Blocks transcription by blocking the binding of RNA polymerase.

Genetic code

It consists of 64 triplet codons (A, G, C, U) $4^3 = 64$. All codons are used in protein synthesis. It has 20 amino acids. It has 3 termination (stop) codons: UAA, UAG, and UGA. So, AUG (methionine) is the start codon (also used internally). It has multiple codons for a single amino acid = degeneracy. It has 5 amino acids are specified by the first two nucleotides only.

The Genetic Code

UUU	Phe	UCU	Ser	UAU	Tyr	UGU	Cys
UUC		UCC		UAC		UGC	
UUA	Leu	UCA		UAA	Stop	UGA	Stop
UUG		UCG		UAG		UGG	Trp
CUU	Leu	CCU	Pro	CAU	His	CGU	Arg
CUC		CCC		CAC		CGC	
CUA		CCA		CAA	Gln	CGA	
CUG		CCG		CAG		CGG	
AUU	Ile	ACU	Thr	AAU	Asn	AGU	Ser
AUC		ACC		AAC		AGC	
AUA		ACA		AAA	Lys	AGA	Arg
AUG	Met	ACG		AAG		AGG	
GUU	Val	GCU	Ala	GAU	Asp	GGU	Gly
GUC		GCC		GAC		GGC	
GUA		GCA		GAA	Glu	GGA	
GUG		GCG		GAG		GGG	

Codon-anticodon interactions

It has codon-anticodon base-pairing is antiparallel. The third position in the codon is frequently degenerate. One tRNA can interact with more than one codon (therefore 50 tRNAs). The Wobble rules.

- C with G or I (inosine).
- A with U or I.
- G with C or U.
- U with A, G, or I.
- I with C, U, or A.

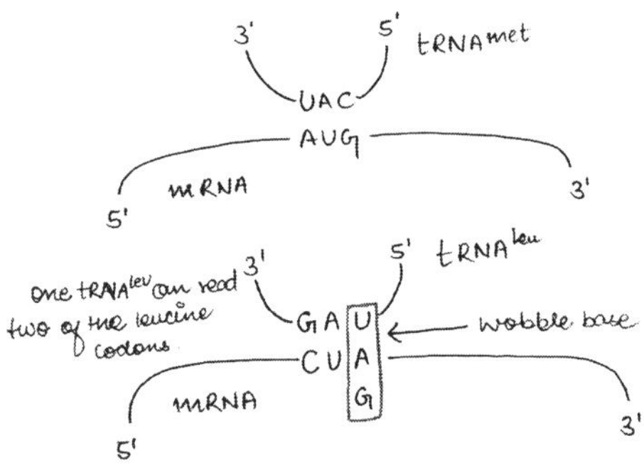

Termination Point

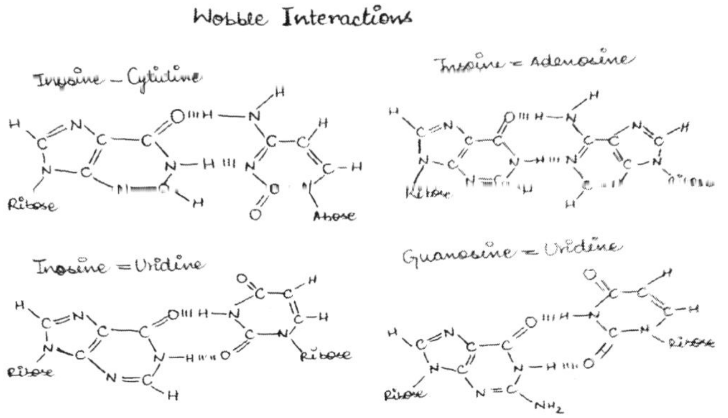

Initiation in Prokaryotes and Eukaryotes

Initiation can occur at internal AUG codons in prokaryotic mRNA. Initiation in eukaryotes occurs only at the first AUG codon. Lac operon in *E. coli* is transcribed as a polycistronic mRNA with multiple AUG codons.

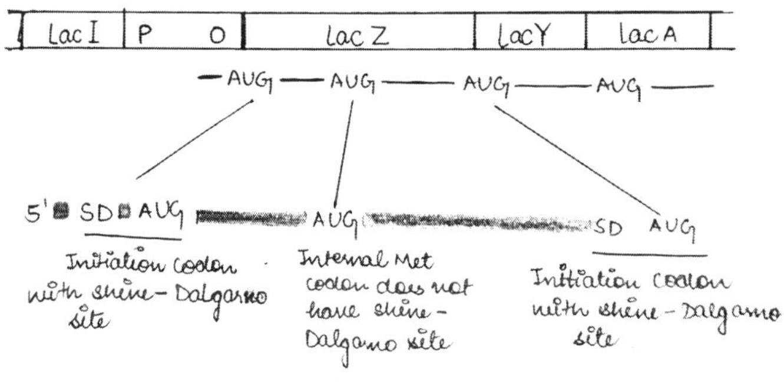

Reading frame

The reading frame is determined by the AUG initiation codon. Every subsequent triplet is read as a codon until reaching a stop codon.

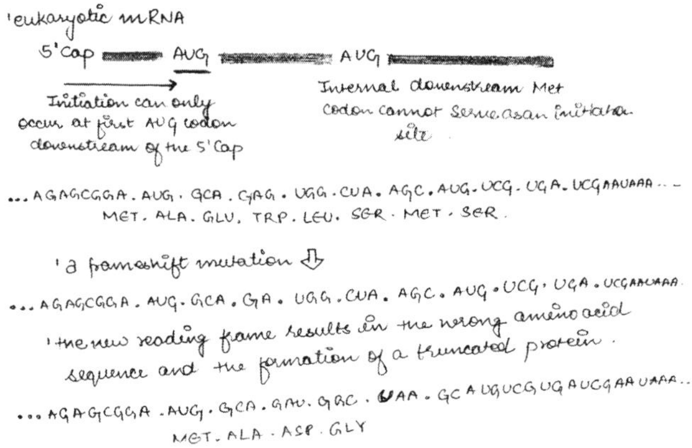

Mutations Affecting Translation

Hemoglobin wayne (3' terminal frameshift mutation). Missense mutations are (e.g., AGC Ser to AGA Arg). Nonsense mutations are (e.g., UGG Trp to UGA Stop). Read through, reverse terminator, or sense mutations (e.g., UAA Stop to CAA Gln) as in hemoglobin constant spring. Silent mutations are (e.g., CUA Leu to CUG Leu) do not affect translation.

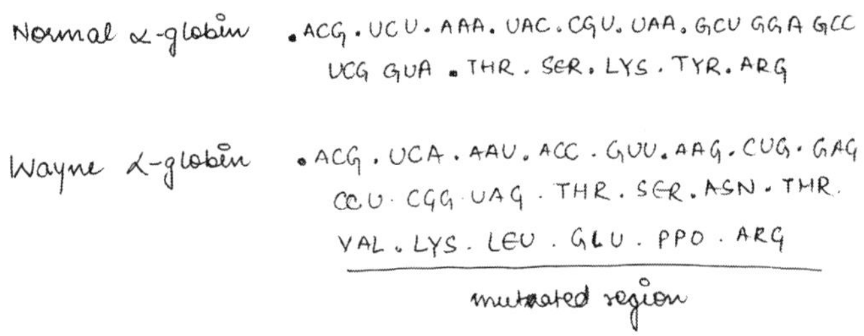

Post-translational modifications

- Cleavage of signal peptides
- Phosphorylation
- Amidation
- Glycosylation
- Hydroxylation
- Ubiquitination
- Addition of prosthetic groups

- Iodination
- Adenylation
- Sulfonation
- Prenylation
- Myristoylation
- Acylation
- Acetylation
- Methylation
- Oxidative crosslinking
- N-Glutamyl cyclization
- Carboxylation

N-terminal modifications

- N-formyl Met (by-product of translation process and incomplete cleavage by deformylase).
- Aminopeptidase cleavage of Met.
- N-terminal acetylation.
- N-terminal myristoylation.
- N-terminal glutamyl cyclization.

C-terminal modification

- C-terminal prenylation
- Farnesylation
- Geranylation/geranyl-geranylation
- C-terminal amidation

Phosphorylation

- Hydroxyl groups - Ser, Thr, and Tyr.
- Amide groups of His, Lys & Carboxy group Asp.
- It often causes changes in protein conformation and/or protein-protein interactions.

C-terminal amidation

- Neuropeptides and hormones
- C-terminal glycine is hydroxylated alpha-hydroxy-glycine.
- Glyoxylate. C-terminal amidation is essential to the biological activity of many neuropeptides and hormones.

Hydroxylation and carboxylation

- Hydroxylation (previous lecture).
- Lys and Pro (e.g. collagen).
- Requires vitamin C.
- Carboxylation
- Requires vitamin K.

Iodination

- Special case: Tyr → thyroxin.

Protein acylation

- Methylation and acetylation.
- Acetylation of histone Lys.
- S-acylation (Cys).
- Prenylation (C-terminus).
- Myristoylation (N-terminus).

Acetylation

N-terminal acetylation

- No simple sequence consensus.
- It usually accompanied by cleavage of the N-terminal Met.

Methylation

- It can occur at carboxyls, amides, sulfhydryl, etc.
- S-Adenosyl Methionine (S-AM) is the activated carbon source.

Histones

- DNA binding proteins.
- The chromatin structure.
- It affect gene expression.

Operon

A unit of bacterial gene expression and regulation, including structural genes and control elements in DNA recognized by regulator gene product(s). Trans-acting – A product that can function on any copy of its target DNA. This implies that it is a diffusible protein or RNA. Cis-acting – A site that affects the activity only of sequences on its own molecule of DNA (or RNA); this property usually implies that the site does not code for protein. Regulator gene – A gene that codes for a product (typically a protein) that controls the expression of other genes (usually at the level of transcription). Structural gene – A gene that codes for any RNA or protein product other than a regulator. In negative regulation, a repressor protein binds to an operator to prevent a gene from being expressed. In positive regulation, a transcription factor is required to bind at the promoter in order to enable RNA polymerase to initiate transcription.

- In inducible regulation, the gene is regulated by the presence of its substrate (the inducer).
- In repressible regulation, the gene is regulated by the product of its enzyme pathway (the corepressor).

Gene regulation *in vivo* can utilize any of these mechanisms, resulting in all four combinations: negative inducible, negative repressible, positive inducible and positive repressible.

Structural gene clusters are coordinately controlled

Genes coding for proteins that function in the same pathway may be located adjacent to one another and controlled as a single unit that is transcribed into a polycistronic mRNA.

The lac operon is negative inducible

Transcription of the *lacZYA* operon is controlled by a repressor protein (the *lac* repressor) that binds to an operator that overlaps the promoter at the start of the cluster. It constitutive expression-A state in which a gene is expressed continuously. In the absence of β-galactosides, the *lac* operon is expressed only at a very low (basal) level. The repressor protein is a tetramer of identical subunits coded by the *lacI* gene. β-galactoside sugars, the substrates of the *lac* operon, are its inducer. The addition of specific β-galactosides induces transcription of all three genes of the *lac* operon. The *lac* mRNA is extremely unstable; as a result, induction can be rapidly reversed.

Lac repressor is controlled by a small-molecule inducer

An inducer functions by converting the repressor protein into a form with lower operator affinity. Repressor has two binding sites, one for the operator DNA and another for the inducer. Gratuitous inducer – Inducers that resemble authentic inducers of transcription, but are not substrates for the induced enzymes.

The repressor is inactivated by an allosteric interaction in which binding of inducer at its site changes the properties of the DNA-binding site (allosteric control). The true inducer is allolactose, not the actual substrate of β-galactosidase.

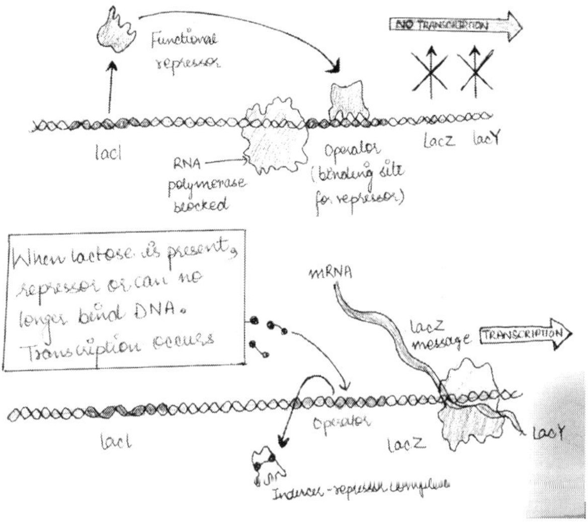

Lac Operon

cis-Acting Constitutive Mutations Identify The Operator

Mutations in the operator cause constitutive expression of all three *lac* structural genes. These mutations are *cis*-acting and affect only those genes on the contiguous stretch of DNA. Mutations in the promoter prevent the expression of *lacZYA* and are uninducible and *cis*-acting. Cis-dominant – A site or mutation that affects the properties only of its own molecule of DNA, often indicating that a site does not code for a diffusible product.

Trans-acting Mutations Identify the Regulator Gene

Mutations in the *lacI* gene are Trans-acting and affect the expression of all *lacZYA* clusters in the bacterium. Mutations that eliminate *lacI* function cause constitutive expression and are recessive (*lacI*$^{-}$). Mutations in the DNA-binding site of the repressor are constitutive because the repressor cannot bind the operator. Mutations in the inducer-binding site of the repressor prevent it from being inactivated and cause uninducibility. When mutant and wild-type subunits are present, a single *lacI*$^{-d}$ mutant subunit can inactivate a tetramer whose other subunits are wild-type. It is dominant negative. Interallelic complementation – The change in the properties of a heteromultimeric protein brought about by the interaction of subunits coded by two different mutant alleles. The mixed protein may be more or less active than the protein consisting of subunits of only one or the other type. Negative complementation – This occurs when interallelic complementation allows a mutant subunit to suppress the activity of a wild-type subunit in a multimeric protein. *lacI*$^{-d}$ mutations occur in the DNA-binding site. Their effect is explained by the fact that repressor activity requires all DNA-binding sites in the tetramer to be active.

Positive Gene Regulation

If glucose levels are low (along with overall energy levels), then cyclic AMP (cAMP) binds to cAMP receptor protein (CRP) which activates transcription. If glucose levels are sufficient and cAMP levels are low (lots of ATP), then the CRP protein has an inactive shape and cannot bind upstream of the *lac* promoter.

Points of control of gene regulation

The control of gene expression can occur at any step in the pathway from gene to a functional protein.

1. Packing/unpacking DNA
2. Transcription
3. mRNA processing
4. mRNA transport

5. Translation
6. Protein processing
7. Protein degradation

34. THE DYNAMIC GENOME: MOBILE GENETIC ELEMENTS IN EUKARYOTES (JUMPING GENES)

Introduction

DNA sequence that can move in the genome. It is also called a mobile DNA element or transposable element. It has "selfish DNA" that exist only to maintain themselves. **Transposition:** The process by which these sequences are copied and inserted into a new site in the genome. Probably had a significant influence on evolution. In the 1940s, Barbara McClintock discovered the first transposable element in maize, earned a Nobel prize in 1983. In the late 1960s, transposition was also found in bacteria.

Retrotransposons

Most mobile elements in bacteria are DNA transposons. In contrast, most mobile elements in eukaryotes are retrotransposons, but eukaryotic DNA transposons also occur.

DNA transposons

- Bacterial insertion sequences (IS element).
- P element in *Drosophila.*

Retrotransposons

LTR retrotransposons.

Non-LTR retrotransposons: the most common type of transposons in mammals.

Non-LTR retrotransposons

- Long interspersed elements (LINEs) ≈6 kb in human account for 21% of the genome.
- Short interspersed elements (SINEs) ≈300 bp in human account for 13% of the genome.

Transposable Elements

Transposable elements are called mobile genetic elements. It comprise 45% of human chromosomal DNA "middle repetitive DNA". It contribute to spontaneous mutation, genetic rearrangements, and horizontal transfer of genetic material. The aid speciation and genomic change (in bacteria transposons are often associated with antibiotic resistance genes). The cells must depress transposition to insure genetic stability.

Types of Transposable Elements

Composite bacterial transposons

It has repeated ends, usually inverted, sometimes direct. The repeated ends themselves are IS elements and can independently transpose. The ends mobilize all intervening DNA. It often antibiotic resistance genes (examples Tn3 (ampicillin), Tn5 (kanamycin), Tn10 (tetracycline). It often reside on plasmids.

Types of DNA Transposons

Excisive mechanism examples: Tn5, Tn10, P elements.

Replicative mechanism examples: Tn3, bacteriophage Mu.

Excisive transposons

Cut-and-paste mechanism are called excisive transposons. It has cut them out of the original site, producing double-strand break. Cut target site and ligate to element ends, thereby inserting at the new site. The original site break repaired usually with sister chromosome, restoring transposon at the original site and sometimes end healed without transposon, can also be associated with deletion at the excision site.

Replicative transposons

The orignal cut of the transposon is only nick and only one strand is initially ligated. Element replicates through itself. Produces as intermediate a "co-integrate" structure. Co-integrate is resolved by resolvase (as TnpR of Tn3) and at a specific site (as *res* of Tn3).

"Degenerate" transposons

Many naturally occurring transposable elements have suffered from mutations and are no longer active. Some of these may have cis-acting end mutations and cannot be mobilized. Others may have intact ends but no transposase: these can be mobilized by an element that is tnp+ ("autonomous" element). Ac Ds system is an example of the latter: Ac (activator) can mobilize Ds (dissociator).

- MITEs (minature inverted repeat transposable elements) are nonautonomous DNA elements.
- SINEs are retrotransposon version (LINEs).

Classification of Retroelements

It has an obligate RNA intermediate, use reverse transcriptase (RT, RNA-dependent DNA polymerase).

- LTR-retroelements: long terminal repeatsTy1/copia, Ty3/gypsy and retroviruses.
- Non-LTR-retroelements "retroposons" LINES.

Characteristics of LTR Retroelements

Long terminal repeats are required for the replication cycle. Genes: gag, pol, (viruses also have env). Pol is polyprotein that gives rise to RT (reverse transcriptase), IN (integrase) RH (RNase H), PR (protease). Forms VLPs virus-like particles. Integrase is functionally and structurally similar to transposase of DNA transposons, DDE motif. The integration gives characteristic TSD.

Characteristics of non-LTR retroelements

It contains 2 ORFs, orf1, orf2. The variable TSD. ORF2 gives rise to EN, endonuclease, (similar to APE) and RT. It uses target primed reverse transcription TPRT. It can transduce 3' downstream non-element segments.

Human L1 (LINE-1) retroelement

It presents 15% of human DNA. It contains 520,000 copies, only 3-5,000 are full-length. It is associated with human disease loci. Transpose specifically in germline.

CHAPTER 6

TECHNIQUES IN BIOCHEMISTRY AND MOLECULAR BIOLOGY

35. CENTRIFUGATION TECHNIQUES-BASIC PRINCIPLES, ANALYTICAL AND PREPARATIVE CENTRIFUGATION, THEIR APPLICATIONS.

Introduction

Centrifugation is a process used to separate or concentrate materials suspended in a liquid medium. The theoretical basis of this technique is the effect of gravity on particles (including macromolecules) in suspension. The two particles of different masses will settle in a tube at different rates in response to gravity. It is a process that involves the use of the centrifugal force for the sedimentation of heterogeneous mixtures with a centrifuge used in industry and in laboratory settings. This process is used to separate two immiscible liquids. More dense components of the mixture migrate away from the axis of the centrifuge, while less-dense components of the mixture migrate towards the axis.

Principle

It is a device for separating particles from a solution according to their size shape, density, viscosity of the medium and rotor speed. In a solution particles whose density is higher than that of the solvent sink sediment and the particles that are lighter than it floats to the top. The greater the difference in density in the faster they move. If there is no difference in density isopyknic conditions, the particles stay steady.

- Routine centrifuge rotors
- Calculation of *g*-force
- Differential centrifugation
- Density gradient theory

Centrifuge rotor

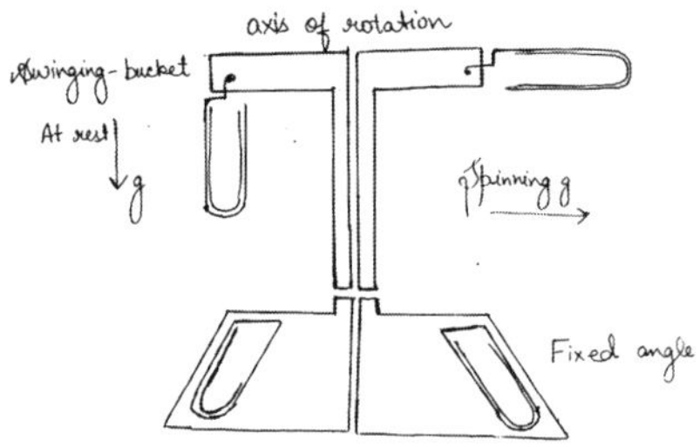

Geometry of rotors

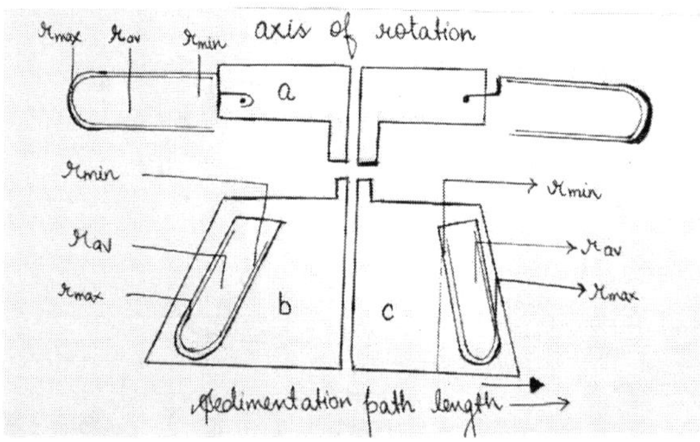

k'-factor of rotors

The *k*'-factor is a measure of the time taken for a particle to sediment through a sucrose gradient

- The most efficient rotors which operate at a high RCF and have a low sedimentation path length therefore have the lowest *k*'-factors

- The centrifugation times (t) and k'-factors for two different rotors (1 and 2) are related by;

$$t_1 = \frac{k_1 t_2}{k_2}$$

Calculation of RCF and Q

- RCF in swinging-bucket and fixed-angle rotors at 40,000 rpm
- Beckman SW41 swinging-bucket (13 ml)
- g_{min} = 119,850g; g_{av} = 196,770g;
- g_{max} = 273,690g
- Beckman 70.1Ti fixed-angle rotor (13 ml)
- g_{min} = 72,450g; g_{av} = 109,120g;
- g_{max} = 146,680g

$$RCF = 11.18 x r \left(\frac{Q}{1000}\right)^2$$

$$Q = 299\sqrt{\frac{RCF}{r}}$$

RCF = Relative Centrifugal Force (*g*-force)
Q = rpm; *r* = radius in cm

Velocity of sedimentation of a particle

$$v = \frac{d^2(\rho_p - \rho_l)}{18\mu} g$$

v = velocity of sedimentation d = diameter of particle
ρ_p = density of particle ρ_l = density of liquid
g = centrifugal force μ = viscosity of liquid

Differential centrifugation

The density of a liquid is uniform. The density of liquid << Density of particles. The viscosity of the liquid is low. Consequence: The rate of particle sedimentation depends mainly on its size and the applied *g*-force.

Size of major cell organelles

• Nucleus	4-12 μm
• Plasma membrane sheets	3-20 μm
• Golgi tubules	1-2 μm
• Mitochondria	0.4-2.5 μm
• Lysosomes/peroxisomes	0.4-0.8 μm
• Microsomal vesicles	0.05-0.3μm

Differential centrifugation of a tissue homogenate

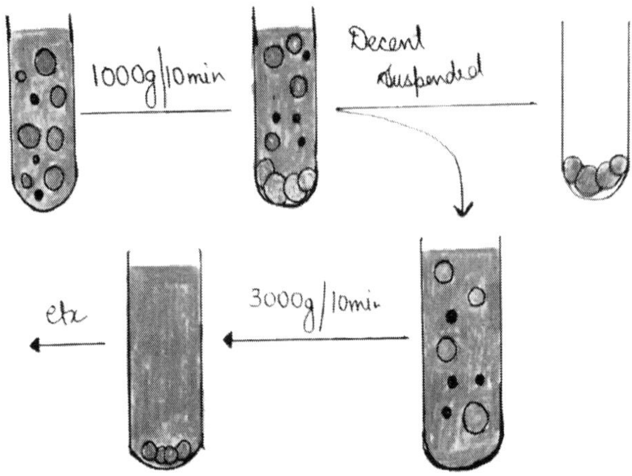

Differential centrifugation of a tissue homogenate

1. Homogenate – 1000g for 10 min
2. Supernatant from 1 – 3000g for 10 min
3. Supernatant from 2 – 15,000g for 15 min
4. Supernatant from 3 – 100,000g for 45 min
   - Pellet 1 – nuclear
   - Pellet 2 – "heavy" mitochondrial
   - Pellet 3 – "light" mitochondrial
   - Pellet 4 – microsomal

Differential centrifugation (III) expected content of pellets

- 1000g pellet: nuclei, plasma membrane sheets
- 3000g pellet: large mitochondria, Golgi tubules
- 15,000g pellet: small mitochondria, lysosomes, peroxisomes
- 100,000g pellet: microsomes

Differential centrifugation

- Poor resolution and recovery because of particle size heterogeneity.
- Particles starting out at r_{min} have the furthest to travel but initially experience lowest RCF.
- Smaller particles close to r_{max} have only a short distance to travel and experience the highest RCF.

Differential centrifugation

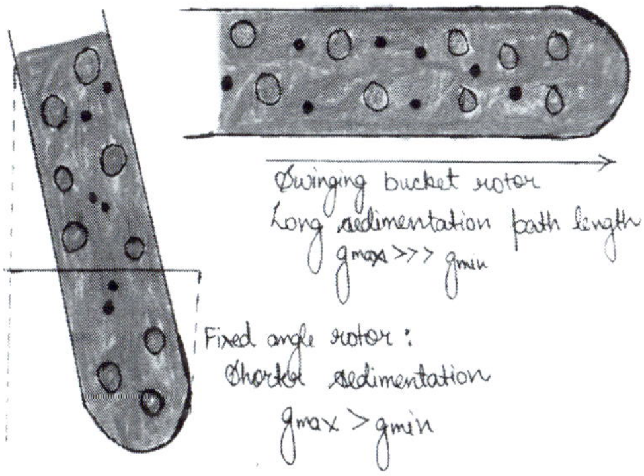

Differential centrifugation

The rate of sedimentation can be modulated by particle density. The nuclei have an unusually rapid sedimentation rate because of their size and high density. Golgi tubules do not sediment at 3000g, in spite of their size: they have an unusually low sedimentation rate because of their very low density: $(\rho_p - \rho_l)$ becomes rate-limiting.

Density gradient centrifugation

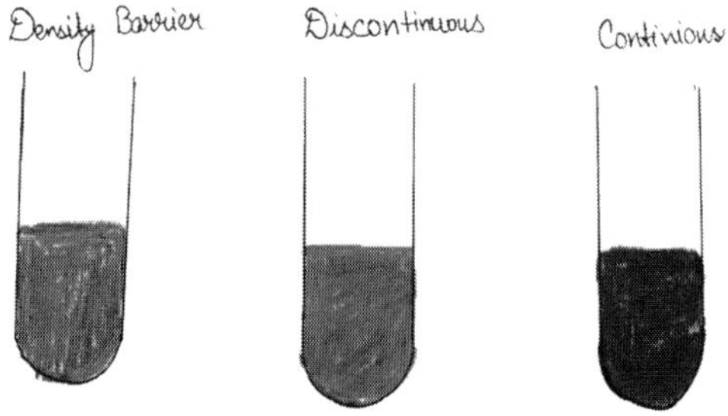

A particle will sediment through a solution if particle density > solution density. If particle density < solution density, the particle will float through the solution. When particle density = solution density the particle stop sedimenting or floating.

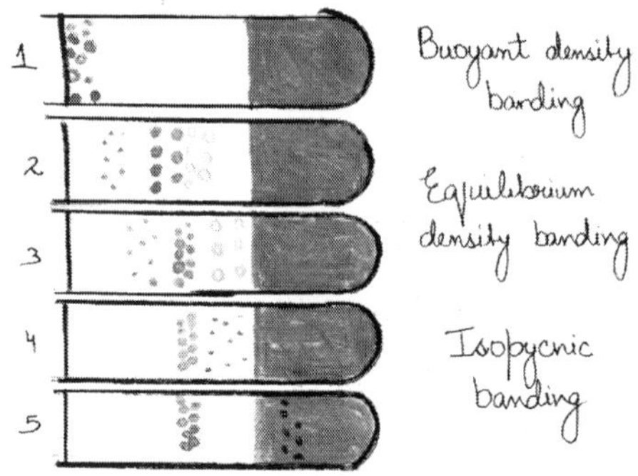

It has formats for separation of particles according to their density.

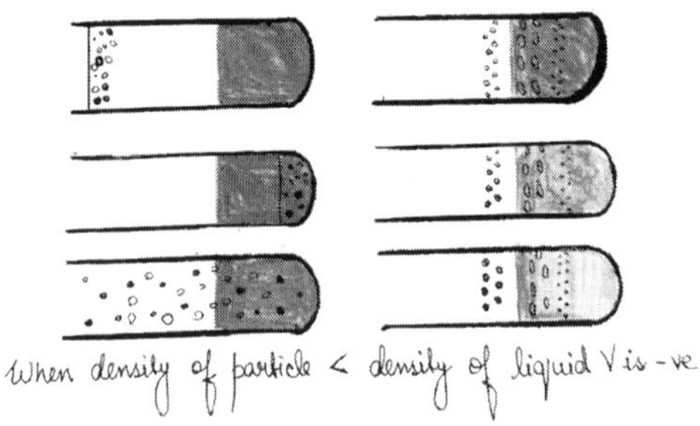

Resolution of density gradients

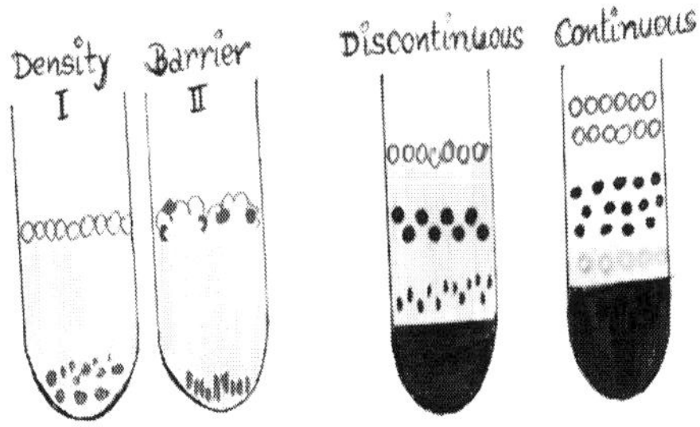

Separation of particles according to size

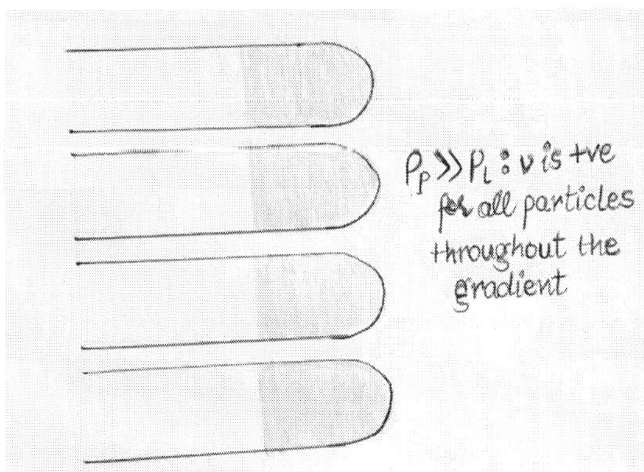

Application of Centrifugation

- Centrifugation can be employed to separate a mixture of two different miscible liquids.
- This technique can also be used in order to study and analyze macromolecules and their hydrodynamic properties.
- Mammalian cells can be purified with the help of a special type of centrifuge.
- Centrifugation is known to have a vital role in the fractionation of many subcellular organelles. Furthermore, centrifugation is also useful in the fractionation of membrane fractions and membranes.
- Centrifugation also has applications in the fractionation of membrane vesicles.

- Chalk can be separated from water with the help of a centrifuge.
- Skimmed milk is a form of milk that has a lower amount of dissolved fats. Skimmed milk can be obtained from regular milk with the help of the process of centrifugation. Here, the centrifuge serves to separate the fat from the milk, leaving the required skimmed milk behind.
- Cyclonic separation is an important process that has vital applications in the separation of particles from air flows.
- Another important application of this technique is in the stabilization and clarification of wine.
- This technique, in combination with other purification techniques, is extremely helpful while separating proteins.
- Other techniques that are used include salting out techniques such as ammonium sulfate precipitation.
- Centrifuges are widely used in the field of forensic chemistry.
- In this field, the technique is employed for the separation of blood components from blood samples.
- Furthermore, the technique is also employed in certain laboratories for the separation of urine components from urine samples.
- Differential centrifugation, a distinct type of centrifugation, is known to have applications in the identification of organelles.

36. SPECTROPHOTOMETRY: UV-VISIBLE SPECTROPHOTOMETER.

Introduction

Photometry refers to the science of light measurement in terms of the light's brightness that the human eye can perceive. As a basic principle, photometers examine how light interacts with reflective materials. The array of photometers allow for photometry to apply in different fields of study. Some instruments are used to observe how light absorbs and/or reflects wavelengths. Other instruments measure light by converting light into an electrical current and then measuring the electrical current produced by the light. Photometry is often used in the study of liquids and solutions in chemistry. Photometers can help measure masses of organic or inorganic materials in a solution or liquid.

Colorimetry

An analytical technique in which the concentration of an analyte is measured by its ability to produce or change the color of a solution.

- Changes the solution's ability to absorb light.

Spectrophotometry

Any technique that uses light to measure chemical concentrations. A colorimetric method where an instrument is used to determine the amount of analyte in a sample by the sample's ability or inability to absorb light at a certain wavelength.

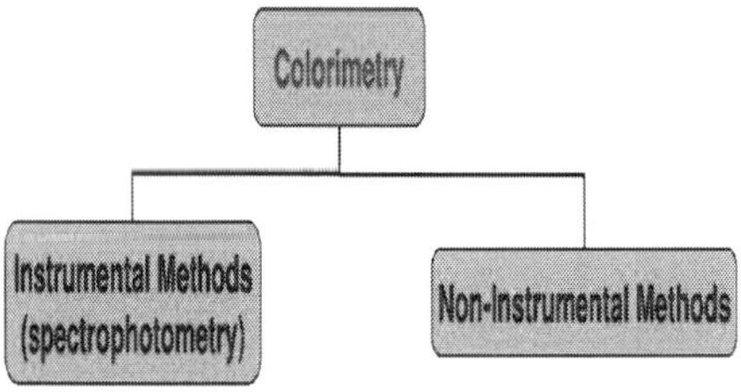

Measurement of Ozone (O_3) Above the South Pole

- O_3 provides protection from ultraviolet radiation
- Seasonal depletion due to chlorofluorocarbons

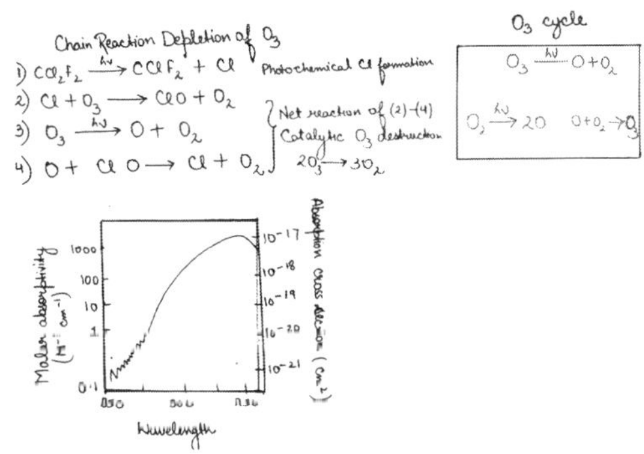

Properties of Light

Particles and waves

Light waves are consist of perpendicular, oscillating electric and magnetic fields.

- Parameters used to describe light.
- Amplitude (A): height of wave's electric vector.
- Wavelength (l): distance (nm, cm, m) from peak to peak
- Frequency (n): number of complete oscillations that the waves make each second

Hertz (Hz): unit of frequency, second^{-1} (s^{-1})

1 megahertz (MHz) = $10^6 s^{-1}$ = 10^6Hz

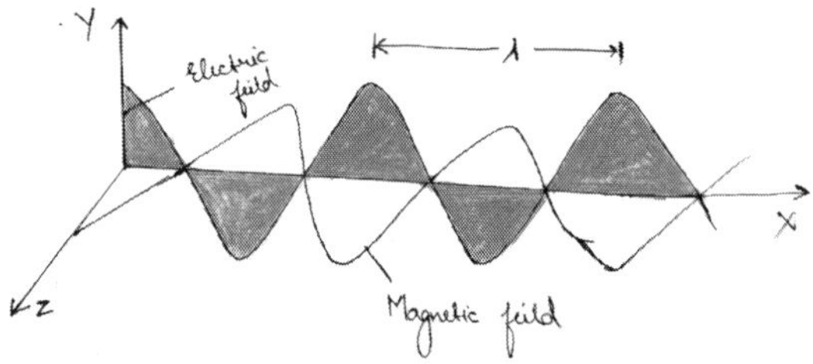

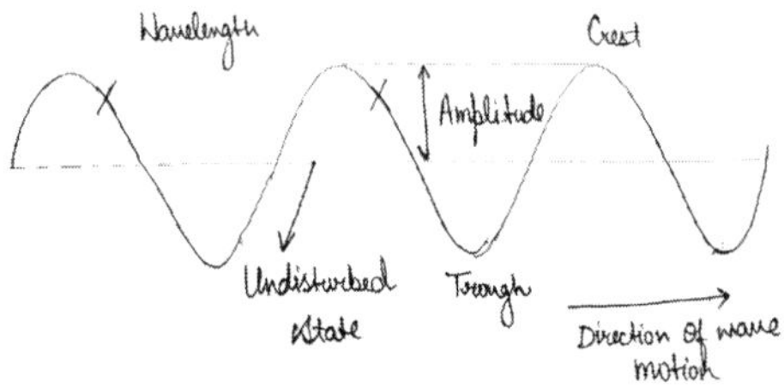

Properties of light

Particles and waves

- Parameters used to describe light.
- **Energy (E):** the energy of one particle of light (photon) is proportional to its frequency

$E = hv$

where: E = photon energy (Joules)

n = frequency (sec^{-1})

h = Planck's constant ($6.626 x 10^{-34}$J-s)

- As frequency (n) increases, energy (E) of light increases

Relationship between frequency and wavelength

$\lambda v = c = V$

$V = c/\lambda$

where: c = speed of light (3.0×10^8 m/s in vacuum))

n = frequency (sec^{-1})

l = wavelength (m)

Relationship between energy and wavelength

$E = hc/\lambda = hcv$

where: = (1/l) = wavenumber

- As frequency (l) decreases, energy (E) of light increases.

Types of light: The electromagnetic spectrum

- Note again, energy (E) of light increase as frequency (n) increases or wavelength (l) decreases

Absorption of Light

1.) Colors of visible light

Many types of chemicals absorb various forms of light. The color of light absorbed and observed passing through the compound are complimentary.

Wavelength of maximum absorption (nm)	Color absorbed	Color observed
380—420	Violet	Green-yellow
420—440	Violet-blue	Yellow
440—470	Blue	Orange
470—500	Blue-green	Red
500 520	Green	Purple
520—550	Yellow-green	Violet
550—580	Yellow	Violet-blue
580—620	Orange	Blue
620—680	Red	Blue-green
680—780	Red	Green

2.) Ground and excited state

When a chemical absorbs light, it goes from a low energy state (ground state) to a higher energy state (excited state). Only photons with energies exactly equal to the energy difference between the two electron states will be absorbed. Since different chemicals have different electron shells that are filled, they will each absorb their own particular type of light. Different electron ground states and excited states

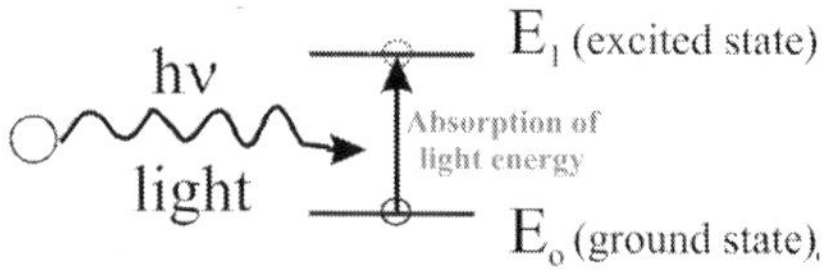

The energy required by a photon to give this transition:

$DE = E_1 - E_o$

3.) Beer's Law

The relative amount of a certain wavelength of light absorbed (*A*) that passes through a sample is dependent as

- The distance the light must pass through the sample (cell path length - b)
- Amount of absorbing chemicals in the sample (analyte concentration – c).
- The ability of the sample to absorb light (molar absorptivity - e).
- The relative amount of light making it through the sample (*P/P$_o$*) is known as the transmittance (T)

$T = P/P_0$

Percent transmittance $\%T = 100 * P/P_0$

T has a range of 0 to 1, %T has a range of 0 to 100%

Absorbance (A) is the relative amount of light absorbed by the sample and is related to transmittance (T). Absorbance is sometimes called optical density (OD).

$$A = -\log\left[\frac{P}{P_0}\right] = -\log(T) = -\log(\%T/100)$$

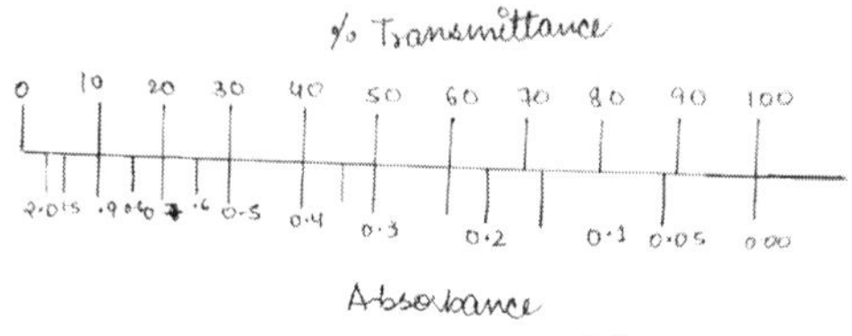

A has a range of 0 to infinity

Absorbance is useful since it is directly related to the analyte concentration, cell pathlength and molar absorptivity. This relationship is known as Beer's Law

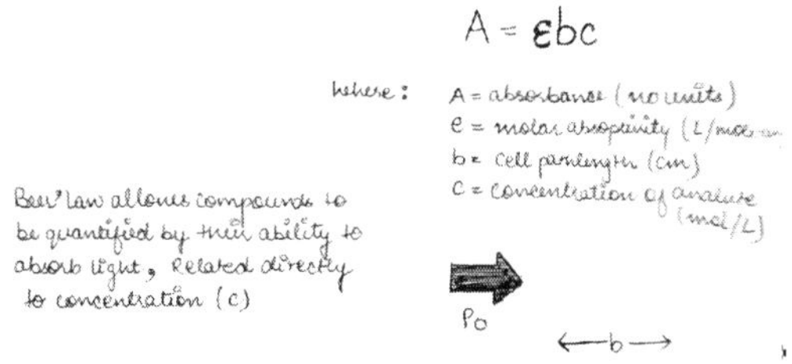

Different chemicals have different energy levels.

- The different ground vs. excited electron states.
- It will have different abilities to absorb light at any given wavelength.

Absorption Spectrum – the plot of absorbance (or e) vs. wavelength for a compound. The greater the absorbance of a compound at a given wavelength (high e), the easier it will be to detect at low concentrations. By choosing different wavelengths of light (1_A vs. 1_B) different compounds can be measured.

Spectrophotometer is an instrument used to make absorbance or transmittance measurements is known as a spectrophotometer.

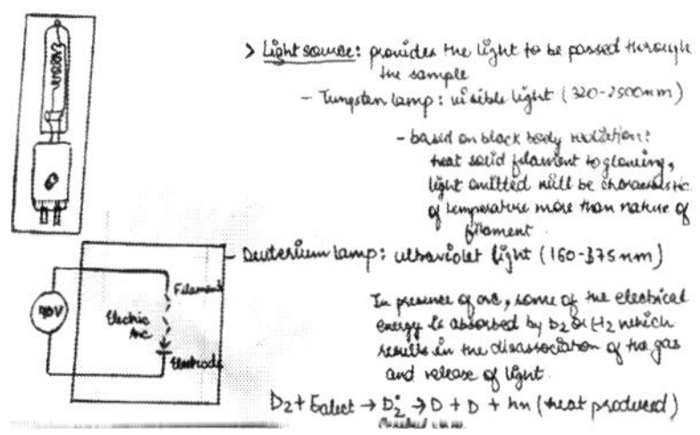

Wavelength selector (monochromator): used to select a given wavelength of light from the light source.

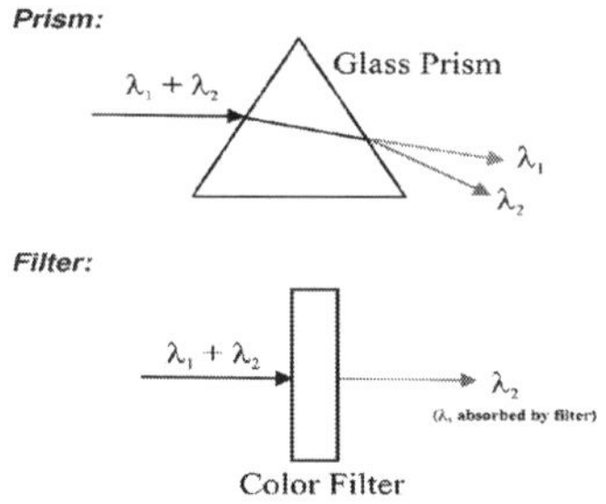

Wavelength selector (monochromator): used to select a given wavelength of light from the light source.

Reflection or diffraction grating

- ***Sample cell***: sample container of fixed length (b).
- It usually a round or square cuvet.
- Made of material that does not absorb light in the wavelength range of interest.
  - Glass – visible region
  - Quartz – ultraviolet
  - NaCl, KBr – Infrared region
- Light Detector: measures the amount of light passing through the sample.
- It usually works by converting the light signal into an electrical signal.

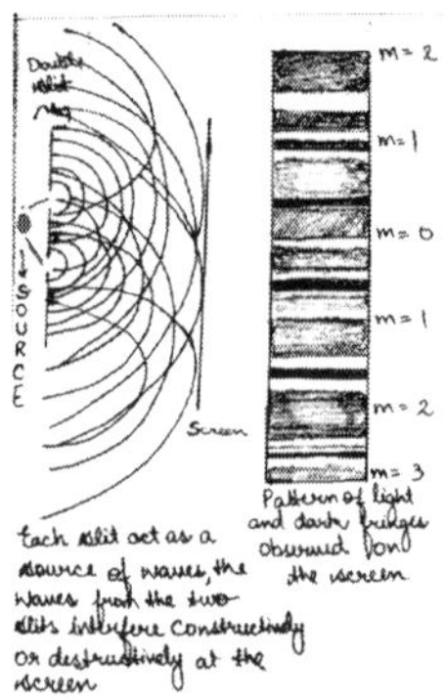

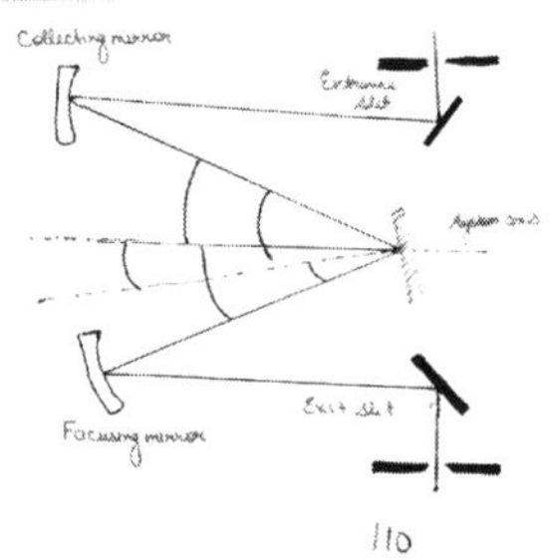

Process

Light hits photoemissive cathode and e^- is emitted. An emitted e^- is attracted to electrode #1 (dynode 1), which is 90V more positive. It causes several more e^- to be emitted. This e^- is attracted to dynode 2, which is 90V more positive than dynode 1, emitting more e^-. The process continues until e^- is collected at anode after amplification at 9 dynodes. The overall voltage between anode and cathode is 900V. One photon produces $10^6 - 10^7$ electrons. The current is amplified and measured.

Types of Spectrophotometers

***Single-beam instrument*:** Sample and blank are alternatively measured in the same sample chamber.

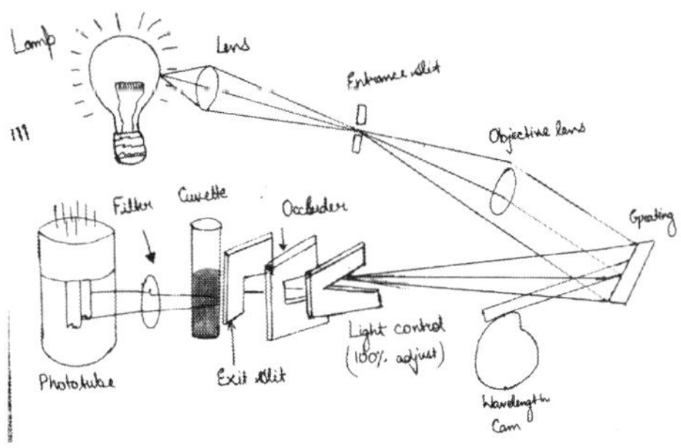

Double-beam instrument

It continuously compares the sample and blank. It automatically corrects for changes in electronic signal or light intensity of the source.

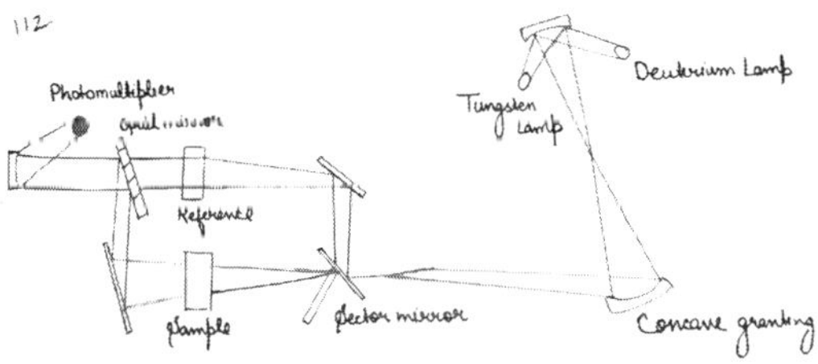

Chemical Analysis

1.) ***Calibration***

- To measure the absorbance of a sample, it is necessary to measure P_o and P ratio

- P_o – the amount of light passing through the system with no sample present
- P – the intensity of light when the sample is present
- P_o is measured with a blank cuvet
- Cuvet contains all components in the sample solution except the analyte of interest
- P is measured by placing the sample in the cuvet.
- To accurately measure an unknown concentration, obtain a calibration curve using a range of known concentrations for the analyte.

2.) Limitations in Beer's Law

The results in a non-linear calibration curve. At high concentrations, solute molecules influence one another because of their proximity.

- Molar absorptivity changes
- Affect on equilibrium, (HA and A^- have difference absorption)

Analyte properties change in different solvents. Errors in reproducible positioning of cuvet. It also problems with dirt & fingerprints.The instrument has electrical noise. Keep A in a range of 0.1 – 1.5 absorbance units (80 -3%T).

Precautions in quantitative absorbance measurements

Choice of wavelength

Choose a wavelength at an absorption maximum. Minimizes deviations from Beer's law, which assumes e- is constant. Pick peak in the absorption spectrum where the analyte is only compound absorbing light. Choose a wavelength where the analyte has the largest difference in its absorbance relative to other sample components.

37. CHROMATOGRAPHIC TECHNIQUES – BASIC PRINCIPLES

Introduction

Mikhail Tswett invented chromatography in 1901 during his research on plant pigments. He used the technique to separate various plant pigments such as chlorophylls, xanthophylls and carotenoids.

Original chromatography experiment

- ***Start***: A glass column is filled with powdered limestone ($CaCO_3$).
- ***End***: A series of colored bands is seen to form, corresponding to the different pigments in the original plant extract. These bands were later determined to be chlorophylls, xanthophylls and carotenoids.

Chromatography

Tswett named this new technique chromatography (Greek = chroma “color” and graphein “writing”) based on the fact that it separated the components of a solution by color.

Types of Chromatography

Tswett’s technique is based on liquid chromatography. There are now several common chromatographic methods. These include

- Paper Chromatography
- Thin Layer Chromatography (TLC)
- Liquid Chromatography (LC)
- High Pressure Liquid Chromatography (HPLC)
- Ion Chromatography
- Gas Chromatography (GC)

Paper and thin layer chromatography

The solvent moves up paper by capillary action, carrying mixture components at different rates.

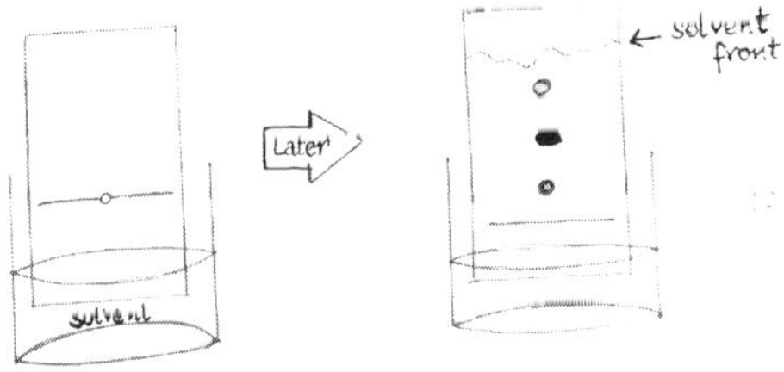

How does Chromatography work?

In all chromatographic separations and the sample is transported in a mobile phase. The mobile phase can be a gas, a liquid, or a supercritical fluid. The mobile phase is then forced through a stationary phase held in a column or on a solid surface. The stationary phase needs to be something that does not react with the mobile phase or the sample. The sample has the opportunity to interact with the stationary phase as it moves past it. Samples that interact greatly then appear to move more slowly. Samples that interact weakly then appear to move more quickly. Because of this difference in rates, the samples can then be separated into their components. It is based on a physical quilibrium that results when a solute is transferred between the mobile and a stationary phase.

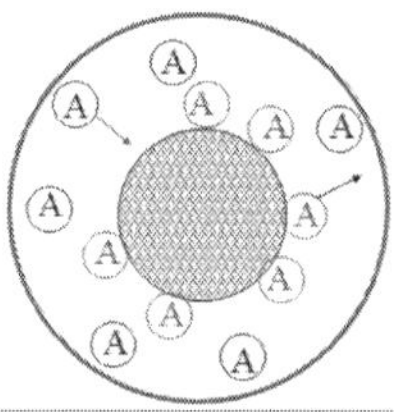

Cross Section of Equilibrium in a column. "A" are adsorbed to the stationary phase. "A" are traveling in the mobile phase.

K = distribution coefficient *or* partition ratio

$$K = \frac{C_S}{C_M}$$

Where C_S is the molar concentration of the solute in the stationary phase and C_M is the molar concentration in the mobile phase.

In a chromatography column, flowing gas or liquid continuously replaces the saturated mobile phase and results in the movement of A through the column. As material travels through the column, it assumes a Gaussian concentration profile as it distributes between the stationary packing phase and the flowing mobile gas or liquid carrier phase.

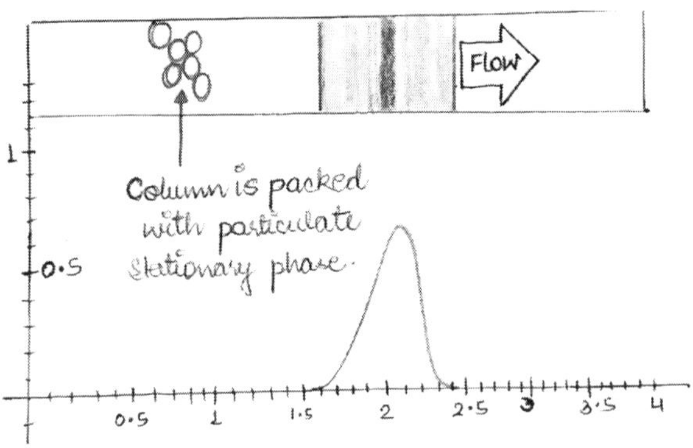

In a mixture, each component has a different distribution coefficient, and thus spends a different amount of time absorbed on the solid packing phase vs being carried along with the flowing gas. More volatile materials are carried through the column more rapidly than less volatile materials, which results in separation.

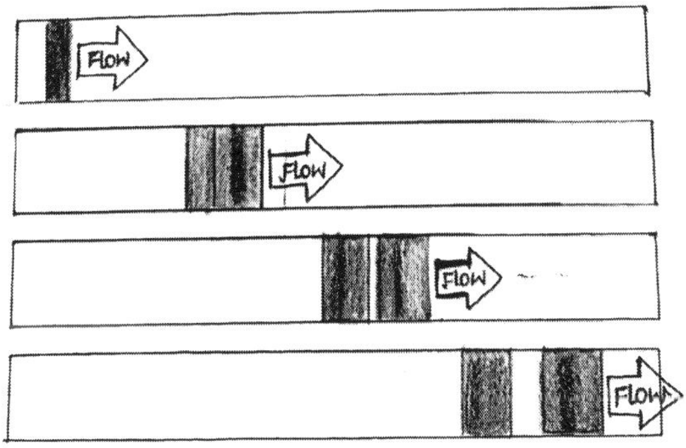

If a detector is used to determine when the components elute from the column, a series of Gaussian peaks are obtained, one for each component in the mixture that was separated by the column.

***Note*:** The first two components were not completely separated.

Peaks in general tend to become shorter and wider with time.

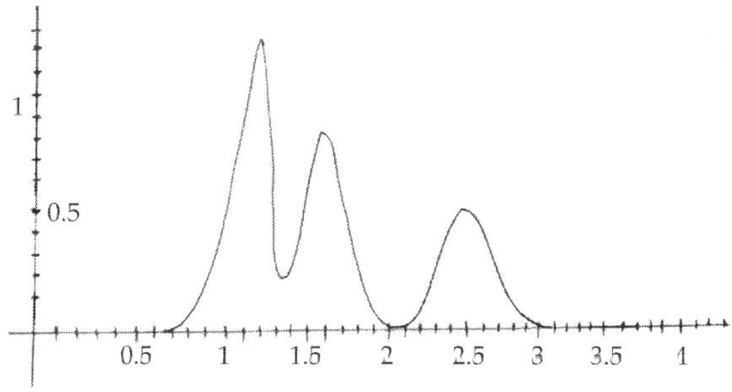

The theoretical plate is a term coined by Martin & Synge. It is based on a study in which they imagined that chromatographic columns were analogous to distillation columns and made up or numerous discrete but connected narrow layers or plates. The movement of the solute down the column then could be treated as a stepwise transfer. Theoretical plates (N) measure how efficiently a column can separate a mixture into its components. This efficiency is based on the retention time of the components and the width of the peaks.

N = Number of theoretical plates (a measure of efficiency)

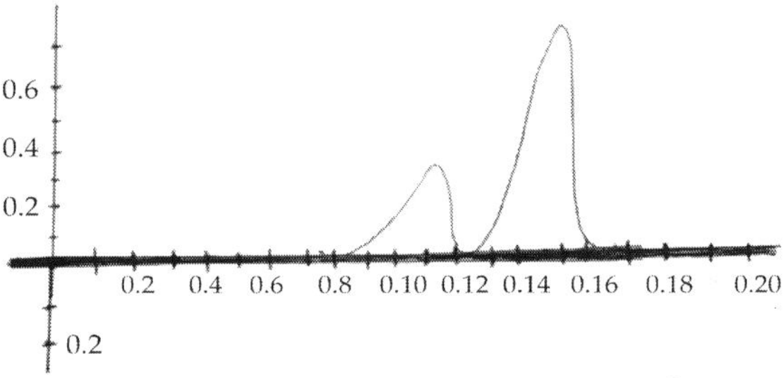

$$N = 16\left(\frac{t_R}{Wb}\right)^2$$

t_R is the retention time; it is measured from the injection peak (or zero) to the intersection of the tangents. w_b is the width of the base of the triangle; it is measured at the intersection of the tangents with the baseline. When the retention time, t_R, is held constant, the column that produces peaks with narrower bases, w_b, will be more efficient have a greater N value. Likewise, a column that produces wider peaks will be less efficient have a smaller N value. This is because a smaller denominator, w_b, will yield a larger overall number and a larger denominator will yield a smaller number.

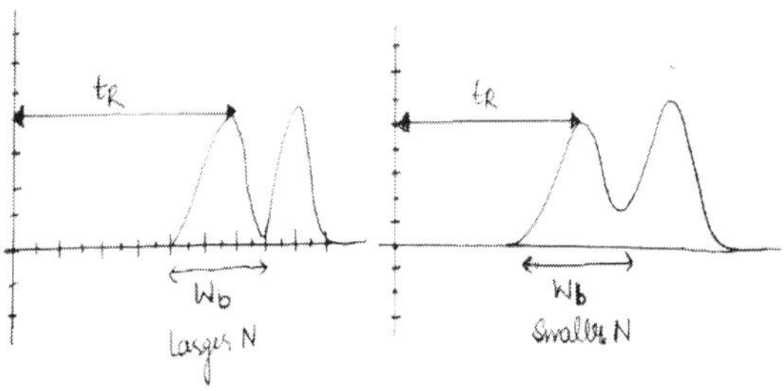

Liquid Chromatography

Liquid Chromatography (LC) is a chromatographic technique in which the mobile phase is a liquid. LC is a much older technique than GC but was overshadowed by the rapid development of GC in the 1950s and 1960s.It is currently the dominate type of chromatography and is even replacing GC in its more traditional applications. In these methods that use large, non-rigid support material

- Particles > 40 mm in diameter
- Poor system efficiencies and large plate heights
- Such systems have the following characteristics:
- Broad peaks
- Poor limits of detection
- Long separation times
- Columns can only tolerate low operating pressures
- Gravity flow or peristaltic pump to apply mobile phase to column

Column chromatography – An Example Of The Equipment Used In Low-Performance Liquid

The sample is usually applied directly to the top of the column. Detection is by fraction collection with later analysis of each fraction.

Advantages

- Simple system requirements.
- Low cost.
- Popular in sample purification.
- Used in the removal of interferences from samples.
- Used in some analytical applications not common due to low efficiency, long analysis times and poor limits of detection.

High-Performance Liquid Chromatography (HPLC)

LC methods that use small, uniform, rigid support material.Particles < 40 mm in diameter.Usually 3-10 mm in practice.Good system efficiencies. Small plate heights. Such systems have the following Characteristics

- Narrow peaks
- Low limits of detection
- Short separation times
- Columns can only tolerate high operating pressures and faster flow-rates

Higher operating pressures need for mobile phase delivery requires special pumps and other system components. The sample applied using a closed system (*i.e.*, injection valve). The detection uses a flowthrough detector.

Advantages

- Fast analysis time.
- Ease of automation.
- Good limits of detection.
- Preferred choice for analytical applications.
- Popular for purification work.

Disadvantages

- Greater expense
- Lower sample capacities

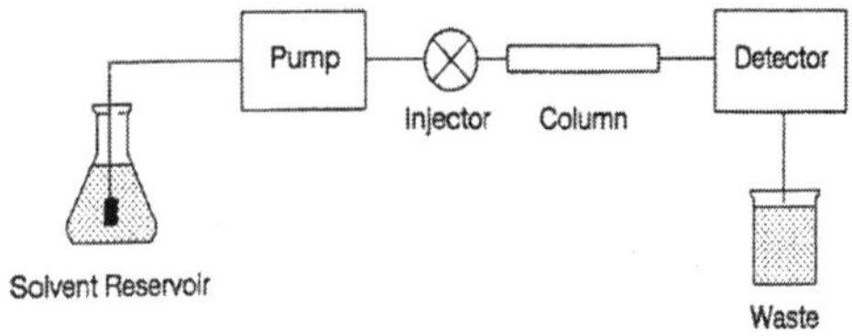

Ion Exchange Chromatography

Ion exchange chromatography (or ion chromatography) is a process that allows the separation of ions and polar molecules based on their affinity to ion exchangers. The principle of separation is thus by a reversible exchange of ions between the target ions present in the sample solution to the ions present on ion exchangers. In this process, two types of exchangers i.e., cationic and anionic exchangers can be used.

- ***Cationic exchangers***: They possess negatively charged groups, and these will attract positively charged cations. These exchangers are also called "Acidic ion exchange" materials, because their negative charges result from the ionization of the acidic group.
- ***Anionic exchangers***: They have positively charged groups that will attract negatively charged anions. These are also called "Basic ion exchange" materials.

Ion exchange chromatography is most often performed in the form of column chromatography. However, there are also thin-layer chromatographic methods that work basically based on the principle of ion exchange his form of chromatography relies on the attraction between the oppositely charged stationary phase, known as an ion exchanger, and analyte. The ion exchangers basically contain charged groups covalently linked to the surface of an insoluble matrix. The charged groups of the matrix can be positively or negatively charged.When suspended in an aqueous solution, the charged groups of the matrix will be surrounded by ions of the opposite charge. In this "ion cloud", ions can be reversibly exchanged without changing the nature and the properties of the matrix.

On exchange Material

Ion exchange material is available naturally and they also can be synthesis.

- ***Natural***: The naturally obtained ion exchangers are different. They are,

1. Cation exchanger like zeolite
2. Anion exchangers like dolomite

B. ***Synthetic***: The synthetic Ion exchange resins arc based on cross-linked polystyrene prepared by co- polystyrene prepared by co-polymersition of styrene and divinyl benzene.

- Other synthetic ion exchange resin used are :-
- Co-polymer of acrylic acid derivative and divinyl benzene
- Cross-linked dextrane.
- Cellulose fibers with introduced charged group.

Gas chromatography

It is an analytical separations technique useful for separating volatile organic compounds consists of : Flowing mobile phase (inert gas - Ar, Ne, N). Injection port (rubber septum - syringe injects sample). kept at a higher temperature than the boiling point. Separation due to differences in partitioning behavior selective retardation. organic compounds separated due to differences in their participating behavior between the mobile gas phase and the stationary phase in the column. in contrast to other types of chromatography, the mobile phase does not interact with molecules of the analyte; its only function is to transport the analyte through the column. The good for volatile samples (up to about 250 °C). 0.1-1.0 microliter of liquid or 1-10 ml vapor. It can detect <1 ppm with certain detectors. It can be easily automated for injection and data analysis.

Components of a gas chromatography

- Gas Supply: (usually N_2 or He).
- Sample Injector: (syringe / septum).
- Column: 1/8" or 1/4" x 6-50' tubing packed with small uniform size, inert support coated with a thin film of nonvolatile liquid
- Detector: TC - thermal conductivity and FID - flame ionization detector

38. ELECTROPHORESIS: THEORY AND DIFFERENT TYPES – PAGE, SDS-PAGE, C APILLARY ELECTROPHORESIS, AND IEF

Introduction

Electrophoresis is a method whereby charged molecules in solution, chiefly proteins and nucleic acids, migrate in response to an electrical field. Their rate of migration through the electrical field depends on the strength of the field, on the net charge, size, and shape of the molecules, and also on the ionic strength, viscosity, and temperature of the medium in which the molecules are moving. As an analytical tool, electrophoresis is simple, rapid and highly sensitive.It can be used analytically to study the properties of a single charged species or mixtures of molecules. It can also be used preparatively as a separating technique. Electrophoresis is usually done with gels formed in tubes, slabs, or on a flat bed. In many electrophoresis units, the gel is mounted between two buffer chambers containing separate electrodes, so that the only electrical connection between the two chambers is through the gel. In most electrophoresis units, the gel is mounted between two buffer chambers containing separate electrodes so that the only electrical connection between the two chambers is through the gel.

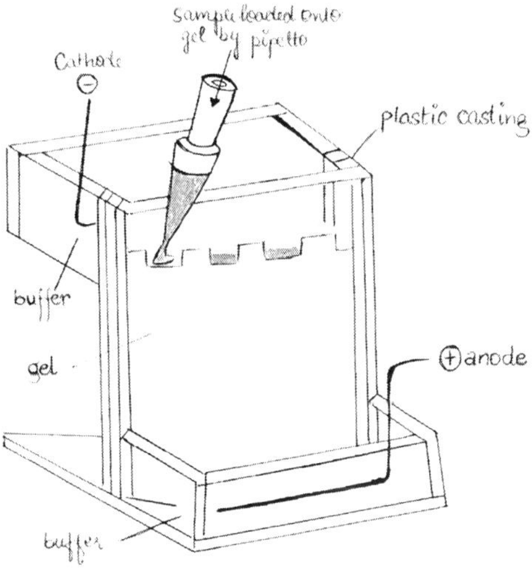

The interrelation of resistance, voltage, current and power. Two basic electrical equations are important in electrophoresis.

- The first is Ohm's Law, $I = E/R$
- The second is $P = EI$
- This can also be expressed as $P = I^2R$

In electrophoresis, one electrical parameter, either current, voltage, or power, is always held constant. Under constant current conditions (velocity is directly proportional to current), the velocity of the molecules is maintained, but heat is generated. Under constant voltage conditions, the velocity slows, but no additional heat is generated during the course of the run. Under constant power conditions, the velocity slows but heating is kept constant. The net charge is determined by the pH of the medium. Proteins are amphoteric compounds, that is, they contain both acidic and basic residues.Each protein has its own characteristic charge properties depending on the number and kinds of amino acids carrying amino or carboxyl groups. Nucleic acids, unlike proteins, are not amphoteric. They remain negative at any pH used for electrophoresis.

During Polymerization

- Exothermic reaction
- Gel irregularities
- Pore size

During electrophoresis

- Denaturation of proteins
- Smile effect
- Temperature regulation of buffers

Role of the Solid Support Matrix

It inhibits convection and diffusion, which would otherwise impede the separation of molecules. It allows a permanent record of results through staining after the run. It can provide additional separation through molecular sieving.

Agarose and Polyacrylamide

Although agarose and polyacrylamide differ greatly in their physical and chemical structures, they both make porous gels. A porous gel acts as a sieve by retarding or, in some cases, completely obstructing the movement of macromolecules while allowing smaller molecules to migrate freely. By preparing a gel with restrictive pore size, the operator can take advantage of molecular size differences among proteins. Because the pores of an agarose gel are large, agarose is used to separate macromolecules such as nucleic acids, large proteins and protein complexes. Polyacrylamide, which makes a small pore gel, is used to separate most proteins and small oligonucleotides. Both are relatively electrically neutral.

Agarose Gels

Agarose is a highly purified uncharged polysaccharide derived from agar. Agarose dissolves when added to boiling liquid. It remains in a liquid state until the temperature is lowered to about 40° C at which point it gels. The pore size may be predetermined by adjusting the concentration of agarose in the gel. Agarose gels are fragile, however, they are actually hydrocolloids, and they are held together by the formation of weak hydrogen and hydrophobic bonds.

Polyacrylamide Gels

Polyacrylamide gels are tougher than agarose gels. Acrylamide monomers polymerize into long chains that are covalently linked by a crosslinker. Polyacrylamide is chemically complex, as is the production and use of the gel.

SDS Gel Electrophoresis

In SDS separations, migration is determined not by intrinsic electrical charge of polypeptides but by molecular weight. Sodium dodecyl sulfate (SDS) is an anionic detergent that denatures secondary and non-disulfide–linked tertiary structures by wrapping around the polypeptide backbone. SDS confers a net negative charge to the polypeptide in proportion to its length. When treated with SDS and a reducing agent, the polypeptides become rods of negative charges with equal "charge densities" or charge per unit length.

Continuous and Discontinuous Buffer Systems

A continuous system has only a single separating gel and uses the same buffer in thetanks and the gel. In a discontinuous system a non-restrictive large pore gel, called a stacking gel, is layered on the top of a separating gel. The resolution obtainable in a discontinuous system is much greater than that obtainable in a continuous one. However, the continuous system is a little easier to set up. Determining molecular weights of proteins by SDS-PAGE. Run a gel with standard proteins of known molecular weights along with the polypeptide to be characterized. A linear relationship exists between the $\log_{10}$ of the molecular weight of a polypeptide and its R_f. Rf = ratio of the distance migrated by the molecule to that migrated by a marker dye-front. The R_f of the polypeptide to be characterized is determined in the same way, and the $\log_{10}$ of its molecular weight is read directly from the standard curve

Blotting

Blotting is used to transfer proteins or nucleic acids from a slab gel to a membrane such as nitrocellulose, nylon, DEAE, or CM paper. The transfer of the sample can be done by capillary or Southern blotting for nucleic acids (Southern, 1975) or by electrophoresis for proteins or nucleic acids.

Isoelectric Point

There is a pH at which there is no net charge on a protein and this is the isoelectric point (pI). Above its isoelectric point, a protein has a net negative charge and migrates toward the anode in an electrical field. Below its isoelectric point, the protein is positive and migrates toward the cathode.

Isoelectric Focusing

Isoelectric focusing is a method in which proteins are separated in a pH gradient according to their isoelectric points. Focusing occurs in two stages; first, the pH gradient is formed. In the second stage, the proteins begin their migrations toward the anode if their net charge is negative or toward the cathode if their net charge is positive. When a protein reaches its isoelectric point (pI) in the pH gradient, it carries a net charge of zero and will stop migrating.

Two-dimensional gel electrophoresis

Two-dimensional gel electrophoresis is widely used to separate complex mixtures of proteins into many more components than is possible in conventional one-dimensional electrophoresis. Each dimension separates proteins according to different properties.

39. MOLECULAR BIOLOGY TECHNIQUES-SOUTHERN BLOTTING, NORTHERN BLOTTING, WESTERN BLOTTING, MICROARRAY TECHNOLOGY

Introduction

DNA, RNA and proteins are separated by blotting techniques. These are described below:

Southern Blotting

A method, developed genes in a DNA restriction fragment is called as Southern blotting technique. Southern blots can easily provide a physical map of restriction sites within a gene located normally and reveal the number of copies of the gene in the genome, and the degree of similarity of the gene when compared with the other complementary genes. A molecular biologist

E.M. Southern (1975) for analyzing the related on a chromosome. The procedure starts with the digestion of the DNA population by one or many restriction enzymes.Consequently, DNA fragments of unequal length are produced. This preparation is passed through agarose gel electrophoresis which results in the separation of DNA molecules based on their size. DNA restriction fragments present in gel are denatured by alkali treatment. Gel is then put on top of the buffer saturated filter paper. The upper surface of the gel is covered with nitrocellulose filter and overlaid with dry filter paper. The dry filter paper draws the buffer through the gel. The buffer contains single-stranded DNA. Nitrocellulose filter binds DNA fragments strongly. After baking at 80 ° C, DNA fragments are permanent and fixed to the when come in contact of the nitrocellulose filter. Then the filter is placed in a solution containing radio-labeled RNA or denatures DNA probe of known sequences. These are complementary in sequence to the blot-transferred DNA. The radiolabeled nucleic acid probe hybridizes the DNA on the nitrocellulose filter. The filter is thoroughly washed to remove the probe. The hybridized regions are autoradiographically detected by placing the nitrocellulose filter in contact with a photographic film.

Northern Blotting

The southern blotting technique could not be directly applied to the blot transfer of mRNA separated by gel electrophoresis, because RNA was found not to bind with nitrocellulose filter **Alwines** et al (1979) devised a technique in which RNA bands are blot transferred from the gel to chemical reactive paper. An aminobenzyloxymethyl cellulose paper. Prepared from Whatman filter paper No. 540 after a series of uncomplicated reactions is diazotized and rendered into the reactive paper and, therefore, becomes available hybridization with radiolabeled DNA probes. The hybridized bands are found by autoradiography.Thus, Alwine's method extends that of the Southern method and tor this reason it has been grunt the jargon Northern blotting. "There is not this northern or western like Southern." These blot transfers are reusable because of the firm covalent bonding of RNA to the reactive paper. Effective in binding the DNA as well Small fragments of DNA can most effectively be transferred to the diazotized paper derivative to nitrocellulose. These techniques were more advanced and demonstrated that mRNA bands can also be blotted directly on nitrocellulose paper and appropriate conditions.

Western Blotting

Towbin et al. (1979) developed the western blotting technique to find out the newly encoded protein by a transformed cell. Its working principle lies on antigen-antibody reaction: hence is an immuno detection technique. In this method, radiolabeled nucleic acid probes are not used. This technique follows the following steps: (Extraction of protein from transformed cells. (I) Separation of protein by using SDS-PAGE (sodium dodecyl sulfate polyacrylamide electrophoresis) where SDS acts as a solvent for electrophoresis. Transfer of electrophoresed gel in a buffer at low temperatmre (40° C) for half an hour (h) Blotting of proteins on nitrocellulose filter paper of nitrocellulose filter, Whaiman filter and coarse filter in transfer buffer. So a kind of nitrocellulose filters Whatman filter and coarse filter in transfer buffer. Placing of Whatman filter paper on a cathode plate followed by stack of coarse filter Whatman filter, electrophoresed gel, nitrecellulose filter, Whatman filter paper, coarse fil- ter stack, Whatman filter and anode plate transferred. Putting the complete set up in transfer tank containing sufficient transfer buffer. Application of an electric field (30 V overnight for 5 hours) to cause the migration of proteins from the gel to nitrocellulose filter and binding on its surface. (The nitrocellulose filter has exact image of pattern of proteins as present in the gel. This type of blotting is called western blotting). Hybridization of proteins by using radiolabeled antibodies (I123- antibodies) of known structure, isolated from the rabbit. Washing of nitrocellulose filter with a wash solution (Tris-buffered saline + tween 20) to facilitate the removal of unhybridised antibodies. Detection of hybridized sequences by autoradiography. The dots of diagram show the presence of desired protein.

Microarray

Allows simultaneous measurement of the level of transcription for every gene in a genome (gene expression). Differential expression, changes over time. Single microarray can test ~10k genes. Data obtained faster than can be processed. Want to find genes that behave similarly. A pattern discovery problem (Green = repressed and Red = induced).

40. COMPLEMENTATION TECHNIQUES; POLYMERASE CHAIN REACTION (PCR), MOLECULAR MARKERS AND PLANT DIVERSITY

Introduction

During the last few years, with the improvement of PCR protocols and due to the availability of automatic thermal cyclers commercially the applications of PCR have increased manifold. However, for the application of PCR, we need a pair of primers which should be based on the knowledge of the nucleotide sequence of DNA to be amplified. Therefore, the non-availability of this information about the DNA segment or gene to be amplified becomes a limitation in the applications of PCR, although this difficulty has been overcome for the study of DNA polymorphism. Some of the applications of PCR technology of random DNA primer are described in this section.

Polymerase chain reaction (PCR)

We discussed the techniques of recombinant DNA and gene cloning, which permit us to obtain an unlimited supply of identical copies of a gene sequence or DNA segment, which is cloned in a prokaryotic a eukaryotic cell with the help of a vector. This technique, discovered around 1975, proved extremely useful in experiments of molecular biology and genetic engineering and thus became an essential tool in all molecular biology laboratories. In 1985, yet another remarkable tool in molecular biology was discovered, which is known as polymerase chain reaction (PCR) nicknamed now as people's choice reaction.

- The methodology is so important that the prestigious journal Science published from the USA considered PCR as the major scientific development of the year 1989, and had chosen **Taq DNA polymerase**, the enzyme used in PCR, in the year 1989. **Kary Mullis** who discovered this Scientific American, how he got the basic idea of this reaction, while driving through the redwood mountains of California, USA.
- The polymerase chain reaction is such a powerful technique that may replace completely the gene cloning with vectors in due course of time.
- The DNA replication involves polymerization of nucleotides using a template DNA strand with the help of an enzyme DNA

polymerase, but this reaction invariably requires a primer strand to which further nucleotides can be added using the DNA polymerase enzyme. If the primer strand is not available, the reaction cannot proceed.

- In the living cells, this primer strand is not a DNA strand but is a small single-stranded RNA molecule synthesized with the help of RNA polymerase enzyme.
- In PCR, a similar reaction takes place in an 'Eppendorf tube', where the primer strand is added from outside in the form of a deoxyoligonucleotide, and DNA polymerase enzyme is added to help in polymerization.
- An unlimited supply of amplified NA is obtained by repeating the reaction, which is made possible by regular denaturation of freshly synthesized double-stranded DNA molecules by heating it to 90-98°C.
- At this high temperature the two strands separate. Once the double-stranded DNA is made single-stranded by heating up to 90-98°C, the mixture with two primers, recognizing the two strands and bordering the sequence to be amplified is cooled to 40-60□ C.
- This allows the primers (which are in excess) to bind to their complementary strands through renaturation.
- The presence of **Taq** DNA polymerase enzyme and all the four essential nucleoside triphosphates allow the synthesis of complementary strands in the usual manner.
- In a thermal cycler this process is automatically repeated 20 and 30 times so that in a single afternoon, a billion copies of the sequence flanked by the left and right primers can be produced.
- In order to continue the synthesis, the temperature of the mixture is alternately increased (for denaturation) and decreased (for renaturation) once every 1-3 minutes (as fixed by the computer device).
- Therefore during temperature rise, the enzymatic activity of DNA polymerase should not be destroyed, otherwise one may have to add a fresh aliquot of the enzyme in each cycle of amplification.

- This became possible only by the discovery of the thermostable enzyme Taq DNA polymerase, isolated from ***Thermus aquaticus*** growing in hot springs.
- This enzyme acts best at 72°C and the need denaturation temperature of 90°C does not destroy its enzymatic activity.
- Later, other thermostable enzymes like ***Pflu*** DNA polymerase isolated from *Pyrococcus furiosus* and Vent polymerase isolated from ***Thermococcus litoralis*** were discovered and were found more efficient.
- These enzymes allowed automation of the entire process and automatic PCR thermal cyclers are now available, which can amplify DNA sequences at a fast speed unattended: **Figure 40.1.**
- **As pointed** out earlier, PCR may eventually replace the gene cloning technique. This technology is much easier and requires a much smaller quantity nanogram or ng 10g of DNA. For cloning, DNA is needed, as the starting material, in microgram (ug) quantities.

Different schemes of PCR

- In the previous section, we discussed the basic outline of the PCR technique, outlining how billions of copies of a DNA sequence can be obtained within a few hours, without the use of a vector and a host as done in gene cloning.
- However, there are several variations to this basic PCR technique, which allow varied applications of this technique.

Inverse PCR

- In this technique, the amplification of those DNA sequences takes place, which is away from the primers and not those flanked by the primers.
- For instance, if the border sequences of a DNA segment are not known and those of a vector are known, there the sequence to be amplified may be cloned in the vector and sequences of the vector may be used as polymerization sequence flanked by the primers and towards the DNA inserted segment.

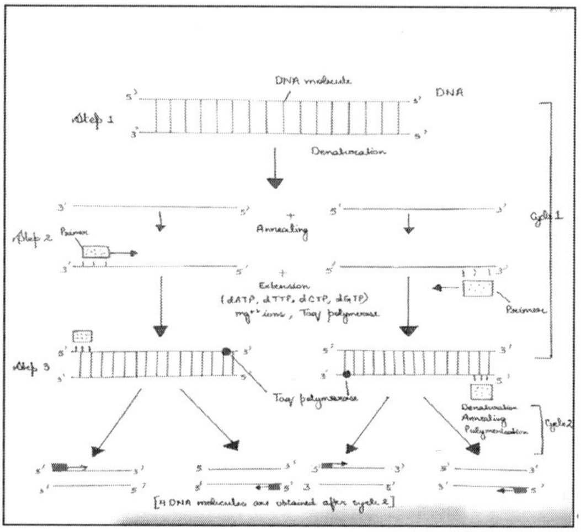

Figure 40.1. Polymerase chain reactions

Anchored PCR

- In the basic PCR technique and the inverse PCR, one has to use two primers representing the sequences lying at the ends of sequences to be amplified.
- But sometimes, we may have knowledge about sequence at only one of the two ends of the DNA sequence to be amplified.
- In such cases, anchored PCR may be used, which will utilize only one primer instead of two primers. In this technique, due to the use of one primer, only one strand will be copied first, after which a poly G tail will be attached at the end of the newly synthesized strand.
- This newly synthesized strand with poly G tail at its 3' end will then become a template for the daughter strand synthesis utilizing an anchor primer with which a poly C sequence is linked to complement with poly G of the template.
- In the next cycle, both the original primer and anchored primer will be used for gene amplification.

Cloning and expression of the PCR product

- The amplified PCR product can be cloned if restriction sites for specific enzymes are introduced at the 5' end of each primer.

- These restriction sites do not interfere with amplification and can be used for cloning with a variety of vectors in the usual manner.
- Similarly, incorporation of a T7 promoter at the 5' end of one primer will allow copies of PCR product on the use of RNA polymerase enzyme to get RNA copies of the PCR product.

Asymmetric PCR for DNA sequencing

- PCR products can also be used as templates for direct DNA sequencing (ligation-mediated PCR - LMPCR).
- Single-stranded DNA can be produced for this purpose using asymmetric PCR, in which the two primers are used in a 100:1 ratio, so that after 20-25 cycles of amplification, one primer is exhausted so that single-stranded DNA is produced in the next 5-10 cycles.
- The morphological markers generally correspond to the qualitative traits that can be scored visually.
- They have been found in nature or as the result of mutagenesis experiments.
- Morphological markers are usually dominant or recessive. Until recently, however, the genetic markers are used to develop maps in plants have been those affecting morphological characters, including genes for dwarfism, albinism and altered leaf morphology etc.
- Scientists have long theorized about the use of genetic maps and markers to speed up the process of plant and animal breeding. In 1923, **Sax** proposed the identification and selection of minor genes of interest by linkage with major genes, which could be scored more easily.
- There are constrainers on the use of morphological markers.
- It causes such large effects on phenotypes that they are undesirable in breeding programs.
- The effects of linked minor genes, making it nearly impossible to identify desirable linkages for selection.
- It is highly influenced by the environment. The morphology phenotype is the result of the genetic constitution and its interaction with the environment.

- Due to varying levels of G × E interaction, it is not appropriate to compare the morphological data of varieties that have been collected across different years and or locations.
- It exhibits dominant intragenic interactions. Therefore homozygous dominant and heterozygous individuals cannot be distinguished.
- The traits show intergenic interactions leading to deviated dihybrid segregation ratios that are not useful for linkage analysis.
- It covers only a limited fraction of the genome of an organism.

There are two types of markers

- Biochemical markers
- Molecular Markers

Biochemical markers

- There are proteins produced by gene expression.
- These proteins can be isolated and identified by electrophoresis and staining.
- Isozymes the different molecular forms of the same enzyme that catalyze the same reaction and proteins.
- They are revealed on electrophoresis through a colored reaction associated with enzymatic activity.
- They are the products of various alleles of one or several genes.
- Monomer and dimer isozymes are most often used because analysis of their segregation is easier. Isozymes are generally codominant.

Moelecular Markers

- The molecular markers are DNA sequences that are readily detected and whose inheritance can be easily monitored.
- The use of molecular markers is based on naturally occurring DNA polymorphism, which forms the basis for designing strategies to exploit for applied purposes.
- A marker must be polymorphic, that chromosome carrying the mutant gene can be distinguished from the chromosome with the normal gene by markers it also carries.

- List of molecular markers such as (Table 40.1).

Table 40.1. Molecular markers

Acronym	Nomenclature
RFLP	Restriction Fragment Length Polymorphism
VNTR	Variable Number Tandem Repeat
SSCP	Single-stranded Conformational Polymorphism
STS	Sequence Tagged Sites
STMS	Sequence Tagged Microsatellite sites
RAPD	Random Amplified Polymorphic DNA
AP-PCR	Arbitrarily Primed Polymerase Chain Reaction
DAF	DNA Amplification Finger Printing
SSR	Simple Sequenced Repeats
CAPS	Cleaved Amplified Polymorphic Sequence
MAAP	Multiple Arbitrary Amplicon Prefilling
SCAR	Sequenced Characterized Amplified Region
SNP	Single Nucleotide Polymorphism
SAMPL	Selective Amplification of Microsatellite Polymorphic Loci
ISSR	Inter-Simple Sequence Repeat
ALFP	Amplified Fragment Length Polymorphism
S-SAP	Sequence specific Amplified Polymorphism
RBIP	Retrotransposon-Based Insertional
IRAP	Inter-Retrotranspososon Amplified Polymorphism
REMAP	Retrotransposon-Microsatellite Amplified Polymorphism

The moleculars markers have some desirable properties:

- It must be polymorphic as it is the polymorphism that is measured for genetic diversity studies.
- *Codominant inheritance:* The different forms of a marker should be detectable in diploid organisms to allow discrimination of homo- and heterozygous.
- The marker should be evenly and frequently distributed throughout the genome.
- It should be easy, fast cheap to detect.
- It should be reproducible.

- It is high exchange of data between laboratories. Unfortunately, there is no single molecular marker that meets all these requirements.

Non-PCR based approaches

- Those are based on DNA-DNA hybridization between a DNA or RNA probe and total genomic DNA, e.g., RFLP (Restriction Fragment Length Polymorphism).

PCR based approaches

- Those based on the PCR amplification of genomic DNA fragments: Random amplified polymorphic DNA (PARD), microsatellites or simples sequences repeat polymorphic (SSRP), amplified fragment length polymorphism (AFLP), arbitrarily primed PCR (AP-PCR), SCAR, SNP etc.
- However some assays combine features from both (e.g., retrotransposon based insertional polymorphism-RBIP).

Targetted PCR and Sequencing

Those based on sequencing: sequenced tagged sites (STS), sequence characterized amplified region (SCARs) sequence tagged microsatellites (STMs), cleaved amplified polymorphic sequences (CAPS) etc.

NON-PCR based Approaches

Restriction fragment length polymorphism (RFLP)

Procedure

- Isolate the DNA.
- Cut the DNA into smaller fragments using restriction enzymes (s).
- Separation of DNA fragments by gel electrophoresis.
- Transfering DNA fragments to a nylon or nitrocellulose membrane filter.
- Visualization of specific DNA fragments using labeled probes (Incubation with clined probe and allow for hybridization).

- Southern Hybridization (Autoradiography)
- Analysis of results (RFLP pattern with positive bands): Figure 40.2.

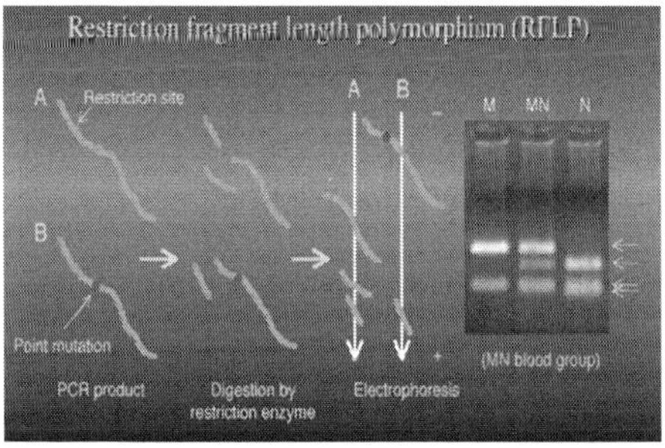

Figure 40.2. Restriction fragment length polymorphism

Uses of RFLP

- It permits direct identification of a genotype or cultivar in any tissue at any developmental stage in an environment independent manner.
- RFLPs are codominant markers enabling heterozygotes to be distinguished from homozygous.
- It has discriminating power that can be at the species/population (single locus probes).
- The method is simple as no sequence-specific information is required.
- Indirect selection using qualitative traits.
- Indirect selection using quantitative trait loci.

PCR Based Approaches

Random amplified polymorphic DNA (RAPD) markers

Procedure

- Isolation of DNA.
- Keep the tubes in PCR thermocycler (Taq DNA polymerase, primer, dNTPs and buffer).
- Denature the DNA (94°C, 1 min).

- DNA strands separated.
- Annealing of primer (36°, 2 min).
- Primer annealed to template DNA strands (DNA synthesis 72°C, 1.5 min).
- Complementary strand synthesis (35 to 45 cycles).
- Amplified products are separated by gel electrophoresis.

Application of RAPD

- Construction of genetic maps.
- Mapping of traits.
- Analysis of the genetic structure of populations.
- Fingerprinting of individuals.
- Targeting markers to specific regions of the genome.
- RAPDs have been utilized for the identification of somatic hybrids.
- RAPDs have been recommended as a useful system for the evaluation and characterization of genetic resources.
- Some success has already been achieved in the characterization of collections from tree genera Gliricidia and Thenobroma.

Limitations of RAPD

RAPD polymorphisms are inherited as dominant-recessive characters. This causes a loss of information relative to markers that show codominance.

RAPD primers are relatively short, a mismatch of even a single nucleotide can often prevent the primer from annealing, hence there is a loss of band.

The production of non-parental bands in the offspring of known pedigrees warrants its use with caution and extreme care.

RAPD is sensitive to changes in PCR conditions, resulting in changes to some of the amplified fragments.

Amplified Fragment Length Polymorphism (AFLP).

Procedure

High quality genomic DNA is cut with restriction enzymes (generally by two enzymes) and double-stranded (ds) oligonucleotide adaptors are ligated to the ends of the DNA fragments.

Selective amplification of sets of restriction fragments is usually carried out with 32P labeled primers designed according to the sequence of adaptors plus 1-3 additional nucleotides. Only fragments containing the restriction site sequence plus the additional nucleotides will be amplified.

Gel analysis of the amplified fragments. The amplification products are separated on highly resolving sequencing gels and visualized using autoradiography.

Fluorescent or silver staining techniques can be used to visualize the products in cases where radiolabeled nucleotides are not used in the PCR.

Advantages of AFLP

This technique is extremely sensitive.

It has high rcproducibility, rendering it superior to RAPD.

It has wide-scale applicability, proving extremely proficient in revealing diversity.

It discriminates heterozygotes from homozygotes when a gel scanner is used.

It permits the detection of restriction fragments in any background or complexity, including pooled DNA samples and cloned DNA segments.

It is not a single fingerprinting technique but it can be used for mapping also.

Disadvantages

It is highly expensive and requires more DNA that is needed in RAPD (1 µg per reaction).

It is technically more demanding than RAPDs as it requires the experience in sequencing gels.

AFLPs are expensive to generate as silver staining, fluorescent dye or radioactivity detects the bands.

DNA Polymorphism using PCR

Using PCR, DNA polymorphism can be studied sequence at loci with (using primers based on known sequence) or at random site primers having unknown but random DNA sequences these are called random primers). In either case, DNA may be amplified from one or more genomic DNAs and subjected to digestion with one or mo to electrophoresis and the restriction patterns may be photographed t reveal polymorphism. Sequences of the prolamin gene and phytochrome gene have been used for the study of polymorphism for these genes in different species of Oryza (wild and cultivated rice). Similarly, storage protein genes in wheat and many genes in other animal and plant species have been analyzed by specific restriction endonuclcases. The digested DNA may be subjected used, for the study of DNA polymorphism. The non-availability of sequences of specific genes for designing primers for DNA amplification has been overcome by a variety of approaches including the following (i) In the case of multigene families like those for RNA, tRNA and heat shock proteins, there are conserved consensus sequences known, which are used for designing universal primers. Similarly for amplification of antibody genes also, universal primers have been used in PCR (ii) Gene bank Sequence Data have also been used for designing primers for amplification of microsatellites like poly (A), poly (G), pol (TC) and poly (TG), etc.

RAPD Markers

The most important use of PCR for e study of DNA polymorphism, however, involves the use of random prime giving Random Amplified Polymorphic DNA. This technique is considered simpler and offers several advantages over more commonly used techniques known as RFLP and RAPDs have actually been used for a study of polymorphism chromosome mapping in maize, soybean, mouse, humans, etc.

VNTR loci

Alec Jeffreys and his colleagues in the U.K suggested the use of PCR for the study of DNA polymorphism at VNTR (variable number tandem repeats, also known as 'minisatellites loci in humans. Primers were developed for the conserved flanking regions of VNTR loci so that all available VNTR loci could be amplified. The PCR products differ according to the number of repeat units present in the different VNTR loci and this variation is visualized in the positions occupied by different bands after electrophoresis.

SSR

The VNTR loci discussed above consist of repeat electrophoresis units in the range of 11-60 base pairs in length. A much simpler type of repeat units exists in the form of dinucleotides, trinucleotides or tetra nucleotides, e.g. (CA)n, (GT)n (AAT)n or (AGAT)n. These repeat units were described as microsatellites, but later also called short tandem repeats or simple sequence repeats (SSR). Some of these SSR loci, such as (CA)n or (GT) n occur in the human with n varying from 10 to 60. The DNA sequences flanking SSRs are conserved, allowing the selection of PCR primers that will amplify the intervening SSR in all genotypes of the target species PCR reaction includes a small amount of 3P labeled nucleotide (only one of the four nucleotides is labeled) or it may include primers, one or both of them being end-labeled. This allows visualization of PCR products using autoradiography after electrophoresis on a standard sequencing gel. Variation in the length of products is a function of the number of SSR units. The principle underlying is detection polymorphism using two parents and their F1 hybrid. Markers resulting from SSR length polymorphisms are placed on genetic maps relation RFLP's, RAPD's, VNTR's, other SSR's and even the morphological markers.

A comparison of very limited sequence data from plants (including algae) those of vertebrates (including humans) indicated that SSR's plants are found with the same frequency as in vertebrates. Recently DNA sequence data for five tree species and Zea mays have been screened for the presence of (AC)n and (AG)n and the frequency in these six species was found to be close to one microsatellite (SSR) per 5x3x 10 bp. Other plant and animal species have also been examined for the presence of a number of SSR's and a wide occurrence of these markers in plants and animals preparing saturated was suggesting their utility in preparing saturated molecular maps. Molecular mapping using PCR Genetic and physical chromosome maps have also been prepared, both in plants and animals, using PCR. Once RAPD's are detected, recombinant inbreeds (RI's) or F2/backcross segregating populations, or doubled haploids (DH) derived from haploids (in plants) can be used to detect linkage and recombination frequencies. The molecular marker which cosegregates RI's or F,'s or DH's suggests linkage and the frequency of cosegregation among RIs or Fa/backcross plants will be a measure of the strength of linkage.

- The molecular markers identified through PCR can also be assigned to specific chromosomes using aneuploids or alien addition lines e.g. wheat-barley additions -21 pairs of wheat chromosomes +1 pair of a specific barley chromosome). A number of molecular markers have been mapped using PCR in maize, soybean, barley, mouse, humans, etc.

Sequence Tagged Microsatellite Sites (STMS) or SSR loci

It has also been examined and mapped using PCR. The gene sequence databases are screened for a microsatellite and the flanking sequences are used for designing primers for PCR. Several STMS or SSR loci using this technique have been mapped in mice, where as many as 200 microsatellites have been detected in the genome database. Besides SSR's or STMS's, other markers including sequence tagged sites or STS's (a short DNA sequence in a molecular genomic DNA marker) and expression sequence tags or ESTs (a short DNA sequence from a cDNA clone representing an expressed gene) have also been for genetic mapping.In both these cases, PCR can be profitably utilized for screening RI's (mouse), somatic hybrid cell lines (humans) or F2 individuals (in plants).This will allow detection of linkage and calculation of recombination frequencies leading to the preparation of genetic maps. Complete molecular maps of several human chromosomes could be prepared using STS's and PCR.

Gene Tagging Using PCR

PCR has also been utilized for developing molecular m to specific genes of economic importance. For instance in tomato 144 random primers were developed to produce 625 PCR products from a set at markers closely link instance in tomato, near isogenic lines (NIL's, which differ only for the presence and absence of a specific gene) for *Pseudomonas* resistance. At least three PCR products were present in one and not in the other of a pair NIR's showing linkage with *Pseudomonas* resistance gene.

Plant diversity

Variation in the genome and epigenetic activities such as gene silencing and altered chromatin structure within or across a crop species is referred to as "plant genetic diversity". This diversity provides the reason why for eg., some crop varieties are tall and some are short, why some species can survive insect attacks while other closely related ones cannot, etc. For millennia farmers have taken advantage of genetic diversity by selecting the most favorable material for the following reason. Crop breeders need plant genetic diversity in their breeding programs to achieve genetic gain. They can exploit the allelic diversity present in the breeding materials by chromosome reassortment and/or recombination, and they can import new alleles by hybridizing elite germplasm with exotic materials. The amount of genetic diversity present in a particular collection of individuals can be divided from comparisons between DNA fingerprints. Such estimates of the genetic diversity present can be used to define the phylogenetic relationship among germplasm is particularly useful as it indicates the combinations of breeding parents, which is most likely to offer a high degree of allelic complementary contributing to a large potential for genetic gain in the offspring. Thus one of the most powerful applications of markers in breeding is the guiding of parental combination before making crosses. Although most fingerprinting assays are based on anonymous markers whose function is not known. The increasing pace of gene diversity and elaboration of gene function is driving a shift towards markers associated with specific genes. The advantages of such markers is that they can allow the selection of parental lines based on specific traits. Fingerprinting is also valuable for establishing varietal identity and purity and variety protection which are real issues for the breeder both in the commercial and the public sector.

CHAPTER 7

PLANT CELL AND TISSUE CULTURE

41. HISTORICAL PERSPECTIVE OF PLANT CELL/ TISSUE CULTURE.

Introduction

The beginning of plant biotechnology or plant tissue culture was made as early as 1898, when a German botanist, G. Haberlandt cultured successfully differentiated individual plant cells, which were isolated from different plant species. Further progress in cell culture research was made although embryo culture, roots and other tissues were successfully achieved in this period. The discovery of the importance of auxins and B-vitamins, the foundation of plant tissue culture was laid down by three scientists (Nobecourt, White and Gautheret) even though small pieces of tissue and not individual's differentiated cells could be grown in cultures during 1934-1939. To boost up these areas ICGEB in a workshop was held in New Delhi from September 18 to 20 1985, recommended the need for more research in developing countries on plant cell culture, differentiation, regeneration and transformation in tropical grain legumes, woody legumes and cereals. This was led to improve growth under stress condition, pest and disease resistance, improved nutritional quality, nitrogen fixation and the control of partitioning within plants. Totipotency: The living cells of multicellular organisms are capable of independent development, when provided with suitable conditions, this statement given by White, 1963. According to Morgan, 1901 coined the term totipotency to denote this capacity of cells to develop into an organism by regeneration. However the concept of totipotency is important for tissue culture.

History

The experiments on wound healing in plants through spontaneous callus formation on the decorticated region of elm plants was coined by historically Henri-Louis Duhamel du Monceau in 1756. But the science is advance related to cell and tissue culture after propounding the cell theory by Schleiden and Schwann (1839). Trecul in 1853 observed callus formation in a number of plants. Hamberlandt is regarded as the father of tissue culture. The organ culture was made attempts from 1902-1930.

- The isolated embryos of some crucifers and successfully grew on mineral salts and sugar solutions by Hanning in 1904. The successfully regenerated a bulk callus, buds, roots from poplar trees on the surface of medium containing IAA which proliferated cell division by Simon 1908.
- The cultured plant material maintains its morphological identity, more or less with the same anatomy and physiology as *in vitro* of the parent plants by Doods and Roberts, 1985.
- The cultured tobacco tumor tissue from the hybrid Nicotiana Glauca and N. Langsdorffi by White 1939.
- R.P White (USA), Gautheret (France) and Nobercourt (France) independently cultured tissues excised from several plants on the defined nutrient media for a long period in 1930.
- The cultured cambium tissue of carrot on Knop's solution was supplemented with other chemicals in trace amount in 1939 by Gautheret.
- The Van Overbeek in 1941 and coworkers used coconut milk for embryo development and callus formation in Datura.
- Miller in 1955 was eventually a potent cell division factor from degraded DNA preparations was isolated and identified named as Kinetin.
- Miller and Skoog in 1957 advanced the hypothesis of organogenesis in cultured callus by varying the ratio of auxin

and cytokinin in the growth medium. The shoot was formed by keeping the ratio of kinetin higher and root developed when the ratio was lower.

- A cell can be propagated by subculturing by Muir in 1953 developed a successful technique for the culture of single isolated cells which is commonly known as the paper-raft nurse technique.
- Pfizer Inc. New York USA in 1952 got the US patent and started producing industrially the secondary metabolites of plants. The first natural production of the natural product shikonin by cell suspension culture was obtained.
- According to Guha and Maheshwari in 1966 developed techniques for the production of the vast numbers of embryos from the culture of pollens and sporogenous tissues of anthers. The methods to double the chromosome number in microspores of Nicotiana and Datura and collected seeds from the homozygous diploid plants within 5 months by Nitsch in 1974.
- To develop the following techniques of cell culture and somatic genetics of mainly grasses and cereals has emphasized by Vasil in 1982.
- The tissue culture was started in India during the mid-1950s at the Department of Botany (University of Delhi) by Panchanan Maheshwari. He is regarded as the father of embryology in India.
- Sipra Guha Mukherjee and S.C Maheshwari in 1964-1967 for the first time developed the haploid through anther and pollen culture at the University of Delhi. The discovery of haploid production was a landmark in the development of plant tissue culture.
- The recombinant DNA technology made it possible to transform artificially cultured plant cells by introducing foreign genes.

Requirements for *in vitro* Cultures

Tissue culture laboratory

The plant cell culture and tissues *in vitro* is not an easy task. It requires all the nutrients and physiochemical factors in maintained in a laboratory. For the plant tissue culture, a good laboratory should have required nutrient medium preparation, sterilization, cleaning and storage of supplies, aseptic condition for working the living materials, controlled environmental conditions for growth and development of culture, observation and evaluation of culture as hoped, recording the observation made during the experiment by White in 1963.

The most basic facilities that an individual's needs for tissue culture such as

- Washing and storage facilities
- Media preparation room
- Transfer Area

Washing and storage facilities

A separate area is required which should have a large sink area with provision for hot and cold running water, distillation apparatus, washing machine, pipette washer, drier and cleaning brushes, keeping and weighing the chemicals and putting glassware.

Media Preparation Room

An area is required for the preparation of media. In such space, there should be provision for bench space for chemicals, labware, culture vessels, closures and miscellaneous equipment required for media prepartion and dispensing

Transfer Area

The transfer of plant tissue was done in an open laboratory bench under clean and dry atmospheric conditions. Later on, a closed plastic box was constructed which consisted of a UV tube. UV light from the tube and the use of 95% ethanol helped to maintain sterile conditions inside the space of the plastic box.

Nutrient Media

White in 1934 observed the unlimited growth of isolated root tissues when provided with nutrient medium containing inorganic salts, sucrose, vitamins, growth hormone and few amino acids.

Inorganic Chemicals

It includes macronutrients such as nitrogen, phosphorus, potassium, calcium, magnesium and sulfur in the form of salts in large amounts and microelements such as boron, molybdenum, copper, zinc, manganese, iron and chloride.

Growth Hormones

There are several hormones that are known to stimulate biological activity in cultured materials.

- Cytokines promote cell division and regulate growth and development similar to kinetin.
- Auxin resembles indole acetic acid (IAA) and stimulates shoot elongation.
- Gibberellins are used in the apical meristem.
- The most widely used cytokines are adenine, kinetin, zeatin, benzyl adenine and auxins are IAA, NAA, 2,4-D.
- For the induction of callus the amount of kinetin should be 0.1 mg/l. The structural formula of cytokines and auxins are given in Figure 41.1.

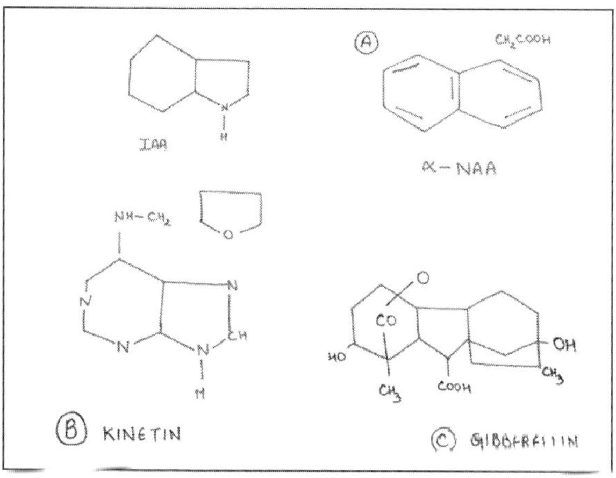

Figure 41.1. Structure of Growth Hormones (a) auxins (b) cytokinin (c) Gibberellin A_3.

Organic constituents

The organic compounds serve as an organic carbon and energy such as Sucrose, D-glucose, peptone, yeast extract, coconut water, malt extract, tomato juice ,etc.

(1) Vitamins

There are a number of vitamins that are required for culture such as Vitamin B_1, Vitamin B_2, Vitamin B_6, Vitamin C, Vitamin H, Vitamin B_{12}.

(2) Amino acids

The most widely amino acids are used such as L-aspartic acid, L-asparagine, L-glutamic acid, L-glutamine and L-arginine.

Solidifying agents

Most commonly agar a polysaccharide obtained from a seaweed i.e a red alga, *Gelidium amansii* is used as a solidifying or gelling agent.

pH

The pH affects the uptake ions optimum pH between 5.0 to 6.0 is required for the growth and development of cultured tissues. The pH should be maintained before the sterilization process or medium.

Maintenance of Aseptic Environment

The contaminants produced toxic metabolites that inhibit the growth of cultured plant tissues. Therefore each stem must be handled aseptically and with great care.

Sterilization of glassware

Glassware should be sterilized in hot air oven at 160-180° C for 2-4 hours.

Sterilization of instruments

The metallic instruments should be sterilized such as scalpels, forceps, needles, spatulas, etc dipping them in 75% ethanol followed by flaming and cooling. It is called incineration.

Sterilization of culture rooms

The floor and room are washed first with detergent then 2% sodium hypochlorite or 955 ethanol. Large surface area is sterilized by exposure to UV light.

Sterilization of nutrient media

The nutrient media are properly dispensed in a glass container plugged with cotton or sealed with plastic closures and sterilized by autoclaving at 15 psi for 30 minutes that gives 121° C for 30 minutes. During autoclaving plant extract, vitamins, amino acids and hormones are denatured so therefore use Millipore filter paper which has 0.2 μm pore diameter for sterilization.

Sterilization of plant materials

The plant materials have microbial contaminates so wash with sodium hypochlorite, hydrogen peroxide, mercuric chloride or ethanol should be used to make plant material sterilized.

Methods of plant cell, tissue organ culture basic Steps

There are the basic steps :

- Preparation of suitable nutrient medium.
- Sterilization of Explants
- Inoculation
- Incubations
- Regeneration
- Hardening
- Plantlet Transfer

42. TOTIPOTENCY AND MORPHOGENEIS, ORGANOGENESIS, RHIZOGENEISIS, EMBRYOGENESIS, NUTRITIONAL REQUIREMENT OF IN VITRO CULTURES, DIFFERENT TECHNIQUES OF IN VITRO CULTURE- MICRO-PRPOAGATION, ANTHER CULTURE, POLLEN CULTURE, OVULE CULTURE, EMBRYO CULTURE, IN VITRO POLLINATION, IN VITRO FERTILIZATION, ENDOSPERM CULTURE., FACTORS AFFECTING IN VITRO CULTURE. ACHIEVEMENTS, SOMACLONAL VARIATION, TYPES, CAUSES, SOMATIC EMBRYOGENESIS AND SYNTHETIC SEED PRODUCTION. PROTOPLAST ISOLATION, CULTURE MANIPULATION AND FUSION. CYBRIDS, PRODUCTS OF SOMATIC HYBRIDIZATION, CRYOPRESERVATION OF GERMPLASM AND SECONDARY METABOLITES.

Introduction

There are different types of culture which are produced from plant material such as Explant culture, organ culture, cell or suspension culture, callus culture, protoplast culture Figure 42.1

Explant Culture

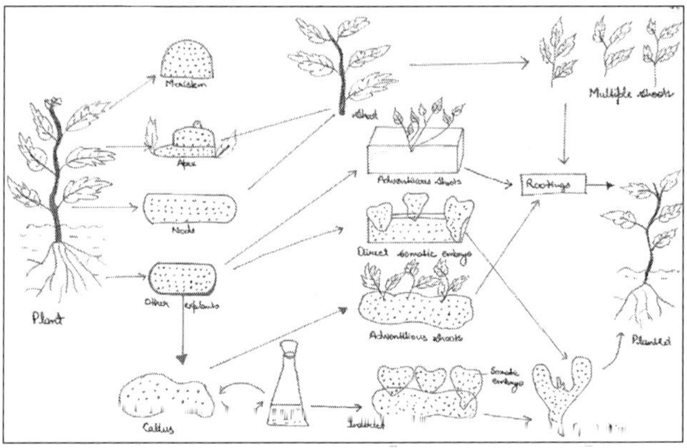

Figure 42.1. Regeneration of plant through plant tissue culture

There are different varieties of seeds plants such as trees, herbs, grasses which exhibit the basic morphological units such as root, stem and leaves. There are various stages and differences in cells, tissues and their topography. Parenchyma is the most versatile of all types of tissue. They are capable of division and growth. However, the development of tissue is characterized by three types of cell growth: cell division, cell elongation and cell differentiation. The healthy part of the plant should be used for culture purposes. Explant cultures are the cultures of plant material. Any part of the plant may be explant such as young and healthy pieces of the stem, leaf, stem hypocotyl, cotyledons, etc. Explant cultures are generally used for induction of callus or regeneration of the plant.

Callus Formation

The callus develops by an infection of microorganisms from wounds due to stimulation by endogenous growth hormones, the auxins and cytokinins. However, it has been artificially developed by adopting tissue culture techniques. It is governed by the source of the explant, nutritional composition of medium and environmental factors. Explants of meristematic tissues developed cells more rapidly than thin three-walled and lignified cells of tissue. Callus is formed through three developmental stages: induction, cell division and differentiation.

Organogenesis

Roots, shoots and leaves are the organs that are induced in plant tissue culture. Since the embryo is an independent structure and does not have vascular supply, it is not supposed to be the plant organ. It starts with stimulation caused by the chemicals of the medium, substances carried over from the original explants and endogenous compounds produced by the culture by Tomas and Davey, 1975. Skoog and co-workers gave the concept of regulation of organogenesis by a balance between cytokinin and auxin. High ration of auxins and cytokinin stimulated the formation of the root in tobacco callus, but a low ratio of the same induced shoot formation.

Root Culture

The root is cultured in a liquid medium. It has several advantages over solid media. The techniques of root culture give certain important information's such as nutritional requirements, infection by Rhizobium and nodulation, physiological activities for eg, production of alkaloids, nicotine ,etc by given Dodds and Roberts, 1985. According to White in 1934 reported first time successful organ culture e.g potentially unlimited growth of excised tomato roots. Subsequently, the roots of several species of gymnosperms and angiosperms have been successfully cultured.

Shoot culture

In the shoot culture, apical meristem is cultured (the region of shoot apex laying to leaf primordium). It is clearly differ from shoot apex by having shoot apex and a few leaves primordial. The culture of relatively large tips (5-10 mm long sections), this technique is also known as meristem culture, Meristemming and mericlones (Murashige, 1974).

Micropropagation

According to Murashige in 1974 has described the procedure of development of micropropagation into three different development stages such as

- Establishment of explant aseptically
- Multiplication of propagules by repeated subcultures on a specific nutrient medium.
- Rooting and hardening of plantlets and planting into soil.

Cell-Suspension Culture

It is prepared by transferring a fragment of callus (500 mg) to the liquid medium (500 ml) and agitating them aseptically to make them cells free. It is difficult to have a suspension of single cells. However the suspension includes single cell, cell aggregates residual inoculum and dead cells (Dodds and Roberts, 1985). King (1980) has described that a good suspension consists of a high proportion of single cells than a small cluster of cells. It is more difficult to have a good suspension than to find out optimum environmental factors for cell separation.

Benefits of Cell Culture

- Benefits for cell culture such as:
- Suspension can be pipetted.
- They are less heterogeneous and cell differentiation is less pronounced.
- They can be cultured in volumes up to 1.500 liters.

- They can be subjected to more stringent environmental controls.
- They can be manipulated for the production of natural products by feeding precursors by Kurz and Constable, 1979.

Somatic Embryogenesis

It can be initiated in two ways Inducing embryogenic cells within the preformed callus. Directly from pre-embryonic determined cell (without callus). Which are ready to differentiate into embryoids by Sharp et al., 1980. In the first case, embryoids are initiated in callus from superficial cell aggregates where cells contain a large vacuole, dense cytoplasm, large starch granules and nucleus by Mc Willian et al., 1974. Two nutritional media of different compositions are required to obtain embryoids. First medium contains auxins to initiate embryogenic cells. Second medium lacks auxin or reduced level of auxin is needed for subsequent development of the embryonic cells into embryoids and plantlets. In both cases reduced amount of nitrogen is required (Ammirato, 1983). The embryogenic cells pass through 3 different stages such as globular heart-shaped and torpedo-shaped to form embryoids. These embryoids can be separated and isolated mechanically by using glassbeads. When embryoids reach torpedo stage they are transferred to filter paper bridge and sucrose on which Whatman no.1 filter paper is placed to make a bridge by Dodds and Roberts, 1985.

Somaclonal Variation

The genetic heterogeneity of cells in a population represents the continuity of genotypes, whereas phenotypically the population is represented as a discrete sum of subclones. After cloning a single cell, from the population of the strain of *Dioscorea deltoidea*, there developed subpopulation that defined in their growth rate and sapogenin content. The mass occurrence somaclonal variants, increase in the resistance, productivity and vital force of the plant have been explained by Carlson, 1983. This would be due to the dominance of nonlethal mutations that lead to heterozygosis with a wild-type allele which is thus phenotypically expressed as hybrid heterosis. There are some somaclonal variations in some plants such as *Brassica spp., Nicotiana tabacum, Oryza sativa, Saccharaum officinarum and Zea mays.*

Protoplast Culture

Protoplast is the biologically active and most significant material of cells. When the cell wall is mechanically or enzymatically removed the isolated protoplast is known as naked plant cell. The plant cell wall acts as a physical barrier and protects the cytoplasm from microbial invasion and environmental stress. It consists of a complex mixture of cellulose, hemicellulose, pectin, lignin, lipids, protein ,etc. For dissolution of different components of the cell wall it is essential to have the respective enzymes. There are a number of stages like

Isolation of Protoplasts

- Surface sterilization of samples (Leaf)
- Rinsing in suitable osmoticum
- Plasmolysis of cells
- Peeling of the lower epidermis
- Isolation and purification of protoplasts

Regeneration

The protoplast regenerates a cell wall, undergo cell division and form a callus.The callus can also be subcultured. Embryogenesis begins from the callus when it is placed on a nutrient medium lacking mannitol and auxin. The embryo developed into seedlings and finally, mature plants; see Figure 42.2.

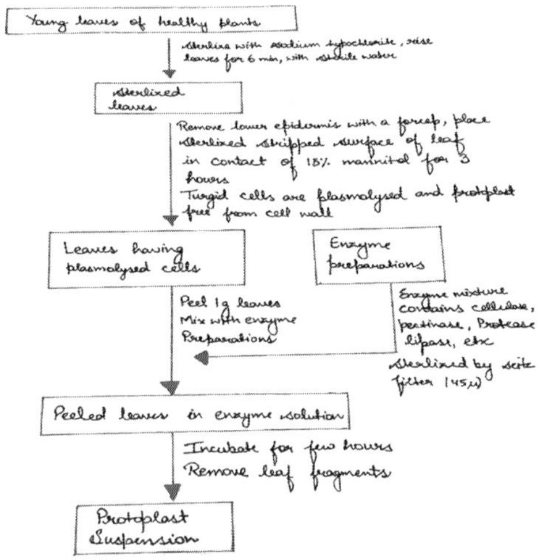

Figure 42.2. Isolation of protoplasts

Protoplast Fusion and Somatic Hybridization

According to Schieder in 1982 discussed four major aspects of protoplast fusion as

- Production of fertile amphidiploid somatic hybrids of sexually incompatible species is achieved. Induced fusion of protoplasts from two genetically different lines of species must result in a variety of homo as well as heterokarytic fusion products.
- Selection of a few two somatic hybrid colonies from the mixed population of regeneration protoplast is a key step in a successful somatic hybridization technique.
- Production of heterozygous lines within one plant species which normally will be propagated only vegetatively e.g potato.
- Transfer of only a part of genetic information from one species to another using the phenomenon of chromosome elimination
- The transfer of cytoplasmic genetic information from one to a second line or species. It has been possible to transfer useful genes e.g nif genes, disease resistance genes, rapid growth genes from one species to another, thereby widening the genetic base for plant breeding: Figure 42.3.

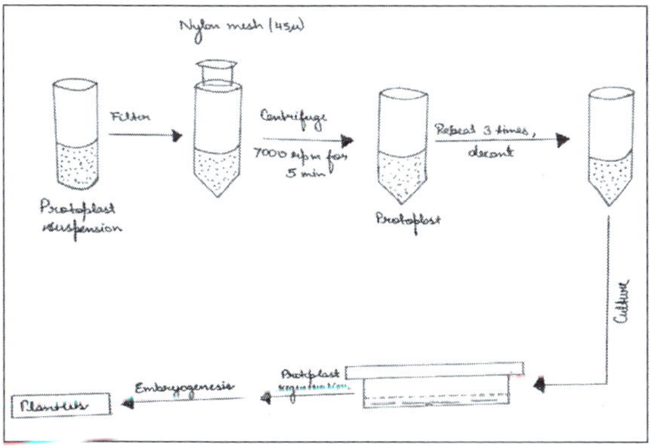

Figure 42.3. Purification, culture and regeneration of protoplasts

Fusion Product

Fusion of cytoplasm of two protoplasts results in coalescence of cytoplasms. The nuclei of two protoplasts may or may not use together even after the fusion of cytoplasms.

The binucleate cells are known as heterokaryon or heterocyte when nuclei are fused the cells are known as hybrid or synkaryocyte and when only cytoplasm fuse and genetic information from one of the two nuclei is lost is known as cybrid i.e cytoplasmic hybrid or heteroplast (Doods and Roberts, 1985) Figure 42.4.

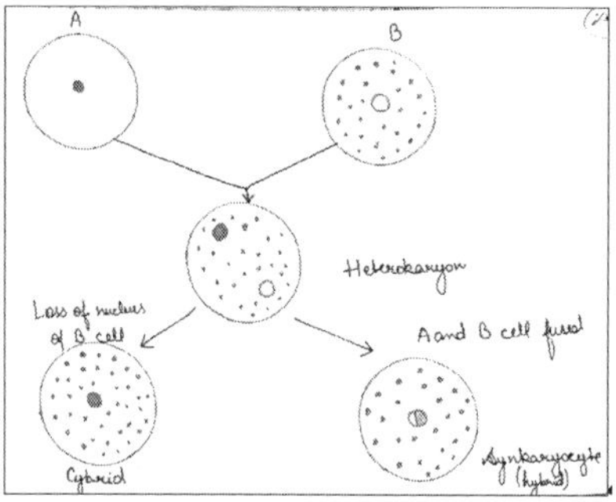

Figure 42.4. Cybrid/Hybrid productions

Methods of Somatic Hybridization

- Isolation of protoplast from suitable plants.
- Mixing of protoplast in a centrifuge tube containing fusogenic chemicals i.e chemicals promoting protoplast fusion such as polyethylene glycol (PEG) 20% W/V, sodium nitrate ($NaNO_3$), maintenance of high pH 10.5 and temperature 37° C.
- Wall regeneration by heterokaryotic cells.
- The fusion of nuclei of heterokaryon to produce hybrid cells.
- Plating and production of colonies of hybrid cells.
- Selection of hybrid, subculture and induction of organogenesis in the hybrid colonies.
- Transfer of mature plants from the regenerated callus.

Selection of Somatic Hybrids and Cybrids

After the fusion protoplasts of media regenerate cell walls and undergo mitosis. This results in a mixed population of parent cells, homokaryotic fusion products and hybrids. Hybrid cells should be differentiated from other cells. There are various selection methods used for the selection of fusion products. Selection methods are dependent on

- Physical properties of fused cells.
- Biological properties of fused cells.
- Biological properties of colonies formed from fused cells.

The somatic hybrids cannot be identified. The first time P.S Carlson and co-workers produced first interspecies somatic hybrids between ***Nicotiana glauca and N. langsdorfii*** in 1972. The most somatic hybridization experiment, selection procedure includes full of chlorophyll-deficient (non-green) chloroplasts of one parent with the protoplasts of the other parents. This help in the visual selection of heterokaryon under the microscope.

Anther and Pollen Culture

A male reproductive organ is diploid in chromosome numbers. As a result of microsporogenesis, tetrads of microspores are formed from a single spore mother cell. They are known as pollen grains after release from tetrads by Bhojwani and Bhatnagar, 1974. Haploid plants are very useful in the direct screening of recessive mutation because in diploid or polyploid screening of recessive mutation is not possible, development of homozygous diploid plants following chromosome doubling of haploid plant cells **Figure 42.5.**In China, the most widely grown wheat is a doubled haploid produced through homozygous diploid lines. Anther cultures of rice are also successfully grown. Haploid plants have been produced in tobacco, wheat and rice through pollen culture.

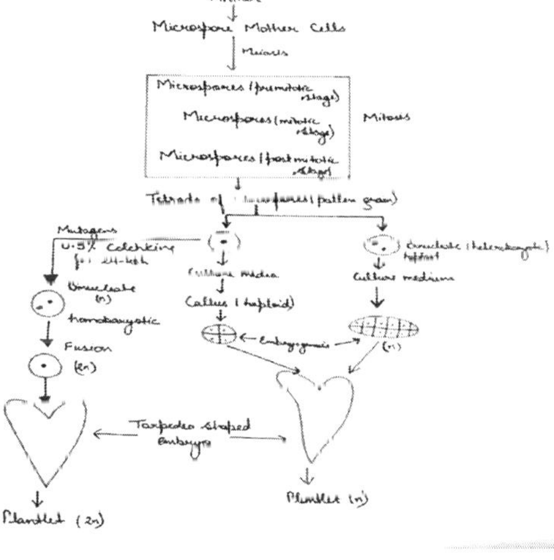

Figure 42.5. Anther Culture

At present more than 247 plant species and hybrids belonging to 38 genera and 34 families of dicots and monocots have been regenerated using anther culture technique. They include economically important crops and trees such as rice, wheat, coconut, rubber trees, etc. by Maheswari et al., 1983.

Culturing Techniques

Anthers are superficially sterilized and washed with double distilled sterile water. They are excised from the flower buds and their proper developmental stages are determined under a microscope. On confirmation of stages such as (a) anthers are directly transfered on nutrient agar or liquid medium where induction of embryogenesis occurs (b) the pollen grains are aseptically removed from the anthers and cultured on liquid medium (Reinert and Bajaj, 1977). Methods of anther culture.

In vitro androgenesis

- *In vitro* androgenesis is the formation of sporophyte from the male gametophyte on artificial medium.
- It is most commonly found in the family ***Solanaceae and Poaceae*** (Graminae).
- Success of *in vitro* androgenesis is based on adjustment of the development stage of pollen, minerals of culture medium and growth regulators as well as thermal shock or other treatments.
- Pollen grains are isolated from excised anther and extracted in a beaker.
- Well collected pollens are washed properly and centrifuged and decanted. They are inoculated in a liquid medium, then subcultured in solid MS medium.
- In culture, pollen grains can be induced to produce callus or embryo from which whole plants are regenerated in one month.
- This technique is most successful in Brassica, Datura, petunia ,etc. Mainly there are two ways of *in vitro* morphogenesis of pollen grains, direct and indirect.

Direct Androgenesis

- It is also called pollen derived embryogenesis. Here pollen directly acts as a zygote and therefore passes through various embryogenic stages similar to zygotic embryogenesis.
- When pollen grains has reached globular stage of embryo, then the plants.
- Direct androgenesis is very common in many plants of the family Solanacease and Brassicaseae by Prakash and Giles, 1987.

Indirect androgenesis

- In indirect androgenesis, the pollen grains, instead of normal embryogenesis, divide erratically to develop a callus. Indirect androgenesis has been found in barley, wheat, Vitis, Coffea, etc.
- Possibility of pollen morphology from the dividing pollen varies.
- The haploid callus, embryo or plantlets may originate from the continued division of vegetative cell of the pollen, multiple division of generative cells, division products of both generative and vegetative cells by Sangwan, 1981.

Mentor pollen technology

- Pollination is a natural phenomenon that occurs in many plant species.
- However, there are certain plants where pollen grains remain viable but unable to set seed on their own pistil i.e, they are self incompatible.
- Similarly, in several other plants, besides being viable they even fail to germinate on stigmas of another species i.e they are interspecific in compatible.
- After the birth of recent mentor pollen technology, the incompatibility problem of pollen grains can be overcome by altering the fertilization ability.
- The pollen which has been purposefully treated is called mentor pollen
- Hence the mentor pollen is defined as the compatible pollen which has been treated in many ways to inhibit its fertilization

ability and retain its power to stimulate incompatibility pollen to accomplish fertilization.

- The mentor pollen itself may or may not germinate on stigmas.
- As a result of this treatment, hybridization between closely related species or genera and pollination in a cross-pollinating plants could be achieved.

Embryo Culture

- Embryo culture has also been done for the production of haploid plants. It is used for the recovery of plants from distinct crosses.
- It is useful when embryos fail to develop due to the generation of embryonic tissues.
- It is being used extensively in the extraction of haploid barley *Hordeum vulgare* from the crosses *H. vulgare* × *H. bulbosum*. Das and Barman (1992) developed the method of regeneration of tea shoots from embryo callus.
- It produced somatic embryoids within 8 weeks of culture in the second medium which differentiated into buds after 2 weeks. Several shoots with 4-6 leaves developed after 16 weeks of culture.

Embryo Rescue

- Viable hybrids are produced as a result of sexual crosses between two varieties of the species. However, if sexual crosses are done between the species of the same genus or between two different genera, production of hybrid is rather difficult because of several barriers arising either during pollination, fertilization or embryogenesis.
- It has been observed that in some cases in spite of successful pollination and fertilization, the embryo does not develop.
- This due to inherent deficiencies or incompatibility between the developing embryo or endosperms.
- In such cases, immature embryos are dissected out from the fruit (seed) and grown artificially on the medium to differentiate into shoot, root and plantlets.
- This technique of growing an immature embryo is termed as **"embryo rescue"**.

- Embryo rescue technique is very useful in wide hybridization, complete growth of the embryo in plants, breaking dormancy of certain seeds where dormancy period is very long. By using the embryo rescue technique wild varieties can be crossed with cultivars.
- As compared with cultivars, the wild species have greater resistance to pests and pathogens and produce grains of better quality.

Triploid Production

- The triploid primary endosperm (fusion of the second male gamete with two polar nuclei).
- Some examples of triploid plants raised from endosperm cultures such as Rice (*Oryza sativa*), Maize (*Zea mays*) and Barley (*Hordeni vulgare*).
- The triploids of poplar (*Popus tremuloids*) have better quality pulpwood. Therefore, it is important to the forest industry.

Protoplast Fusion In Fungi

- The protoplast fusion in fungi is the improvement of strains to be used for commercial purposes.
  - For example, two strains of *Cephalosporium acremonium* were crossed and Cephalosporin C was improved.
- To improve the citric acid production in A. niger were unsuccessful (Peberdy, 1989).

The protoplast techniques are applied in the following four major areas to get the crosses:

- Between different species and genera.
- Between apparently incompatible strain of the same species.
- Between different strains or isolates of a species primarily for breeding.
- Between isogenic strains which may provide an opportunity for genetic mapping.

Intraspecific Protoplast Fusion

- It is the cross between the same species in an individual which involves the isogenic strains or the non-isogenic ones eg *Absidia glauca*, *Pleurotus ostreatus*, *Fusarium graminearum* and Treesi ,etc.

Interspecific protoplast fusion

- It is the crosses between two different species It is important in the area where new products are to be produced eg., *S. cerevisiae* × *S. fermentali*, *Pleurotus ostreatus* × *P. florida* ,etc.

Application in agriculture

- There are number sof application in agriculture which has solved the problem of incompatibility of plants and scope of production of new verities within a short period of time.
- Improvement of Hybrids
- Production of Encapsulated Seeds
- Production of Diseases Resistant Plants
- Transfer of *nif* gene to Eukaryotes

Bioethics in plant genetic engineering

- There are new varieties of production of a plant due to mutation. Cultivation of genetically modified crops by farmers is increasing fast thought out the world. It will support healthcare and industry and provide food, feed and fiber security at a global basis. The major concern about genetically modified plant or crops foods is given below such as:
- Genetically modified crops may bring about changes in evolutionary patterns.
- Effect of genetically modified crops on biodiversity and the environment.
- Transfer of transgene from genetically modified food to pathogenic microbes.
- Risk of change in fundamental vegetable nature of plants.
- Effect of genetically modified on non-target and beneficial insects and microbes.

- Pollen transfer from genetically modified plants.
- The risk of transfer of allergics.

Some Important contributions have been described below.

Year	Author
1902	Plant Tissue culture (Haberlandt)
1904	Embryo culture of selected crucifers (Hannig)
1922	Asymbiotic germination of orchid seeds invitro (Knudson)
1922	Invitro culture of root tips (Robbins)
1925	Use of embryo culture technique in interspecific crosses of Linum (Laibach).
1934	Successful culture of tomato roots (White).
1939	Successful establishment of continuously growing callus cultures (Gautheret, Nobecourt and White).
1941	Use of coconut milk containing a cell division factor for the first time in Datura (Van Overbeek).
1944	*In vitro* adventitious shoot formation in tobacco (Skoog).
1946	Rising of whole plants of Lupinus and Tropaeolum by shoot tip culture (Ball).
1952	Use of meristem culture to obtain virus-free dahlias (Morel and Martin).
1952	First application of micrografting (Morel and Martin).
1952	Got the US Patent on their claim that plant cells could be grown in a liquid nutrient medium and had the potential to produce useful compounds (Routien and Nickel of Pfizer and Co. Inc., New York).
1953	Production of haploid callus of the gymnosperm Ginkgo biloba from pollen (Tulecke).
1954	First plant from a single cell (Muir et al.,).
1955	Discovery of kinetin a cell division hormone (Miller et al.,).
1957	Discovery of the regulation of organ formation by changing the ratio of auxin: cytokinin (Skoog and Miller).
1958	Regeneration of somatic embryos *in vitro* from the nucellus of citrus ovules (Maheshwari and Rangaswamy).
1959	Regeneration of embryos from callus clumps and cell suspensions of Daucus carota (Reinert, Steward).
1959	Publication of the first handbook on plant tissue culture (Gautheret).
1960	First successful test tube fertilization in Papaver rhoeas (Kanta).

1960	Use of the microculture method for growing single cells in hanging drops in a conditioned medium (Jones et al.).
1960	Enzymatic degradation of cell walls to obtain large number of protoplasts (Cocking).
1960	Filtration of cell suspensions and isolation of single cells by plating (Bergmann).
1962	Development of Murashige and Skoog nutrition medium (Murashige and Skoog).
1964	Production of first haploid embryos from pollen grains of *Datura innoxia* (Guha and Maheshwari).
1967	Androgenic haploid plants of Nicotiana were produced (Bourgin and Nitsch).
1968	Restriction endonuclease term coined to a class of enzymes involved in cleaving DNA (Meselson and Yuan).
1970	Selection of biochemical mutants *in vitro* by the use of tissue culture-derived variation (Carlson).
1970	The first achievement of protoplast fusion (Power et al.).
1970	Discovery of first restriction endonuclease from *Haemophilus inflenzae* Rd. It was later purified and named HindII (Kelly and Smith).
1970	In certain cancer-causing animal viruses, the genetic information flow in reverse form. Discovery of reverse transcriptase (Temin).
1971	The first restriction map using HindII enzyme to cut circular DNA of SV 40 into 11 specific fragments was prepared (Nathans).
1971	Regeneration of first plants from protoplasts (Takebe et al.)
1972	Interspecific hybridization through protoplast fusion in two species of Nicotiana (Carlson et al.).
1972	First recombinant DNA molecule produced using restriction enzymes (Jackson et al.).
1972	Joining of two restriction fragments regardless of their origin produced by the same restriction enzyme by the action DNA ligase (Mertz and Davis).
1973	Use of Lobban and Kaiser technique to develop hybrid plasmid-insertion of EcoRI fragment of DNA molecule into circular plasmid DNA of bacteria using DNA ligase. Gene from African clawed toad inserted into plasmid DNA of bacteria (Cohen et al.).
1974	Ti plasmid is the tumor inducing principle of *Agrobacterium* (Zaenen et al., Larebeke et al.).
1975	Positive *in vitro* selection of maize callus cultures resistant to T toxin of *Helminthosporium maydis* (Gengenbach and Green).

1975	High resolution 2 Dimensional gen electrophoresis procedure developed that led to the development of proteomics (O'Farrel).
1976	Shoot initiation from cryopreserved shoot apices of carnation (Seibert).
1977	Successful integration of the Ti plasmid DNA from *A. tumefaciens* in plants (Chilton et al.,).
1977	Gene sequencing based on the degradation of DNA chain (Maxam and Gilbert).
1977	Discovery of split genes (Sharp and Roberts)
1978	Somatic hybridization of tomato and potato resulting in potato (Melchers et al.).
1979	Co-cultivation procedure developed for the transformation of plant protoplasts with *Agrobacterium* (Marton et al.)
1980	Commercial production of human insulin through genetic engineering in bacterial cells (Eli Lilly and Co).
1981	Somaclonal variation (Larkin and Scowcroft).
1982	Incorporation of naked DNA by protoplasts resulting in the transformation with isolated DNA (Krens et al.)
1984	Transformation of tobacco with Agrobacterium transgenic plants developed (De Block et al., Horsch et al.).
1986	The genetic fingerprinting technique developed for identifying individuals by analyzing polymorphism at DNA Sequence level (Jeffreys).
1990	Random amplified polymorphic DNA technique developed (Williams et al., Welsh and McClelland).
1991	DNA microarray system using light-directed chemical synthesis system developed (Fodor).
1995	DNA fingerprinting by amplified fragment length polymorphism (AFLP) technique developed (Vos et al.,).
1997	Sequencing of *E.coli* genome
2001	Sequencing of the human genome successfully completed (Consortium and Venter et al.)
2002	Draft sequence of rice genome (*Oryza sativa* L.ssp. japonica)
2003	Draft sequence of rice genome (*Oryza sativa* L.ssp. indica) (Yu et al., 2002).
2005	Map based sequence of the rice genome (international Rice genome sequencing project).

Secondary Metabolites

- The plants' knowledge of healing has been closely linked from the time of man's earliest social and cultural groupings.
- The medicine man was usually an accomplished botanist.
- Even in historical times, botany and medicine continued to be virtually one and the same disciplines until about 1500 A D.
- When the two began to separate from their close association to the advantage of both sciences
- It is well established that industrialization has many direct and indirect effects on the human population.
- Increased stress is the most evident though offset by increased health awareness among the people and better medical facilities.
- Nevertheless increase in diseases (mostly in urban populations) such as coronary heart diseases, diabetes, hyperlipidemia, AIDS and Cancer cannot be denied.
- Data on Cancer show both increased incidence of the disease as well as an increase in percent survival of patients due to better medical facilities.
- This increase in the world population associated with increased health awareness and disease incidence will generate pressure on our natural resources, i.e., medicinal plants for drugs, cosmetics, health supplements, dyes and fragrances.
- Besides medicines, mankind is almost completely dependent on plants for such requirements as food, shelter and clothing.
- However, most of the natural products are compounds biosynthetically derived from primary metabolites such as amino acids, carbohydrates and fatty acids and are generally categorized as secondary metabolites.
- These secondary metabolites are the major source of pharmaceuticals, food additives, fragrances and pesticides.
- It is estimated that the market potential for herbal drugs in the western range billion to 47 billion by the year 2000.
- A similar situation exists for plant- based food additives, fragrances and biopesticides. Natural resources for plant-based chemicals are limited but the consumer preference for natural products is large.

Medicinal Plants

- Plants still remained, however, a great source of therapeutic agents until the beginning of the 20th century.
- With the development of chemistry in the last century, plants have been looked at as sources of New therapeutic agents.
- This investigation still continues and newer drugs of plant origin are being discovered every year. Some examples of drugs of plant origin discovered in the recent past are: muscle relaxants from South American arrow poisons; tumor-inhibiting agents from the family Apocynaceae, Ochrosia, Podophyllum and Taxus; tranquilizers from Rauwolfia serpentina, cortisone and sapogenins from several plants (e.g., Digitalis, Dioscorea); blood agglutinin from the Leguminosae and phyllanthoides from Phyllanthus.
- The importance of plant-derived drugs in modern medicine is usually not fully recognized (Table 1.2).
- Many chemical compounds of medicinal value are still obtained from plants for one or the other reason, such as Morphine, Codeine, Atropine, cocaine, caffeine, Quinine, quinidine, sparteine, ergotamine, vincaleucoblastine, taxol and host others.
- In fact, the complete Indian system of medicines, which is 3000 years old, is based on plants and their extracts.
- Plants are not only the source of pharmaceuticals, they also provide chemicals of unknown and unusual chemical structures.
- These compounds provide new pharmacological properties and may serve as well as a starting material for more complex biologically active compounds.
- But we still don't know all the plants present on the planet Earth
- Many more plants have yet to be evaluated chemically and pharmacologically.
- A large number of plants have been screened in the last three decades for their chemical constituents as well as for pharmacologically active principles.
- The National Cancer Institute, Maryland, USA is actively engaged in screening plants for their important pharmacological properties, and several hundred thousand plants have already been screened for their chemical and pharmacological attributes.

- With the advent of modern sophisticated tools of analytical chemistry, small samples can be evaluated, which has expedited the screening programs.
- The study of medicinal plants based on ancient literature and their investigation in the light of modern knowledge is known as ethnology.
- However, the branch of medicine dealing with history, collection, selection, identification and preservation of crude drugs and raw materials is Pharmacognosy.
- A large majority of these medicinal and aromatic plant raw materials are still collected in the wild.
- Pharmaceutical and perfumery industries throughout the world have been making consistent endeavors to discover newer, more potent and cheaper sources of raw materials and their derived chemicals to broaden the product range in the trade.
- In recent years cheaper natural or synthetic sources of several phytochemicals have been developed, Cinchona and steroidal yam are examples of changing fortunes as plantation crops despite synthetics having captured world trade.
- Cinchona bark and its principal alkaloid Quinine has been the main armory of medical men to fight and control malaria in the last 200 years.
- However, the price of Quinine fell for the first time in 1951 to a low level of US $35 per kg but the discovery of the quinidine alkaloid in cinchona bark sustained its demand and supported the market price for some time; quinidine found use in cardiac medicine.
- Many developments in this field, such as the production of these phytochemicals through chemical conversions in laboratories using cheaper starting chemicals, dampened its profitability as a plantation crop, however, therefore, a large number of drug companies abandoned its cultivation. These examples clearly illustrate the volatile nature of raw materials trade in the agricultural sector of pharmaceuticals.
- Plant cells are highly sophisticated chemical factories in which a large variety of chemical compounds are manufactured with great precision from simple raw materials.

- Plants are thus a very important renewable source of raw materials for the production of a variety of chemicals and drugs. India, being a tropical country and rich in biodiversity, produces and exports raw medicinal plants and their extracts.
- According to one estimate, the present world trade in plant based dry raw materials and phytochemicals is around US $33,000 million.

Biotechnological Approaches Applied in Medicinal Plant

- *In vitro* grown plant cells and tissues have been used extensively for the production of secondary metabolites.
- Depending on the objectives, a system such as callus, cell suspension or root/shoot culture is used.
- Once the *in vitro* culture system is established it is used to achieve objectives through different approaches viz., micropropagation, genetic manipulation or cultivation in a Bioreactor.
- The following applications of biotechnology are being pursued in the subject of secondary metabolites.
- Development of seed material for domestication and micropropagation.
- Continuous improvement of plants for secondary metabolites through somaclonal and genetic variation.
- Understanding metabolic pathways.
- Alternative methods using Bioreactors (biomass); immobilization, organ culture (shoot/hairy root).

Seed Material Used for Domestication for Development

- Plants are collected from their wild habitats for medicinal uses.
- This supply is unable to meet the demand if a plant finds multipurpose uses or applications.
- There are several examples of insufficient availability of planting material necessitating, raising it through biotechnological methods as the only way to get a sufficient amount.
- This approach has been used for the Domestication of such plants as Digitalis species, Dioscorea species, Chlorophytum species, *Commiphora wightii* and many others.

- *Micropropagation:* Micropropagation is the most widely used application of plant tissue culture and micropropagation protocols for a large number of medicinal plants have been reported.
- Micropropagation is routinely used wherever a particular selected or unselected genotype has to multiply at an enormous rate to meet the demand of cultivators to replace the existing stock of material (perhaps having inferior characters such as low productivity, disease susceptibility, poor responsiveness) for introduction to a new state/country for cultivation, e.g., *Atropa belladonna, Rauwolfia serpentina, Dioscorea floribunda, Mentha species*, etc.

Somaclonal Variation and Genetic Engineering Improvement

- Decline of potential for the production of active principle or essential content is a common feature with the cultivation of medicinal plants and a continuous selection is required to maintain the high yield of active constituents for commercially viable programs.
- Therefore, selection through conventional breeding as well as in cell culture is explored for obtaining high-yielding cultivars.
- Genetic manipulation has been employed to obtain hairy root cultures of a large number of solanaceous plants and manipulation of biosynthetic pathways to obtain the desired product, e.g. change in flower color by blocking the chalcone synthase gene responsible for flower color.

Metabolic Pathways

- Plant tissue culture has been extensively used to understand the nutrition of the plant cell and the secondary metabolites it can produce.
- Later, the selection of cell lines for high yield of secondary metabolites and enzymes involved in the biosynthesis were investigated using radioactive precursors wherever considered necessary. The growth of the cell in a totally controlled environment of physical and chemical factors provides an excellent system for studying changes in the production of secondary metabolites, which are always present in small quantities.
- This basic information has provided significant clues for genes

and their functioning, leading to genetic manipulation for biosynthetic pathways so as to obtain desired products by either blocking a pathway or enhancing the metabolic reaction.

New Compounds Production

- Plant cell cultures have produced (a) new compounds previously not known in the intact plant, (b) new derivatives of a known compound and new derivatives by biotransformation of molecules incorporated in the medium.
- Such results provide an unlimited opportunity to screen the cultures for new compounds. It is presumed that the production of new compounds/derivatives might be due to altered gene function in the cultured cells compared to the in-vivo grówn mother plant.

Scaling up Technology Through Bioreactors

- Plant cell cultures grown in a bioreactor enable biomass production of a desired medicinal plant or its active principle through transformation such a system has been used for a long time for the production of antibiotics using fungi.
- The basic technology was derived from microbial systems and has been continuously improved for the growth of plant cells and organ cultures on a large scale.
- Bioreactors up to 20,000 liters have been designed and tested for the growth of plant cell cultures. e.g., shikonin from *Lithospermum erythrorhizon*, taxol from *Taxus brevifolia* and berberine from *Coptis japonica*
- The geography of HIV infections shows that although sub-Saharan Africa still has by far the most infections (6 million adults), other regions are catching up fast.
- By the end of the millennium the number of people infected in Asia could rise to more than 3 million, most of them in India and Thailand The situation is worse when we consider the worldwide occurrence of various types of cancers. With an increase in dreaded diseases associated with luxurious life style, the demand for natural products will increase in the coming two decades.
- This demand will exert tremendous pressure on already Scant Natural Resources.

- Therefore, biotechnological alternative methods hold great promise for meeting the demand and the challenge.
- The demand for natural products is also increasing day by day because of increased health and antipollution awareness as exemplified by the use of natural dyes instead of synthetic dyes for coloring clothes, biofertilizers and biopesticides Health care products for improving skin color and texture, anti-aging and vitalizing drugs, medicines for hyperlipidemia and hypercholesterolemia are already in great demand.

CHAPTER 8

ANIMAL CELL CULTURE

43. HISTORY, TYPES OF TISSUE CULTURE, ADVANTAGES AND LIMITATIONS.

Introduction

The most important areas of today's research which have potential economic value and prospects of commercialization are the cell and tissue culture based on the production of vaccines, monoclonal antibodies, etc.

History of Animal Cell Culture

Alex carrel used tissue and embryo extract as culture media. The fibrin clot of plasma served as an anchor for cell attachment and the extract provided growth factors and nutrients. A major innovation to get cell suspension was the use of trypsin for cell disaggregation from tissue explants. This allowed single cell culture. Thus the technique of cell cultures differs from tissue culture. The use of biological fluids and extracts was a contaminated major problem because of getting contaminated. In 1907, Ross Harrison made the first attempt to culture animal cells and cultivated embryonic nerve cells of a frog by using the hanging drop method. Thereafter, this method was extended and a wide range of mammalian cells was cultured *in vitro*. However, after supplementing with the embryo of chick and plasma, cells proliferated well. These provide nutrients and proliferation factors. Moreover, cell culture method was extracted quite improved after the discovery of antibiotics during the 1940s. Such studies provided an impetus for the culture of animal cells in a large scale. During this period many human carcinoma cell lines (e.g. HeLa cell line) were isolated and grown in culture. Alexis Carrel, the famous physiologist kept the chick embryo heart alive and its beating continued *in vitro* for about three months. Animal cell culture studies resulted in the total stop of using monkeys for multiplication of animal viruses.

Requirements for Animal Cell And Tissue Culture

Animal cell Growth in Culture

Unlike plant and microbial cells, animal cells can grow only to limited generations even in the best nutritive media. This growth also depends on the sources of tissue isolated. The special features of different cell cultures are briefly discussed (i) Neuronal cells constitute the nervous system, In culture, the neuronal cells cannot divide and grow.

The cells that form connective tissue (skin) is called a fibroblast. Fibroblast can divide and grow in culture to some generations. After completing several generations, they die. It means that all normal animal cells are mortal. (iii)In culture, the animal cells divide and grow. Consequently, they fill the surface of the container in which they are growing. Thereafter, they stop further growth. This phenomenon is termed as contact inhibition i.e. inhibition of further cell growth after reaching the wall of the container. The environment of cell growth in culture differs from that of *in vitro*, for example:

- Absence of cell-cell interaction and cell matrix interaction,
- Lack of three-dimensional architectural appearance, and
- Changed hormone and nutritional environment. The way of adherence to glass or plastic containers in which they grow, cell proliferation and shape of cell results in alterations.
- In culture, the cancer cells apparently differ from the normal cells. Due to uncontrolled growth and more rounded shape, they lose contact inhibition. Therefore, they are piled on each other. These features are exploited by cancer specialists i.e. the cancerologists to identify cancer cells from the normal cells.

Requirements for animal cell and tissue culture are the same as described for plant cell, tissue and organ culture Desirable requirements are: (i) air conditioning of a room, (i) hot room with temperature recoder (iii) microscope room for carrying be installed. (iv) darkroom glassware and culture media, and should out microscope work , (v) plastics-shelving (vi) sealed container and drawers, (vii) specialized chemicals.

Substrates for Cell Growth

- There are many types of vertebrate cells that require support for their growth I anchorage-dependent cells e (e.g plastic, glass, palladium types may be used.

Plastic as a substrate

- Disposable plastics are cheaper substrates as they are commonly used *in vitro* condition .

Glass as a substrate

- Glass apparatus etc. These are sterilized by chemicals, radiations, dry heat fin oven) and heat (in autoclaved test tubes, slides, coverslips flasks, rods, moist palladium deposited on agarose was used as For the first time and glia. the efficiency of cultured treated onto the surface

Culture Media

Culture for animal cells and tissue are prepared by certain chemical constituents from inorganic more difficult than that of microorganisms and substances. However , chemical media used for the culture of animal cells and the culture media provide the optimum growth factors (e.g pH, osmotic pressure) (a) Natural Media: Natural media are the natural sources of nutrient and proliferation of animal cells and tissue. It is used since a long time but now available in the market in the form of liquid microbes).

Glassware, equipment and culture media

The thoroughly washed glassware and equipment and carefully prepared culture media are millipore filter paper . The sterilized materials are removed sterilized by heat, steam or when the temperature cools down and used according to the procedure adopted.

Equipment Required for Animal Cell Culture

The equipment required for animal cell culture are given below:

Laminar air flow (LAF)

LAF hood acts as animal cells, Culture manipulation in aseptic conditions protects from contamination by any microorganisms such as bacteria and fungi. The contaminated bacterial/fungal cells grow more rapidly than the cultured animal cells. Therefore, the growth of animal cells fails to occur in the presence of contaminants. The working area of the LAF hood is first made sterile by using 70% ethanol. Manipulation of any cell is done by keeping the LAF in ON position aseptic working table for inoculation of Blower The LAF hood performs two functions: t provides a sterile environment for cell manipulation (ie. protects tissue culture from operator) and (protects the operator from the potential infection risk from the culture. There are different types of LAF hoods, airflow hoods (cabinets) are commercially available in different size and shapes. In this apparatus, sterile air flows inside the space of the cabinet which maintains the sterile conditions required for all transfer work. Different types of laminar airflow hoods are available on the ba- dishes Flask sis of nature of the cells and kinds of organisms and tissue culture as below.

CO_2 Incubators

The CO, incubators provide suitable environmental conditions to the growing animal cells. A silicon gasket is used on the inner door which makes the chamber of the incubator airtight. It separates the internal environment from the external environment. The external environment possesses microbial contaminants, while the internal environment remains always contamination-free. The filtered air i.e. a high efficiency particulate air (HEPA) is injected inside the chamber which maintains the internal environment of the chamber sterile. The relative humidity inside the chamber remains high; therefore, correct osmolarity is maintained and the medium is protected from desiccation.

Centrifuges

There are different types of centrifuges based on speed. A low speed centrifuge is needed for most of the cell cultures. The separated beads of cells are disrupted simply by a gentle breaking action. Commonly cells are centrifuged at 20° C. The motor evolves heat that raises the temperature.

Therefore, the use of low temperature for centrifugation is preferred so that the cells should not be exposed to high temperature.

Inverted Microscope

Use of an inverted microscope is important to observe cell cultures in situ. Because the cells are found on the bottom of the tissue culture vessel (e.g Petri plates). The culture medium remains above the growing cells in plates. If such plates are put over the stage of an ordinary microscope, the growing cells at the bottom cannot be observed. Therefore, the inverted microscope is used for such a purpose.

Culture room

All types of cultured plant tissues are incubated under the conditions of well-controlled temperature, humidity, illumination and air circulation. The culture room should have light and temperature control systems. Generally, the temperature is maintained at 25 ± 2°C and 20 98% relative humidity and uniform air ventilation. Generally cultures darkness each for a period of 12 hours.

Data Collection (Observation)

The cultures are monitored at regular intervals in the culture room for the growth and development of cultured tissues. Observation is also made under the aseptic area in laminar airflow.

Isolation of Animal Material (Tissue)

During the experimental work, attempts should be made that animal materials are not contaminated. When glassware and media are sterilized, animal materials should be handled. Before that, a balanced salt solution (BSS) is required. This solution consists of 1000 units of penicillin and 0.5 mg of streptomycin washed in BSS aseptically to avoid contamination. The tissue to be cultured should be properly sterilized with 70 percent ethanol, and removed surgically under aseptic conditions. Attempts should be made to avoid contamination. Thus, the tissue isolated is either stored in freeze or used immediately.

Establishment of Cell Cultures

There are many types of animal cells that can grow in such as tumor cells, pigmented melanoma cells, neuroblastoma cells, steroid producing adrenal cells, growth hormone prolactin-secreting cells, teratoma cells capable of differentiating in cold trial conditions pigmented or cartilage cells, etc. On the basis of the purpose of the experiment, The immortal cell line can be developed from cultured tissues. Healthy animal tissues are cultured on artificial nutrient media that proliferate and differentiate into able of dividing into a mixture of different types of cells.

Cell Lines

The primary cell culture cannot long time because the cell utilizes all nutrients of the medium. Therefore, sub remains viable for culturing needs to be done on another fresh medium. Hence, the primary culture is removed adherent cells are dissociated enzymatically medium and passed into fresh culture flask or by repeated pipetting. Then the cells are diluted with fresh. This results in multiple copies of proliferating cells. During the course of repeated sub-culture and selection, the cell line gets evolved and properly established consisting of rapidly proliferating cells. Thus the unaltered form of cell fine (only for a limited number of generations) is called continuous cell line which propagates in a single type of cell with a negligible amount of non- logarithmic ways. This mixture of cells is in single-cell suspension which may be used as a primary culture or starter culture. The primary culture (in the form of single-cell suspension is subculture by transferring into culture dishes/flasks containing special growth nutrients t optimal growth conditions Consequently, some cells attach to the surface and proliferate to yield a single cell line, in spite of being damaged in suspension. Therefore, while subculturing the suspension should be diluted with fresh medium at a certain ratio and transferred into a flask (Anon, 1988). Primary Cell Cultures: The first step in establishing cells in culture is to dissociate organs into a single cell suspension. It is done by mechanical or enzyme methods. The cells are transferred into a special glass or plastic containers containing a culture medium. Under these conditions, maintenance of growth of such cells is called primary cell culture

Hybridoma Technology

Blood consists of two major components, various kinds of cells and the fluid called serum serum consists of a heterogeneous population of antibodies.

Different antibodies are different types of B-lymphocytes. Hence, serum consists of different types of antibodies which are often called polyclonal antibodies. The antibodies bind to specific domains of antigens (called antigenic produced by dif determinants or epitopes) and neutralize them.

Monoclonal Antibodies Production of Monoclonal Antibodies (MoAb)

The mouse MOAB has revolutionised the field of biology and more specifically immunology. The mice MoAb have been used in human patients with varying levels of success in leukemia, lymphoma, melanoma and colorectal cancer. Clinical trials have indicated several limiting factors such as: (i) heterogenecity of tumour cells (not all malignant cells carry relevant antigen), circulating free antigens (they bind Fab on antibody molecules and thus block Moab from binding to the target cells), (iii) antigenic modulation (antigen modulated off the cell surface as a consequence of the binding of MoAbs to the cancer cell as in leukemia. This problem can be overcome by applying introduced into hamster kidney cells.

Applications of Monoclonal Antibodies

The monoclonal antibodies are used in the four main ways i.e. disease diagnosis, disease treat- nt, passive immunization and detection and purification of biomolecules. However, their exploitation for various uses is based on two features: (i) exacting the specificity of antibody binding, and (ii) presence of similar structure of all antibody molecules.

Disease Diagnosis

Using MoAb medical conditions and diseases of sufferers can be diagnosed. One of the approaches is the 'antibody- sandwich' strategy which is also called ELISA (enzyme-linked immunosorbent assay). The circulatory antigens in the blood can be assayed using ELISA or radioimmunoassay (RIA) technique. By using ELISA you can test for HIV, hepatitis, typhoid, herpes, etc. For each test, a separate ELISA kit is used; see **Figure 42.1.**

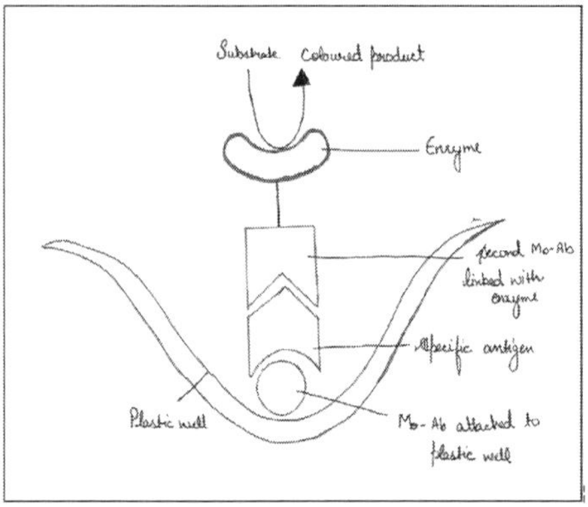

Figure 42.1. ELISA testing

44. BASIC CELL CULTURE, PRIMARY CULTURE, DEVELOPMENT OF MEDIA- PHYSICOCHEMICAL PROPERTIES, COMPLETE MEDIA, SERUM AND SUPPLEMENTS; SERUM-FREE MEDIA, CELL LINES, GROWTH CYCLE AND CULTURE MEDIA

Introduction

Cell culture is the process by which prokaryotic, eukaryotic or plant cells are grown under controlled conditions. But in practice, it refers to the culturing of cells derived from animal cells. Cell culture was first successfully undertaken by Ross Harrison in 1907. Roux in 1885 for the first time maintained embryonic chick cells in a cell culture. Historical events in the development of cell culture

- 1878: Claude Bernard proposed that physiological systems of an organism can be maintained in a living system after the death of an organism.
- 1885: Roux maintained embryonic chick cells in a saline culture.
- 1897: Loeb demonstrated the survival of cells isolated from blood and connective tissue in serum and plasma.
- 1903: Jolly observed cell division of salamander leucocytes *in vitro.*
- 1907: Harrison cultivated frog nerve cells in a lymph clot held by the 'hanging drop' method and observed the growth of nerve

fibers *in vitro*for several weeks. He was considered by some as the father of cell culture.

- 1910: Burrows succeeded in the long term cultivation of chicken embryo cell in plasma clots. He made detailed observations of mitosis.
- 1911: Lewis and Lewis made the first liquid media consisted of seawater, serum, embryo extract, salts and peptones. They observed limited monolayer growth.
- 1913: Carrel introduced strict aseptic techniques so that cells could be cultured for long periods.
- 1916: Rous and Jones introduced proteolytic enzyme trypsin for the subculture of adherent cells.
- 1923: Carrel and Baker developed 'Carrel' or T-flask as the first specifically designed cell culture vessel. They employed the microscopic evaluation of cells in culture.
- 1927: Carrel and Rivera produced the first viral vaccine - Vaccinia.
- 1933: Gey developed the roller tube technique
- The 1940s: The use of the antibiotics penicillin and streptomycin in culture medium decreased the problem of contamination in cell culture.
- 1948: Earle isolated mouse L fibroblasts which formed clones from single cells. Fischer developed a chemically defined medium, CMRL 1066.
- 1952: Gey established a continuous cell line from a human cervical carcinoma known as HeLa (Helen Lane) cells. Dulbecco developed plaque assay for animal viruses using confluent monolayers of cultured cells.
- 1954: Abercrombie observed contact inhibition: motility of diploid cells in monolayer culture ceases when contact is made with adjacent cells.
- 1955: Eagle studied the nutrient requirements of selected cells in culture and established the first widely used chemically defined medium
- 1961: Hayflick and Moorhead isolated human fibroblasts (WI-38) and showed that they have a finite lifespan in culture.
- 1964: Littlefield introduced the HAT medium for cell selection.

- 1965: Ham introduced the first serum-free medium which was able to support the growth of some cells.
- 1965: Harris and Watkins were able to fuse human and mouse cells by the use of a virus.
- 1975: Kohler and Milstein produced the first hybridoma capable of secreting a monoclonal antibody.
- 1978: Sato established the basis for the development of serum-free media from cocktails of hormones and growth factors.
- 1982: Human insulin became the first recombinant protein to be licensed as a therapeutic agent.
- 1985: Human growth hormone produced from recombinant bacteria was accepted for therapeutic use.
- 1986: Lymphoblastoid γIFN licensed.
- 1987: Tissue-type plasminogen activator (tPA) from recombinant animal cells became commercially available.
- 1989: Recombinant erythropoietin in trial.
- 1990: Recombinant products in the clinical trial (HBsAG, factor VIII, HIVgp120, CD4, GM-CSF, EGF, mAbs, IL-2).

Major Development's in Cell Culture Technology

First development was the use of antibiotics which inhibits the growth of contaminants. Second was the use of trypsin to remove adherent cells to subculture further from the culture vessel. Third was the use of a chemically defined culture medium.

Why is cell culture used for?

Areas where cell culture technology is currently playing a major role.

- Model systems for Studying basic cell biology, interactions between disease causing agents and cells, effects of drugs on cells, process and triggering of aging & nutritional studies

Toxicity testing

- Study the effects of new drugs.

Cancer research

- Study the function of various chemicals, viruses & radiation to convert normal cultured cells to cancerous cells.

Virology

Cultivation of virus for vaccine production also used to study their infectious cycle.

Genetic Engineering

Production of commercial proteins, large-scale production of viruses for use in vaccine production e.g. polio, rabies, chickenpox, hepatitis B & measles.

Gene therapy

Cells having a functional gene can be replaced by cells which are having non-functional genes.

Tissue Culture

In vitro cultivation of organs, tissues & cells at defined temperature using an incubator & supplemented with a medium containing cell nutrients & growth factors is collectively known as tissue culture. Different types of cell grown in culture includes connective tissue elements such as fibroblasts, skeletal tissue, cardiac, epithelial tissue (liver, breast, skin, kidney) and many different types of tumor cells.

Primary Culture

Cells when surgically or enzymatically removed from an organism and placed in a suitable culture environment will attach and grow is called as primary culture. Primary cells have a finite life span. Primary culture contains a very heterogeneous population of cells. Subculturing of primary cells leads to the generation of cell lines. Cell lines have a limited life span, they passage several times before they become senescent. Cells such as macrophages and neurons do not divide *in vitro* so can be used as primary cultures. The lineage of cells originating from the primary culture is called a cell strain.

Continous Cell Lines

Most cell lines grow for a limited number of generations after which they cease. Cell lines either occur spontaneously or induced virally or chemically transformed into Continuous cell lines. Characteristics of continuous cell lines

- Smaller, more rounded, less adherent with a higher nucleus / cytoplasm ratio.
- Fast growth and have an aneuploid chromosome number.
- Reduced serum and anchorage dependence and grow more in suspension conditions.
- Ability to grow up to higher cell density.
- Different in phenotypes from donor tissue.
- Stop expressing tissue-specific genes.

Types of cells

- On the basis of morphology (shape & appearance) or on their functional characteristics. They are divided into three.
- Epithelial like- cells attached to a substrate and appears flattened and polygonal in shape.
- Lymphoblast like- cells do not attach remain in suspension with a spherical shape.
- Fibroblast like- cells attached to a substrate appears elongated and bipolar.

Culture media

- Choice of media depends on the type of cell being cultured.
- Commonly used Media are MEM, GMEM, EMEM,DMEM, etc.
- Media is supplemented with antibiotics viz. penicillin, streptomycin, etc.
- Prepared media is filtered and incubated at 4°C.

Why sub culturing

- Once the available substrate surface is covered by cells (a confluent culture) growth slows & ceases.

- Cells to be kept in healthy & in a growing state have to be sub-cultured or passaged.
- It's the passage of cells when they reach 80-90% confluency in flask/dishes/plates.
- Enzymes such as trypsin, dispase, collagenase in combination with EDTA breaks the cellular glue that attached the cells to the surface.

Culturing of cells

- Cells are cultured as anchorage dependent or independent
- Cell lines derived from normal tissues are considered as anchorage-dependent grows only on a suitable substrate e.g. tissue cells
- Suspension cells are anchorage-independent e.g. blood cells
- Transformed cell lines either grow as a monolayer or as suspension

Adherent cells

- Cells that are anchorage dependent.
- Cells are washed with PBS (free of ca & mg) solution.
- Add enough trypsin/EDTA to cover the monolayer.
- Incubate the plate at 37°C for 1-2 min
- Tap the vessel from the sides to dislodge the cells.
- Add a complete medium to dissociate and dislodge the cells with the help of a pipette which is remained to be adherent
- Add complete medium depends on the subculture.
- requirement either to 75 cm or 175 cm flask.

Suspension cells

- Easier to passage as no need to detach them.
- As the suspension cells reach to confluency.
- Aseptically remove 1/3rd of the medium.
- Replaced with the same amount of pre-warmed medium.

Transfection Methods

- Calcium phosphate precipitation.
- DEAE-dextran (dimethylaminoethyl-dextran).
- Lipid mediated lipofection.
- Electroporation.
- Retroviral infection.
- Microinjection.

Cell Toxicity

- Cytotoxicity causes inhibition of cell growth.
- Observed effect on the morphological alteration in the cell layer or cell shape.
- Characteristics of abnormal morphology are the giant cells, multinucleated cells, a granular bumpy appearance, vacuoles in the cytoplasm or nucleus.
- Cytotoxicity is determined by substituting materials such as medium, serum, supplements flasks, etc. at a time.

Working with Cryopreserved Cells

- Vial from liquid nitrogen is placed into 37 C water bath, agitate vial continuously until the medium is thawed.
- Centrifuge the vial for 10 mins at 1000 rpm at RT, wipe top of vial with 70% ethanol and discard the supernatant.
- Resuspend the cell pellet in 1 ml of complete medium with 20% FBS and transfer to properly labeled culture plates containing the appropriate amount of medium.
- Check the cultures after 24 hrs to ensure that they are attached to the plate.
- Change medium as the color changes, use 20% FBS until the cells are established.

Freezing cells for storage

- Remove the growth medium, wash the cells with PBS and remove the PBS by aspiration.
- Dislodge the cells by trypsin-versene.

- Dilute the cells with a growth medium.
- Transfer the cell suspension to a 15 ml conical tube, centrifuge at 200g for 5 mins at RT and remove the growth medium by aspiration.
- Resuspend the cells in 1-2ml of freezing medium.
- Transfer the cells to cryovials, incubate the cryovials at -80° C overnight.
- The next day transfer the cryovials to Liquid nitrogen.

Cell viability

- Cell viability is determined by staining the cells with trypan blue.
- As trypan blue dye is permeable to non-viable cells or death cells whereas it is impermeable to this dye.
- Stain the cells with trypan dye and load to a haemocytometer and calculate % of viable cells
- % of viable cells= Nu. of unstained cells x 100 total nu. of cells

Common cell lines

Human cell lines

- -MCF-7 breast cancer
- HL 60 Leukemia
- HEK-293 Human embryonic kidney
- HeLa Henrietta lacks
- Primate cell lines
- Vero: African green monkey kidney epithelial cells
- Cos-7 African green monkey kidney cells
- And others such as CHO from hamster, sf9 & sf21 from insect cells

Contaminant's of cell culture

- Cell culture contaminants of two types.
- Chemical-difficult to detect caused by endotoxins, plasticizers, metal ions or traces of disinfectants that are invisible.

- Biological-cause visible effects on the culture are mycoplasma, yeast, bacteria or fungus or also from cross-contamination of cells from other cell lines.

Effects Of Biological Contamination's

- They compete for nutrients with host cells.
- Secreted acidic or alkaline by-products causes the growth of the host cells.
- Degraded arginine & purine inhibits the synthesis of histone and nucleic acid.
- They also produce H_2O_2 which is directly toxic to cells.

Detection of Contaminants

In general indicators of contamination are turbid culture media, change in growth rates, abnormally high pH, poor attachment, multi-nucleated cells, graining cellular appearance, vacuolization, inclusion bodies and cell lysis. Yeast, bacteria & fungi usually show visible effects on the culture (changes in medium turbidity or pH). Mycoplasma is detected by direct DNA staining with intercalating fluorescent substances e.g. Hoechst 33258. Mycoplasma is also detected by enzyme immunoassay by specific antisera or monoclonal abs or by PCR amplification of mycoplasmal RNA. The best and the oldest way to eliminate contamination is to discard the infected cell lines directly

Basic Equipment Used in Cell Culture

- Laminar cabinet-Vertical is preferable.
- Incubation facilities- Temperature of 25-30° C for insect & 37° C for mammalian cells, CO2 2-5% & 95% air at 99% relative humidity. To prevent cell death incubators set to cut out at approx. 38.5°C.
- Refrigerators- Liquid media kept at 4° C, enzymes (e.g. trypsin) & media components (e.g. glutamine & serum) at -20°C
- Microscope- An inverted microscope with 10x to 100x magnification
- Tissue culture ware- Culture plastic ware treated with polystyrene

Rules for working with cell culture

- Never use contaminated material within a sterile area.
- Use the correct sequence when working with more than one cell line.
- Diploid cells (Primary cultures, lines for the production of vaccines ,etc.).
- Diploid cells (Laboratory lines).
- Continous, slow growing line.
- Continous, rapidly growing lines.
- Lines that may be contaminated.
- Virus producing lines.

Basic aseptic conditions

- If working on the bench use a Bunsen flame to heat the air surrounding the Bunsen.
- Swab all bottle tops & necks with 70% ethanol.
- Flame all bottlenecks & pipette by passing very quickly through the hottest part of the flame.
- Avoiding placing caps & pipettes down on the bench; practice holding bottle tops with the little finger.
- Work either left to right or vice versa, so that all material goes to one side, once finished.
- Clean up spills immediately & always leave the work place neat & tidy.

Safety aspect in cell culture

- Possibly keep cultures free of antibiotics in order to be able to recognize the contamination.
- Never use the same media bottle for different cell lines. If caps are dropped or bottles touched unconditionally touched, replace them with new ones.
- Necks of glass bottles prefer heat at least for 60 secs at a temperature of 200°C.
- Switch on the laminar flow cabinet 20 mins prior to start working.

- Cell cultures that are frequently used should be subcultured & stored as duplicate strains
- Use actively growing cells that are in their log phase of growth, which are 80-90% viable.
- Keep exposure to trypsin at a minimum.
- Handle the cells gently. Do not centrifuge cells at high speed or roughly re-suspend the cells.
- Feeding & subculturing the cells at more frequent intervals then used with serum-containing conditions may be necessary.
- A lower concentration of 104cells/ml to initiate subculture of rapidly growing cells & a higher concentration of 105cells/mlfor slowing growing cells.

Growth Cycle

- Animal cell growths in culture have a characteristics growth pattern similar to bacteria.
- As cell grow and divide in a monolayer or in suspension, they usually follow a characteristic growth pattern composed of three phases:
- Lag Phase
- Log Phase
- Stationary phase

Log Phase

- It is a period of zero growth when cells are first inoculated into the growth medium.
- The length of this phase depends on the type of cells and their metabolic states during inoculation.
- It is a period of adaptation during which the cells recover from trypsinization, reconstruct their cytoskeleton, secrete matrix to aid attachment and spread out on the substrate.

Log Phase

- The exponential growth phase is a period of continuous cell doubling.

- Animal cells normally exhibit a doubling time of between 15 and 25 hours.
- The length of the log phase depends on the seeding density, the growth rate of the cells and the density at which cell proliferation is inhibited by density.

Stationary Phase

- It is the period after growth when there is no change in the culture cell density.
- The phase occurs when the nutrients have been depleted or inhibitory metabolites have accumulated in the culture.
- All available growth surfaces are occupied and all the cells are in contact with surrounding cells.
- Further growth of cells can be obtained by subculturing the cells in a fresh medium.

Culture Media

The culture medium is a liquid or gel designed to support the growth of cells, tissue or organs. A typical cell culture medium is composed of a complement of amino acid, vitamins, inorganic salts, glucose and serum as a source of growth factors, hormones and attachment factors. There are different types of media used for animal cell culture that may be grouped into two broad categories: Natural media or artificial media.

Natural Media

It consists solely of naturally occurring biological fluids. It is very useful and convenient for a wide range of animal cell cultures. The major disadvantage of natural media is its poor reproducibility due to lack of knowledge of the exact composition of these natural media.

Artificial Media

It contains partly or fully defined components that are prepared artificially by adding several nutrients (organic and inorganic). Artificial media supplemented with serum or with suitable formulations of organic compounds supports prolonged survival of the cell culture. It may be defined in different groups into the following four classes:

- Serum containing media
- Serum-free media
- Chemically defined media
- Protein free media

REFERENCES

1. Abollino, O., Giacomino, A., Malandrino, M., Mentasti, E., Aceto, M., Barberis, R., (2006). Assessment of metal availability in a contaminated soil by sequential extraction. Water Air Soil Pollution, 173(1-4): 315-338.
2. Agarwal, V. K., (2000). Molecular Biology. New Delhi: S. Chand.
3. Agarwal, V. K. and Verma. P. S., (2000*). Concepts of Molecular Biology*. New Delhi: S. Chand.
4. Ahmadpour, P., Ahmadpour, F., Mahmud, T., Abdu, A., Soleimani, M., Tayefeh, F.H., (2014). Phytoremediation of heavy metals: a green technology. *In:* Asrari, E. (Ed), heavy metal contamination of water and soil: Analysis, Assessment, and Remediation strategies. CRC Press, Taylor and Francis, 249.
5. Aldrich, J., and Vaker, R., (1970). Plant Dis. Reptr. 54: 446-448.
6. Alexander, M., (1977). Introduction to Soil Microbiology, John Wiley & Sons, New York.
7. Alfermann, A.W. and Petersen, M., (1995). Natural product formation by plant cell biotechnology. Plant Cell Tiss. Org. Cult, 43: 199-205.
8. Ali, H., Khan, E., Sajad, M.A., (2013). Phytoremediation of heavy metals-concepts and applications. Chemosphere, 91(7): 869-881.

9. Amirato, P.V., (1983). In Handbook of Plant Cell Culture. Vol. I, Techniques for Propagation and Breeding (Eds. D.A., Evans, W.R., Sharp, P.V., Amminrato and Y., Yamada), PP. 82-123, Macmillan., New York.
10. Anand, A., et al., (2004). Physiol. Biol. 10:233-242.
11. Aneja, K.R., (2014). Experiments in Microbiology, Plant Patholoy and Biotechnology. Fourth Edition. New Age International (P) Limited Publisher.
12. Aneja, K.R., (2014). Laboratory Manual of Microbiology and Biotechnology. Scientific International PVT. LTD Medtech.
13. Anonymous (1980 a) Nature (London) 286:110.
14. Anonymous (1980 b). NBRI News letter 7:28.
15. Anonymous (1998). National Cancer Institute, www.http://rex.nci.nih.gov.in_Chadha, K.L. and Gupta, R. (1995). Advances in Horticulture, Vol II. Medicinal and Aromatic Plants. Malhotra Publ. House, New Delhi.
16. Arora, P.N., and Malhan, P.K., Biostatistics.
17. Atkinson, G.F., (1961). Mushrooms, 2nd ed. Hofner Publishing Co., New York.
18. Atkins, F.C., (1983). Mushroom J., 125:168-171.
19. Baker, R., and Nash, S.M., (1965). Phytopath. 55:1381-1382.
20. Baker, K.F., and Snyder W.C., (1965). Ecology of Soil Borne Plant Pathogens-Prelude to Biological Control, John Murray, London, P. 571.
21. Banerjee, P.K., Problems on genetics molecular genetics and evolutionary Genetics.
22. Bhattacharya, D., and Roy, Choudury, Statistcs theory and Practices.
23. Barabasz, A., Wilkowska, A., Tracz, K., Ruszczyñska, A., Bulska, E., Mills, R.F., (2013). Expression of *HvHMA2*in tobacco modifies Zn-Fe-Cd homeostasis. Journal of Plant Physiology, 170: 1176-1186.
24. Barell, B.G., Air, G.M., and Hutchindon, C.A., (1976). Nature. 264:34-41.
25. Becking, J.H., (1982). In Advances in Agricultural Microbiology, eds. N.S. Subba Rao), pp. 89-110, Oxford & IBH Publish, Co.,

New Delhi.

26. Bergersen, F.J., (1971). Ann. Rev. Plant Physiol. 22:121-140.
27. Betsch, David F. "DNA Fingerprinting in Human Health and Society." Retrieved: 12th April 2006 http://www.accessexcellence.org/RC/AB/BA/DNA_Fingerprinting_Basics.html
28. Carlson, P.S., (1973). Science 180: 1366-1368.
29. Bhaskaran, S., (1987). In Biotechnology in Agriculture, (eds. Natesh, S., Chopra, V.L., and Ramachandan, S.,) PP. 101-108, Oxford &IBH Publ. Co., New Delhi.
30. Biello, D., (2006). Tobacco Plant Transformed into Plague Vaccine Factory. Retrieved 10 Jan. http://www.sciam.com/article.cfm?chanID=sa016&articleID=000E3140-EB2E-13C2-AB2E83414B7F0000
31. Binns, A.N., (1981). In Propagation of Higher Plants Through Tissue Culture: Emerging Technoloics and Strategies, (eds. Constantin, M.J., Henke, R.R., Hughes, K.W., and Conger, B.V.,) Enviorn Exp. Bot. 21:325-332.
32. Biotechnology: A Dutch Perspectives (1981). A report by the Netherlands Study Centre for Technology Trends, (ed. J.H.F. Van Apeldoorn).
33. Boffey, S.A., (1987). In Biotechnology: The Biotechnology Principles (eds. Trevan, M.D., Boffey, S., Goulding, K.H., and Stanbury, P.,) pp 11-152. Open University Press: Milton Keynes, England.
34. Bouwer, E.J., and Zehnder, A.J.B., (1993). Trends Biotechnology. 11:287-318.
35. Bouwer, E.J. et al., (1998), In Biological Treatment of Hazardous Wastes (eds. Lewandowaski, G.A., and Deflippi, L.J.) pp 237-270, John Wiley & Sons.
36. Brooks, R., Lee, J., Reeves, R.D., Jaffre, T., (1977). Detection nickeliferous rocks by analysis of herbarium specimens of indicator plants. Journal of Geochemical Exploration, 7: 49-57.
37. Bull, A.T., Holt, G., and Lilly, M.D., (1983). Biotechnology: International Trends and Perspectives, p. 84, Oxford & IBH Publ. Co. PVT. Ltd., New Delhi.
38. Carlson, P.S., (1983). Cell and Tissue Culture Techniques for Cereal Crop Improvement, pp. 407-418, New York.

39. Christov, L., and Van Driessel, B., (2003). Indian J. Biotech. 2:444-450.
40. Clark, D.S., (1962). Biotehnol Bioeng. 4:1721.
41. Cohen, S.N., Chang, A.C.Y., Boyer, H.W., and Helling, R.B., (1973). Proc. Nat. Acad. Sci. (U.S.A) 70:3240-3244.
42. Collins, H.A.,and Watts, M., (1983). In Handbook of Plant Cell Culture: Techniques for Propagation and Breeding, Vol. I (eds. Evans, D.A., Sharp, W.T., Ammirato, P.V., and Yamada, Y.) pp. 729-747. Macmillan, New York.
43. Crueger, W., and Crueger, A., (1984). Biotechnology-A Text Book of Industrial Microbiology. Sinaver Associates Inc., Sunderland, USA, pp 23:207-216.
44. Cummings, Michael R., and Williams S. Klug., (2004).Concepts of Genetics. New Delhi: Pearson Education.
45. Das, D., and Das, A Statistcs in Biology and Psychology.
46. Das, N.G., Statistical Methods.
47. David, L., Nelson, Michael, M., (2017). Lehninger Principles of biochemistry 7th edition.
48. De, S.N., Business Statistics.
49. Dodds, J.H., and Roberts, L.W., (1985). Experiments in Plant Tissue Culture, 2nd Ed. Cambridge University Press, Cambridge.
50. Dubey, R.C., (2009). A textbook of Biotechnology. S.Chand and Company LTD..
51. Ferber, D., (2004). "The New Science of Cell Hacking." Popular Science, June pp: 43-44.
52. Frazier, W. C. and Westhoff, D. C., (1993). Food Microbiology. New Delhi: Tata McGraw-Hill.
53. Fisher, R.A. Statistical methods for Research Workers.
54. Gautam, N., Verma, P.K., Verma, S., Tripathi, R.D., Trivedi, P.K., Adhikari, B., Chakrabarty, D., (2012). Genome wide identification of rice class I metallothionein gene: tissue expression patterns and induction in response to heavy metal stress. Functional and Integrative Genomics, 12: 635-647.
55. Garg, N., Garg, K.L., Mukerjii, K.G., (2013) Laboratory Manual of Food Microbiology. I.K. International Publishing House Pvt. Ltd.

56. Gibbs, W., (2001).“Cybernetic Cells.” Scientific American, Aug. pp: 54-57
57. Greipsson, S., (2011). Phytoremediation, National Education Knowledge, 3(10): 7.
58. Gritfith, A., Gelbart, W.M., Miller, S.H., and Lewontin, R.C., Modern genetic analysis.
59. Gu, C.S., Liu, L.Q., Deng, Y.M., Zhu, X.D., Huang, S.Z., Lu, X.Q., (2015). The heterologus expression of the Iris lacteal var. chinensis type 2 metallothionenin IIMT2b, gene enhances copper tolerance in *Arabidopsis thaliana*. The Bulletin of Environmental Contamination and Toxicology, 94:247-253.
60. Guha, S., Maheswari, S.C., (1964). Nature (London), 204:297.
61. Guha, S., Maheswari, S.C., (1966). Nature 212:97-98.
62. Gupta, C.P., Sharma, A., Dubey, R.C., and Maheshwari, D.K., (1999). Cytobios. 99:183-189.
63. Gupta, C.P., Dubey, R.C., Kang, S.C., Maheshwari, D.K., Current Science. 81:91-94.
64. Haberlandt, G., (1902). Sber Akad. Wiss. Wien. III: 69-92.
65. Hanson, Earl D., (1983) Recombinant DNA Research and the Human Prospect. Washington, DC. American Chemical Society.
66. Henshaw, G.G., (1982). In plant Tissue Culture, pp. 789-792, Tokyo.
67. Huang, G.Y., and Wang, Y.S., (2010). Expression and characterization analysis of type 2 metallothionein from grey mangrove species (*Avicennia marina*) in responses to metal stress. Aquatic Toxicology, 99:86-92.
68. ICGEB. (1998). Activity Report, 1998.
69. Khan, I.A. and Khanum, A., (1998). Role of Biotechnology in Medicinal and Aromatic Plants I Ukaaz Publ., Hyderabad.
70. Khoshoo, T.N., (1988). Enviornmental Concern and Strategies. II Ed. PP. 253-404. Ashish Publ. House, New Delhi.
71. King, P.J., (1980). Int. Rev. Cytol. Supplement. 11A:25-54.
72. Kowles Richard, Solving problems in Genetics.
73. Kumar, H.D., (2003). Modern Concepts of Biotechnology. New Delhi: Vikash Publishing House.
74. Lee, S., (2014). Artifical induction of cadmium tolerance and its further enhancement via heterologus co-expression of SpHMT1 and AtPCS1

in the yeast cells. Journal of Korean Society Applied Biological Chemistry. 57:307-310.

75. Lopez, D. A., R. M. Williams., and Michlke, K., (1994). Enzymes: The Foundation of Life. München: The Neville Press.
76. Lund, P., "What is Bioinformatics?" Bioscience Course Manual, The University of Birmingham. pp: 35.
77. Ma, Y., Rajkumar, M., Luo, Y., Freitas, H., (2011). Inoculation of endophytic bacteria on host and non-host plants-effects on plant growth and Ni uptake. Journal of Hazard Materials, 195: 230-237.
78. Mahajan, B.K., Methods in Biostatistics.
79. Maestri, E., Marmiroli, M., Visioli, G., Marmiroli, N., (2010). Metal tolerance and hyperaccumulation: costs and trade-offs between traits and environment. Environmental and Experimental Botany, 68(1); 1-13.
80. Maheswari, S.C., Rasid, A., and Tyagi, A.K., (1983). IAPTC Newsletter 41:2-9.
81. Mandels, M., and Stenberg, D., (1976). J. Ferment. Technol. 54:267-286.
82. Mandal, P.K., (1988). Gene Therapy. Sci. Reptr. 25:156-157.
83. Mason, H.S., and Arntzen, C.J., (1995). Trends Biotechnol. 13:388-392.
84. Maxam, A.M., and Gilbert, W., (1977). Proc. Natl. Acad. Sci. (U.S.A) 74:560-564.
85. Memon, A.R., Schroder, P., (2009). Implications of metal accumulations mechanisms to phytoremediation. Environmental Science and Pollution Research, 16(2):162-175.
86. Miller, C.O., Skoog, F., Saltz, M. and Strong, F.M., (1955). J. Am. Chem. Soc. 77:1992.
87. Misawa, M., (1997). Plant tissue culture: an alternative for production of useful metabolites. FAO Agric Service Bull, 108: 1-87.
88. "Molecular Biotechnology in Life." Author not mentioned. Retrieved: 5th April 2006 http://www.goldenessays.com/free_essays/4/technology/molecular-biotechnology-in-life.shtml.
89. Morgan, T.H., (1901). Regerantion, Macmillian: London.
90. Nitsch, C., (1974). C.R. Acad. Sci. (Paris), 278:1031-1034.
91. Murashige, T., (1974). Ann. Rev. Plant Physiol 25:135-166.

92. Murashige, T., and Skoog, F., (1962). Physiol Plant, 15:473-497.

93. Paras, Y., Annu, Y., Kumar, P., Arora, J.S., Datta, T.K., De, S., Goswami, S.L., Yadav, M., Jain, S., Nagpal, R., Yadav, H.,. Department of Animal Biotechnology, Animal Biochemistry Division and Dairy Microbiology Division, National Dairy Research Institute, Karnal 132001 (Haryana), India; SOS in Chemistry, Jiwaji University, Gwalior-474011, M.P., India.

94. Peberdy, J.F., (1989). Mycology Research 93:1-20.

95. Peters, P., (1993). Biotechnology- a guide to genetic engineering. Wmc. Brown Publishers, Dubuwue, P. 253.

96. Pillai, R.S.N, & Bagavati, V., Statistics.

97. Prakash, J., and Giles, K.L., (1987). Int. Rev. Cytol. 107:273-292.

98. Prescott, L.M., Harley, J.P., and Kein, D.A., (2003). Microbiology. McGraw-Hill High Education, New York.

99. Purohit, S., (2005) Agricultural Biotechnology. India: Agrobios.

100. Purohit, S., (2004) Biotechnology: Fundamental and Applications. India: Agrobios.

101. Rajkumar, M., Ae, N., Prasad, M.N.V., Freitas, H., (2010). Potential of siderophore-producing bacteria for improve heavy metal phytoextraction. Trends Biotechnology, 28(3): 142-149.

102. Ramawat, K.G., and Merillon, J.M., (2011). Biotechnology Secondary Metabolites. Oxford & IBH publishing CO.PVT.LTD. New Delhi. ISBN 81-204-1387-3.

103. Reinert, J., and Bajaj, Y.P.S., (1977). In Applied and Fundamental Apects of Plant Cell, Tissue and organ ulture (eds. Reinert, J., and Bajaj, Y.P.S) pp. 251-267, Springer-Verlag Berlin.

104. Roberts, R.J., (1983). Nuceilc Acid Res. 11:135-168.

105. Robson, R.L., (1979). Biochem. J. 181:569-575.

106. Ruschel, A.P., and Vose, P.B., (1983). In Hand Book of Plant Cell Culture-Techniques for Propagation and Breeding, Vol. 1 (eds. Evans, D.A; Sharp, W.R; Ammirato, pp. and Yamada, Y.) pp 828-851, Macmillian Publ. Co., New York.

107. Russel, P.J., Genetics (5th edition)

108. Sasson, A., (1984). Biotechnology: Challenges and Promises, UNESCO, paris.

109. Saxena, H.C., Elementary statistics.
110. Schell, J., and Van Montagu, M., (1983). Biotechnology 1: 175-180.
111. Schesinger, H., (1999). "DNA Conductor." Popular Science, Aug.. pp: 47.
112. Schlatmann, J.E, Hoopen, HJ G. and Heijnen, J., (1996). Large scale production of secondary metabolites by plant cell cultures. In: Plant Cell Culture: Secondary Metabolites. F. Dicosmo and M. Misawa (eds.) CRC Press, NY. I
113. Schlegel, H.S., (1986). Genera Microbiology, 6th Ed. (Tranlated by M. Kogut) Camcridge University Press, London.
114. Sharp, W.R., Sondahl, M.R., Caldas, L.S., and Maraffa, S.B., (1980). Hort. Rev. 2:268-310.
115. Shrof, Joannie M. "Miracle Vaccines." *US News & World Report,* 23 Nov. 1998. pp: 157
116. Singh, Mahabir, and Sandeep Ahlawat. "Applied Genetics." MTBiology Today, April 2004: pp: 27-29.
117. Skoog, F., (1944). Am. J. Bot. 31:19-24.
118. Smith, P.H., (1966). Dev. Industrial Microbiol. 7:156-159.
119. Southern, E.M., (1975). J. Mol. Biol. 98:503-517.
120. Snustad, D. Peter, and Simmons, M.J., Principle of genetics (4th edition).
121. Srivastava, S.K., et al., (1995). Indian J. Exp. Biol. 33:962-966.
122. Strick-Berger, M.W., Genetics (3rd edition).
123. Standfield, W.D., Schaums Outline Theory & Problems is Genetics (2rd edition).
124. Swaminathan, M.S., (1980). Second Annual Day Lecture, NBRI, Lucknow.
125. Tangahu, B.V., Sheikh Abdullah, S.R., Basri, H., Idris, M., Anuar, N., Mukhlisin, M., (2011). A review on heavy metals (As, Pb, and Hg) uptake by plants through phytoremediation. International Journal of Chemical Engineering, 31:939161.
126. Thomas, E and Davey, M.R., (1975). From Single Cell to plants, Wyked=ham: London.
127. Verpoorte, R., Heijden, R.V, Schripsema, J, Hoge, JH.C. and Ten-Hoopen, H.J.G., (1993). Plant cell biotechnology for the production of alkaloids; present status and pros-i pects. J. Nat. Products, 56: 186-207.

128. White, P.R., (1934). Plant Physiol. 9:585-600.

129. Willams, J.G., (1981). In Genetic Engineering, Vol. I, (ed. Williamson, R.,), PP. 1-59, AcademicPress London.

130. Wiseman, A., (eds.) (1975). Handbook of Enzyme Technology, Halsted Press, New York.

131. Www.scribd.com., www.slideshare.com., www.coursehero.com., www. biocyclopedia.com., www. molecular-plant-biotechnology. info., www.slideplayer.com., www.documents.mx., www.docplayer. net., www.bmb.psu.edu.,www.archive.org.,www.mafiadoc.com., www.powershow.com., www.bioscience.jbpub.com., www. outhorstream.com., www.docslide.com., www.msluay.com., www. scienceprofonlue.com., www.csus.edu., www.krishikosh.egrouth. ac.in., www.tankonyvtor.hu., www.silo.pub., www.studyblue.com., www.Preservearticles.com., www.biotechuniverse.blogspot.com.

132. Wikipedia Online Encyclopedia, Retrieved 20 Feb. 2006 http://www. wikipedia.org/wiki/Biotechnology.

133. Wikipedia Online Encyclopedia, Retrieved 20 Feb. 2006 http://www. wikipedia.org/wiki/genetic_engineering.

134. Wikipedia Online Encyclopedia, Retrieved 20 Feb. 2006 http://www. wikipedia.org/wiki/genetically_modified_food.

135. Wikipedia Online Encyclopedia, Retrieved 22 Feb. 2006 http://www. wikipedia.org/wiki/bioinformatics

136. Yamada, Y., aand Fujita, Y., (1983). In Handbook of Plant Cell Culture-Technique for Propagation and Breeding. (eds. Evans, D.A; Sharp, W.R., Ammirato and Yamada, Y) pp. 717-728. Macmillian Publ. Co., New York.

137. Yeoman, M.M. and Yeoman, Y., (1996). Manipulating secondary metabolism in cultured plant cells. New Phytol, 134: 552-569.

138. Zar, Jerrold, H., Biostatistical Analysis (4th edition).

Index

C

D

E

F

G

H

I